EARTHQUAKE DATA IN ENGINEERING SEISMOLOGY

GEOTECHNICAL, GEOLOGICAL AND EARTHQUAKE ENGINEERING

Volume 14

For further volumes:
http://www.springer.com/series/6011

Earthquake Data in Engineering Seismology

Predictive Models, Data Management and Networks

edited by

SİNAN AKKAR
Middle East Technical University, Ankara, Turkey

POLAT GÜLKAN
Middle East Technical University, Ankara, Turkey

and

TORILD VAN ECK
Royal Netherlands Meteorological Institute, De Bilt, The Netherlands

Springer

Editors
Sinan Akkar
Middle East Technical University
Earthquake Engineering Research Center
Department of Civil Engineering
06531 Ankara
Turkey
sakkar@metu.edu.tr

Polat Gülkan
Middle East Technical University
Earthquake Engineering Research Center
Department of Civil Engineering
06531 Ankara
Turkey
a03516@metu.edu.tr

Torild van Eck
Royal Netherlands Meteorological Institute
ORFEUS c/o Seismology Division
3730 AE De Bilt
The Netherlands
torild.van.eck@knmi.nl

ISSN 1573-6059
ISBN 978-94-007-3479-1 ISBN 978-94-007-0152-6 (eBook)
DOI 10.1007/978-94-007-0152-6
Springer Dordrecht Heidelberg London New York

Printed on acid-free paper

Springer is part of Springer Science+Business Media (www.springer.com)

Preface and Acknowledgements

Researchers and practitioners who make use of processed strong motion data have access to a multitude of Internet resources. This is far removed from times when such data was exchanged among users on magnetic tapes, and later, on CDs. Today, researchers need instant access to waveforms that must have been processed in uniformly verifiable ways for a wide variety of objectives. The products of this branch of earthquake science must be searchable according to a multitude of different criteria defined by user needs. The increased utilization of strong motion data has made dialogue among network operators, data providers and the user community to become more important than ever. It is clear that uniformly processed waveform recordings that have been archived consistently would increase the confidence of users and encourage a broader exchange of strong-motion data for purposes of seismic safety in countries that live under the earthquake peril.

A strong motion sensor that is triggered into recording during an earthquake provides a waveform that contains the complex interaction among its source, propagation path and local site geology of the station. Each record is unique, but the collective processing of records obtained under different environmental circumstances provides the only reliable means of obtaining parameterized statistics of future earthquakes for engineering design. They also enable a better understanding of the fundamental physics of strong ground motion. Repositories of this type of data thus provide researchers the information to mine for deriving improved ground motion prediction equations. The GMPE development area has been actively developed during about the last decade and a half thanks to improvements in standards and procedures for archived ground motion data.

The Workshop on Accelerometric Data Exchange and Archiving (ADEA) where the contributions in this book have been presented has been made possible by collaboration among several institutions. On the Turkish side it has been the culmination of a 4-year project (award no. 105G016) that had been supported during 2005–2009 by the Scientific and Technological Research Council of Turkey (TUBITAK). The program was executed jointly by Middle East Technical University (METU) and the Ministry of Public Works and Reconstruction (MPWR). On the part of the University, the Earthquake Engineering Research Center (EERC) and for the Ministry the General Directorate for Disaster Affairs (GDDA) served as the designated focal points where the work was performed. The Turkish Atomic Energy Authority (TAEK) was an early supporter of the Workshop. The Network of Research Infrastructures for European Seismology (NERIES), an EU FP6 I3 project (contract RII3-CT-2006-026130), supported the meeting as its co-sponsor. Kinemetrics, Inc. provided funding for some participants. We record our thanks to all meeting sponsors. Articles were peer-reviewed, often at short notice, by many individuals in addition to the Editors. Among these individuals we are particularly indebted to Reşat Ulusay and M. Tolga Yılmaz for their meticulous comments for manuscripts we directed to them.

No workshop or its book is possible without certain individuals surpassing their fair share of the work. ADEA was no different. John Douglas and Zehra Çağnan helped in many ways, not only before and during the Workshop but also during the phase when we brought the book to its final form, so we feel obliged to cite them by name. Our students at METU, Tuba Eroğlu and Emrah Yenier, went beyond the call of duty in handling correspondence, attention to participants and shaping the manuscript for the book. Finally we express gratitude to Petra Steenbergen of Springer whose guidance in editorial matters was of much benefit for us.

Ankara, Turkey — Sinan Akkar
Ankara, Turkey — Polat Gülkan
De Bilt, The Netherlands — Torild van Eck
June, 2010

Contents

Contributors

Sinan Akkar Department of Civil Engineering, Earthquake Engineering Research Center, Middle East Technical University, 06531 Ankara, Turkey, sakkar@metu.edu.tr

G. Ameri Istituto Nazionale di Geofisica e Vulcanologia, Milano, Italy, ameri@mi.ingv.it

S. Aoi National Research Institute for Earth Science and Disaster Prevention, 3-1 Tennodai, Tsukuba, Ibaraki 305-0006, Japan, aoi@bosai.go.jp

D. Bindi Deutsches GeoForschungsZentrum GFZ, Potsdam, Germany, bindi@gfz-potsdam.de

J.J. Bommer Civil and Environmental Engineering Department, Imperial College London, London SW7 2AZ, UK, j.bommer@imperial.ac.uk

D.M. Boore U.S. Geological Survey (MS977), Menlo Park, CA 94025, USA, boore@usgs.gov

I.S. Borcia National Institute for Building Research (INCERC), Bucharest, Romania, isborcia@yahoo.com

Z. Çağnan Department of Civil Engineering, Middle East Technical University Northern Cyprus Campus, Kalkanlı, Güzelyurt, KKTC, Mersin 10, Turkey, cagnan@metu.edu.tr

K.W. Campbell EQECAT, Inc., Beaverton, OR, USA, kcampbell@eqecat.com

C. Cauzzi Swiss Federal Institute of Technology, Zurich, Schweiz Erdbebendienst (SED), Switzerland, carlo.cauzzi@sed.ethz.ch

M. Çelebi U.S. Geological Survey (MS977), Menlo Park, CA 94025, USA, celebi@usgs.gov

F.J. Chávez-García Instituto de Ingeniería, Universidad Nacional Autónoma de México, Ciudad Universitária, 04510 México, D.F. México

J. Clinton Swiss Seismological Service (SED – ETHZ), Sonneggstrasse 5, 8092 Zürich, Switzerland; Swiss Federal Institute of Technology, Zurich, Switzerland, j.clinton@sed.ethz.ch

I.G. Craifaleanu National Institute for Building Research (INCERC), Technical University of Civil Engineering (UTCB), Bucharest, Romania, i.craifaleanu@gmail.com

G. Cua Swiss Seismological Service (SED – ETHZ), Sonneggstrasse 5, 8092 Zürich, Switzerland; Swiss Federal Institute of Technology, Zurich, Switzerland, georgia.cua@sed.ethz.ch

B. Dost Seismology Division, Royal Netherlands Meteorological Institute, 3730 AE De Bilt, The Netherlands, bernard.dost@knmi.nl

J. Douglas BRGM – RNSC/RIS, 3 avenue C. Guillemin, BP 36009, 45060 ORLEANS Cedex 2, France, j.douglas@brgm.fr

D. Fäh Swiss Seismological Service (SED – ETHZ), Sonneggstrasse 5, 8092 Zürich, Switzerland; Swiss Federal Institute of Technology, Zurich, Switzerland, donat.faeh@sed.ethz.ch

L. Frobert EMSC/CSEM -European-Mediterranean Seismological Centre, Bruyères-le Châtel, France, frobert@emsc-csem.org

H. Fujiwara National Research Institute for Earth Science and Disaster Prevention, 3-1 Tennodai, Tsukuba, Ibaraki 305-0006, Japan, fujiwara@bosai.go.jp

D. Giardini Department of Earth Sciences, ETHZ, CH-8092 Zurich, Switzerland; Swiss Seismological Service (SED – ETHZ), Sonneggstrasse 5, 8092 Zürich, Switzerland, domenico.giardini@sed.ethz.ch

S. Godey European Mediterranean Seismological Center (EMSC), 91297 Arpajon, Cedex, France, godey@emcs-csem.org

X. Goula Institut Geològic de Catalunya, Barcelona, Spain, xgoula@igc.cat

P. Guéguen Laboratoire de Géophysique, Interne et Tectonophysique – Université de Grenoble, France, philippe.guegen@obs.ujf-grenoble.fr

Polat Gülkan Department of Civil Engineering, Earthquake Engineering Research Center, Middle East Technical University, 06531 Ankara, Turkey, a03516@metu.edu.tr

F. Haslinger Swiss Seismological Service (SED – ETHZ), Sonneggstrasse 5, 8092 Zürich, Switzerland; Swiss Federal Institute of Technology, Zurich, Switzerland, florian.haslinger@sed.ethz.ch

R. Jacquot LGIT/CNRS/LCPC – Joseph Fourier University, Grenoble, France, raphael.jacquot@obs.ujf-grenoble.fr

I. Kalogeras NOA-GI – National Observatory Athens, Geodynamic Institute, Athens, Greece

S. Koutrakis NOA-GI – National Observatory Athens, Geodynamic Institute, Athens, Greece

T. Kunugi National Research Institute for Earth Science and Disaster Prevention, 3-1 Tennodai, Tsukuba, Ibaraki 305-0006, Japan, kunugi@bosai.go.jp

L. Luzi Istituto Nazionale di Geofisica e Vulcanologia, Milano, Italy, luzi@mi.ingv.it

K. Makra ITSAK – Institute of Engineering Seismology and Earthquake Engineering, 55102 Thessaloniki, Greece

M. Manakou Department of Civil Engineering, Aristotle University of Thessaloniki, GR-54124 Thessaloniki, Greece

B. Margaris ITSAK – Institute of Engineering Seismology and Earthquake Engineering, 55102 Thessaloniki, Greece, margaris@itsak.gr

M. Massa Istituto Nazionale di Geofisica e Vulcanologia, Milano, Italy, massa@mi.ingv.it

C. Michel Swiss Seismological Service (SED – ETHZ), Sonneggstrasse 5, 8092 Zürich, Switzerland; Swiss Federal Institute of Technology, Zurich, Switzerland, clotaire.michel@sed.ethz.ch

H. Nakamura National Research Institute for Earth Science and Disaster Prevention, 3-1 Tennodai, Tsukuba, Ibaraki 305-0006, Japan, manta@bosai.go.jp

C.S. Oliveira Instituto Superior Técnico, Lisboa, Portugal, csoliv@civil.ist.utl.pt

M. Olivieri Swiss Seismological Service (SED – ETHZ), Sonneggstrasse 5, 8092 Zürich, Switzerland; Swiss Federal Institute of Technology, Zurich, Switzerland, marco.olivieri@sed.ethz.ch

F. Pacor Istituto Nazionale di Geofisica e Vulcanologia, Milano, Italy, pacor@mi.ingv.it

R. Paolucci Department of Structural Engineering, Politecnico di Milano, Milano, Italy, paolucci@stru.polimi.it

C. Pappaioanou Institute of Engineering, Seismology and Earthquake Engineering, Thessaloniki, Greece, chpapai@itsak.gr

C. Péquegnat LGIT/CNRS/LCPC – Joseph Fourier University, Grenoble, France, pequegna@obs.ujf-grenoble.fr

K. Pitilakis Department of Civil Engineering, Aristotle University of Thessaloniki, GR-54124 Thessaloniki, Greece, kpitilak@civil.auth.gr

I.C. Praun National Institute for Building Research (INCERC), Bucharest, Romania, praun@incerc.ro

R. Puglia Istituto Nazionale di Geofisica e Vulcanologia, Milano, Italy, puglia@mi.ingv.it

D. Raptakis Department of Civil Engineering, Aristotle University of Thessaloniki, GR-54124 Thessaloniki, Greece, raptakis@auth.gr

A. Roca Institut Geològic de Catalunya, Barcelona, Spain, aroca@igc.cat

E. Şafak Kandilli Observatory and Earthquake Research Institute, Bogazici University, Istanbul, Turkey, erdal.safak@boun.edu.tr

A. Savvaidis ITSAK – Institute of Engineering Seismology and Earthquake Engineering, 55102 Thessaloniki, Greece, alekos@itsak.gr

A. Skarlatoudis GeoLab – Geophysical Laboratory, University of Thessaloniki, Thessaloniki, Greece, askarlat@geo.auth.gr

R. Sleeman ORFEUS c/o Seismology Division, Royal Netherlands Meteorological Institute, 3730 AE De Bilt, The Netherlands, reinoud.sleeman@knmi.nl

T. Susagna Institut Geològic de Catalunya, Barcelona, Spain, tsusagna@igc.cat

N. Theodoulidis ITSAK – Institute of Engineering Seismology and Earthquake Engineering, 55102 Thessaloniki, Greece, ntheo@itsak.gr

Torild van Eck ORFEUS c/o Seismology Division, Royal Netherlands Meteorological Institute, 3730 AE De Bilt, The Netherlands, torild.van.eck@knmi.nl

C. Zülfikar Department of Earthquake Engineering, Kandilli Observatory and Earthquake Research Institute, Istanbul, Turkey, can.zulfikar@boun.edu.fr

P. Zweifel Swiss Seismological Service (SED – ETHZ), Sonneggstrasse 5, 8092 Zürich, Switzerland; Swiss Federal Institute of Technology, Zurich, Switzerland, peter.zweifel@sed.ethz.ch

Principal Contributors

S. Aoi (Tsukuba)

J.J. Bommer (London)

D.M. Boore (Menlo Park)

Z. Çağnan (Güzelyurt)

K.W. Campbell (Oregon)

M. Çelebi (Menlo Park)

J. Clinton (Zurich)

I.G. Craifaleanu (Bucharest)

J. Douglas (Orleans)

P. Gülkan (Ankara)

L. Luzi (Milan)

B. Margaris (Thessaloniki)

R. Paolucci (Milan)

C. Péquegnat (Grenoble)

K. Pitilakis (Thessaloniki)

A. Roca (Barcelona)

E. Şafak (Istanbul)

T. van Eck (De Bilt)

Co-Contributors

S. Akkar (Ankara)

G. Ameri (Milan)

D. Bindi (Potsdam)

I.S. Borcia (Bucharest)

C. Cauzzi (Zurich)

F.J. Chávez-García (México)

G. Cua (Zurich)

B. Dost (De Bilt)

D. Fäh (Zurich)

L. Frobert (Bruyères-le Châtel)

H. Fujiwara (Tsukuba)

D. Giardini (Zurich)

S. Godey (Bruyères-le Châtel)

X. Goula (Barcelona)

P. Guéguen (Grenoble)

F. Haslinger (Zurich)

R. Jacquot (Grenoble)

I. Kalogeras (Athens)

S. Koutrakis (Athens)

T. Kunugi (Tsukuba)

K. Makra (Thessaloniki)

M. Manakou (Thessaloniki)

M. Massa (Milan)

C. Michel (Zurich)

H. Nakamura (Tsukuba)

C.S. Oliveira (Lisbon)

M. Olivieri (Zurich)

F. Pacor (Milan)

C. Pappaioanou (Thessaloniki)

I.C. Praun (Bucharest)

R. Puglia (Milan)

D. Raptakis (Thessaloniki)

A. Savvaidis (Thessaloniki)

A. Skarlatoudis (Thessaloniki)

R. Sleeman (De Bilt)

T. Susagna (Barcelona)

N. Theodulidis (Thessaloniki)

C. Zülfikar (Istanbul)

P. Zweifel (Zurich)

Part I
Ground-Motion Predictive Models

Chapter 1
Ground-Motion Prediction Equations (GMPEs) from a Global Dataset: The PEER NGA Equations

D.M. Boore

Abstract The PEER NGA ground-motion prediction equations (GMPEs) were derived by five developer teams over several years, resulting in five sets of GMPEs. The teams used various subsets of a global database of ground motions and metadata from shallow earthquakes in tectonically active regions in the development of the equations. Since their publication, the predicted motions from these GMPEs have been compared with data from various parts of the world – data that largely were not used in the development of the GMPEs. The comparisons suggest that the NGA GMPEs are applicable globally for shallow earthquakes in tectonically active regions.

1.1 Introduction

The Pacific Earthquake Engineering Research (PEER) Center conducted a multi-year project (the "Next Generation Attenuation (NGA)" project) to derive ground-motion prediction equations (GMPEs, to use a term coined in Appendix A of [4]) from data collected globally for shallow earthquakes in active tectonic regions. Detailed descriptions of the project and the GMPEs are readily available in a special issue of Earthquake Spectra (Vol. 24, No. 1), as well as other papers, such as Campbell et al. [6], and for that reason this article only gives a capsule description of the essential details of the NGA project. New to this article will be a number of figures not contained in the Earthquake Spectra special issue. Because of length limitations, this article will primarily be composed of extended figure captions.

D.M. Boore (✉)
U.S. Geological Survey (MS977), Menlo Park, CA 94025, USA
e-mail: boore@usgs.gov

S. Akkar et al. (eds.), *Earthquake Data in Engineering Seismology*, Geotechnical, Geological, and Earthquake Engineering 14, DOI 10.1007/978-94-007-0152-6_1,

1.2 The NGA-Empirical Project

1.2.1 Personnel

The new GMPEs were derived by five developer teams: Abrahamson and Silva (AS), Boore and Atkinson (BA), Campbell and Bozorgnia (CB), Chiou and Youngs (CY), and Idriss (I). Because the Idriss GMPEs are of limited use (they are only for rock sites), they will not be discussed in this article. A number of working groups performed studies in support of the derivation of the GMPEs (see [10], for details).

1.2.2 Scope

The developer teams were given the task of developing GMPEs for a median measure of ground motion ("GMRotI50", as defined in [5]; in addition [2, 7, 14, 9] also provide equations to convert GMRotI50 to maximum spectral amplitude), including peak acceleration, velocity, and displacement, as well as 5%-damped pseudo spectral acceleration (PSA) for periods from 0.01 to 10s. The equations were to be valid for magnitudes ranging from 5.0 to 8.5 (for strike slip faulting) and 5.0 to 8.0 (for reverse slip faulting) and distances from 0 to 200 km. Models for the aleatory variability were to be included.

1.2.3 Database

A major effort was put into developing the database to be used by the developers (see [8], for details). Generally, the data are from shallow earthquakes located in tectonically active, shallow lithosphere, with reliable earthquake metadata being available.

1.3 Model Development

1.3.1 Dataset Selection

The developers used subsets of the full database, with justifications for the data not used. For example, BA excluded data from aftershocks, records for which metadata were missing or for which only one horizontal component was available, non "free-field" installations, etc. Figure 1.1 shows the magnitude-distance distribution of data for pga and for PSA at T=10.0s period. Note that there are many fewer data at longer periods than at shorter periods, a natural consequence of the low-cut filtering used in processing the data. Also note that there are no normal-fault data for T=10.0s. For these reasons, the GMPEs at longer periods (and for normal faults) will be less certain than at shorter periods.

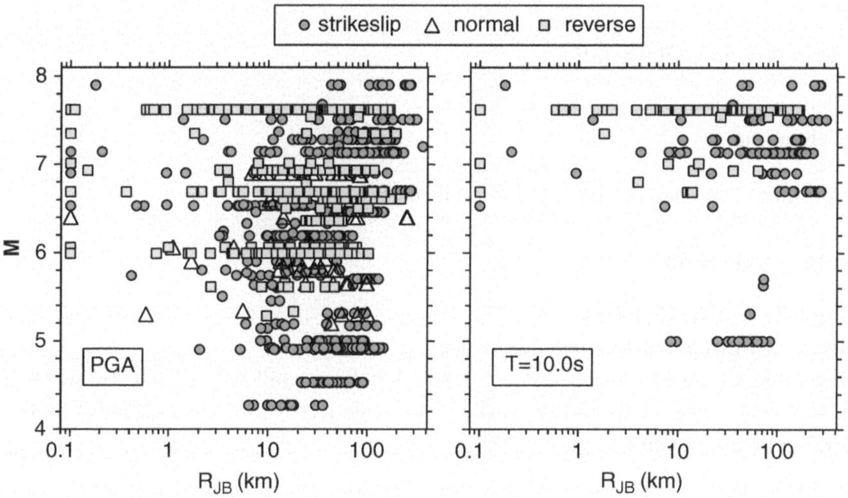

Fig. 1.1 Magnitude and distance distribution of data used by BA. Each symbol represents a recording

1.3.2 Functional Forms

Each developer team chose different functional forms for the GMPEs, in order to capture effects that they thought should be modeled. The functional forms are a tradeoff between simplicity of use and being able to represent the complexity in ground motions, due to many physical effects. Some of the effects captured in the functional forms are given below, as divided into source, path, and site contributions:

Source

- Fault mechanism (all developers)
- Aftershock vs. mainshock (AS, CY; aftershocks not used by BA, CB)
- Depth to top of rupture (Z_{TOR}) (all but BA)
- Magnitude scaling (all, different functions)
- Radiation pattern (not included)
- Directivity (not included)

Path

- Near-source (effect of fault size – CY)
- Far-source
  - Geometric spreading (single or multi-segment (CY), **M** (moment magnitude) dependent (all but CY))
  - Anelastic attenuation (included by all but CB, **M** dependent in CY)
  - Source-site geometry (“hanging wall effect”) (all but BA)

Site

- Near-surface geology (all)
  - Linear
  - Nonlinear
- Sediment thickness ("basin depth"; all but BA)

Aleatory Variability

- Inter (between) event (τ)
- Intra (within) event (σ): usually larger
- Dependence on **M** and/or rock reference-level motion? All but BA, but with different functions. Reduction of variance on soil sites for larger rock input motions makes physical sense.

The effects to be captured by the GMPEs were usually determined by a combination of exploratory data analysis and, for effects for which the recorded data are not sufficient to determine the effects, theoretical considerations. An example of the need for saturation in motions at short periods as magnitude increases is given in Fig. 1.2.

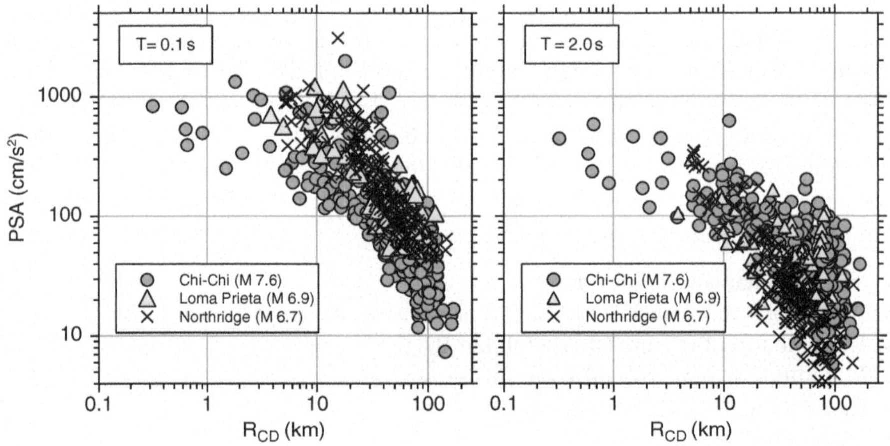

Fig. 1.2 Ground motions for three earthquakes for two oscillator periods. Note that the motions from the M 7.6 Chi–Chi earthquake are smaller and larger than for the smaller earthquakes at short and long period, respectively

In all but the Idriss GMPEs, the site response is considered to be nonlinear. Because the amount of nonlinearity and the amplitude of the site response are functions of both period and the V_{S30} (the variable chosen to characterize the site), the functional forms required to account for these effects are generally complex. Figure 1.3 shows the site response for the BA equations, clearly showing the nonlinear site response.

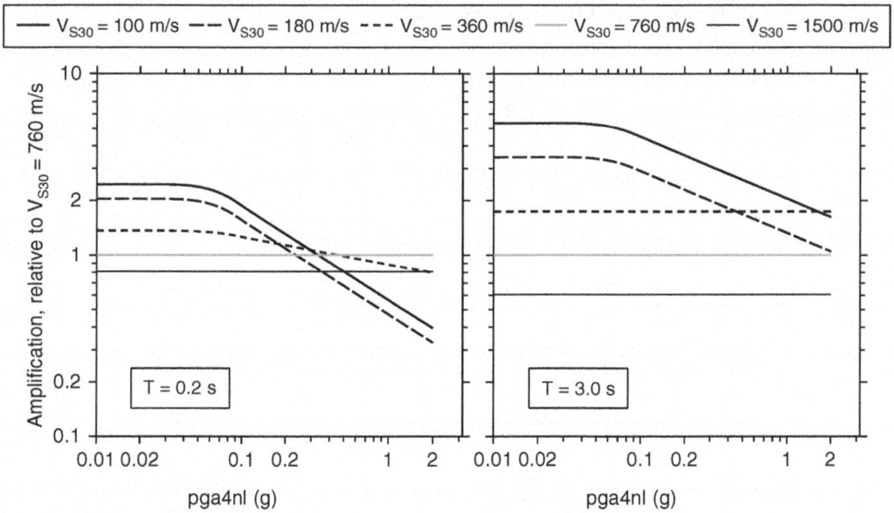

Fig. 1.3 Site response for the BA GMPEs. Note the larger amplifications for long periods and the nonlinear response at both periods, particularly for softer sites

Table 1.1 Effects included in the GMPEs of the NGA developer teams. ("emp"=based on analysis of empirical data; "RV"= reverse-slip fault)

Effect	AS08	BA08	CB08	CY08	I08
Saturation at short distances	X	X	X	X	X
Style-of-faulting	X	X	X	X	X
Rupture depth factor	X		X (RV only)	X	
Hanging wall factor	X		X	X	
Nonlinear site amp	X	X	X	X (emp)	
Sed. depth factor	X		X	X (emp)	
M-dependent σ	X			X	X
Nonlinear effects on σ	σ, τ		σ	σ, τ	

As with site amplification, capturing a number of the physical effects requires complex GMPEs. This is a necessary consequence of going beyond simple magnitude, distance, and scalar site effects as predictor variables. Table 1.1 summarizes the effects included in the GMPEs for the various developer teams.

The predictor variables in the GMPEs include the following:

M	Moment magnitude
R_{RUP}	Closest distance to coseismic rupture (km)
R_{JB}	Closest distance to surface projection of coseismic rupture (km)
R_X	Horizontal distance from top edge of fault perpendicular to strike (km)
Z_{TOR}	Depth to top of coseismic rupture (km)
	Fault type (depends on rake or P, T plunges)

F_{AS}	1 for aftershocks, 0 for mainshocks
Dip	Average dip of rupture plane (degrees)
W	Downdip rupture width (km)
V_{S30}	Average shear-wave velocity in top 30 m of site profile (m/s)
$Z_{1.0}$	Depth to 1.0 km/s shear-wave velocity horizon (m)
$Z_{2.5}$	Basin (Sediment) depth; depth to 2.5 km/s shear-wave velocity horizon (km)
Period	Spectral period for PSA (s); 0 for PGA, –1 for PGV

Table 1.2 gives the predictor variables used by each developer team:

Table 1.2 Predictor variables in the GMPEs for each developer team

Predictor variable	AS	BA	CB	CY
M	X	X	X	X
R_{RUP}	X		X	X
R_{JB}	X	X	X	X
R_X	X			X
Z_{TOR}	X		X	X
Fault type	X	X	X	X
F_{AS}	X			X
Dip	X		X	X
W	X			
V_{S30}	X	X	X	X
$Z_{1.0}$	X			X
$Z_{2.5}$			X	
Period	X	X	X	X

One thing to keep in mind is that the predictor variables can be correlated, and the effect of a variable not included in GMPEs can be captured by a variable in the equations with which it is correlated. A good example is V_{S30} and basin depth. Figure 1.4 shows that the two are correlated:

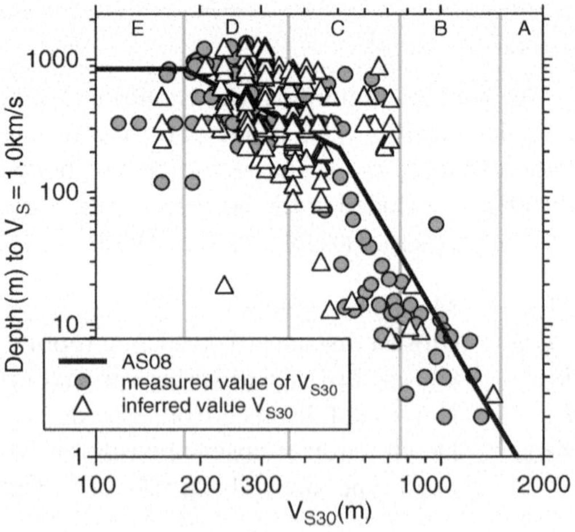

Fig. 1.4 Correlation of depth of basin and VS30

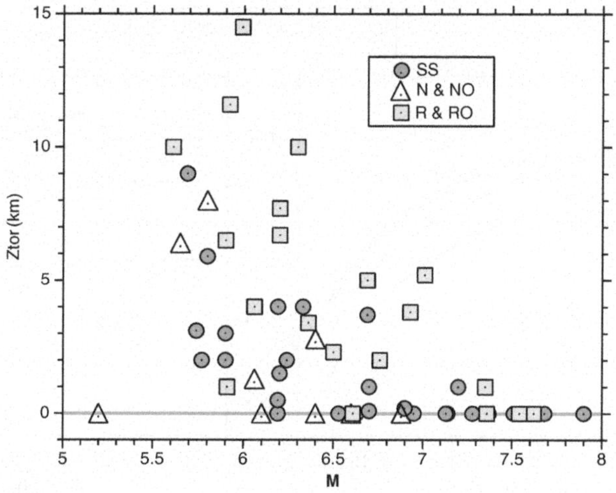

Fig. 1.5 Correlation of fault type and magnitude with depth-to-top of rupture (SS, strike slip; N, normal; NO, normal oblique; R, reverse; RO, reverse oblique)

Another example of correlations is given in Fig. 1.5, which shows that fault type, magnitude, and depth-to-top of the rupture surface are correlated. In particular, note that few reverse-slip faults reach the surface, and that almost all earthquakes with magnitudes larger than 7.0 do reach the surface. Thus it might be difficult to unravel the physical effects associated with fault type, magnitude, or depth-to-top of rupture based on empirical analysis alone.

1.4 Results

Figure 1.6 shows PSA at two periods from the BA GMPEs (the other GMPEs give similar results) as a function of distance for four magnitudes and a rock-like site. This figure shows the saturation with magnitude (which is more pronounced for short periods than for long periods), as well as the stronger magnitude dependence at long periods than short periods.

The saturation with magnitude is also shown in Fig. 1.7, which plots the average motions for each event, reduced to a common distance using the BA GMPEs, versus magnitude. This figure also shows the fault-type dependence of the motions.

The effect of site characterization is shown in Fig. 1.8, again using the BA GMPEs. Note the effect of the nonlinear response, which results in a decrease in short-period motions on softer sites.

A detailed comparison of the GMPEs for all NGA developer teams is given in Abrahamson et al. [1]. Here I show a few sample comparisons. Figure 1.9 compares the spectra as a function of period for strikeslip and reverse slip faults, for **M** 7 and R_{JB} = 10 km. In spite of the differences in datasets and functional forms, the

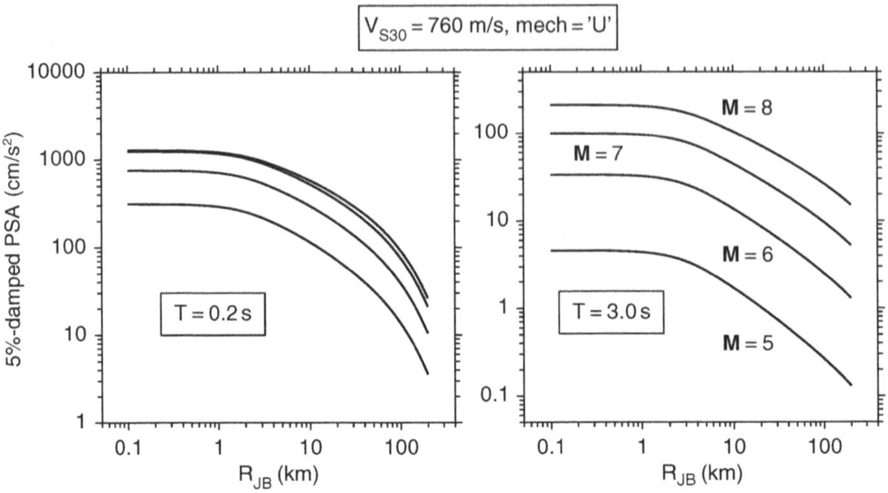

Fig. 1.6 PSA vs. distance, from BA GMPEs

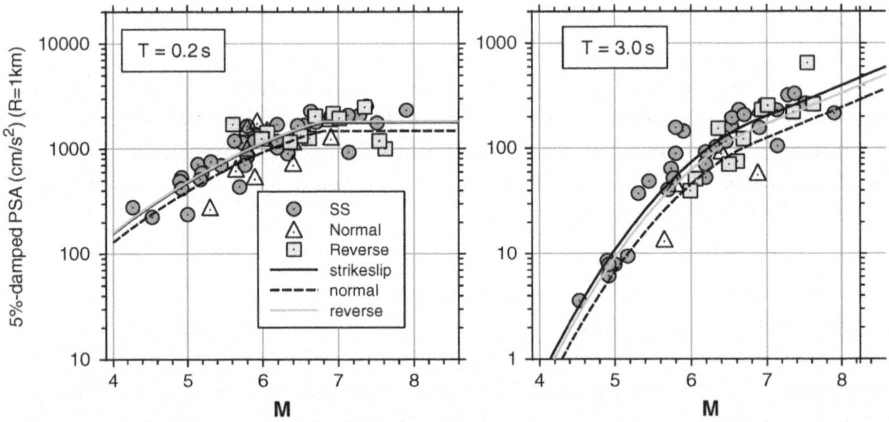

Fig. 1.7 The symbols are average *PSA* for each event reduced to 1 km, using BA GMPEs, as a function of magnitude (one symbol per event). The *curves* show the magnitude dependence in the BA GMPEs

different GMPEs give generally similar values of PSA, at least for the magnitude, distance, and site condition used for this figure (there can be larger differences for other distances, magnitudes, and site conditions).

One of the physical effects included in most of the GMPEs is the effect of being over the hanging wall of a fault (where observations and laboratory models suggest that the motions should be larger than for sites not over the hanging wall).

Figure 1.10 shows the PSA for a 1 s oscillator for a scenario case, in which the fault extends at a 45° dip to the right, from the surface to a depth of 15 km (the fault crops out at 0 km distance).

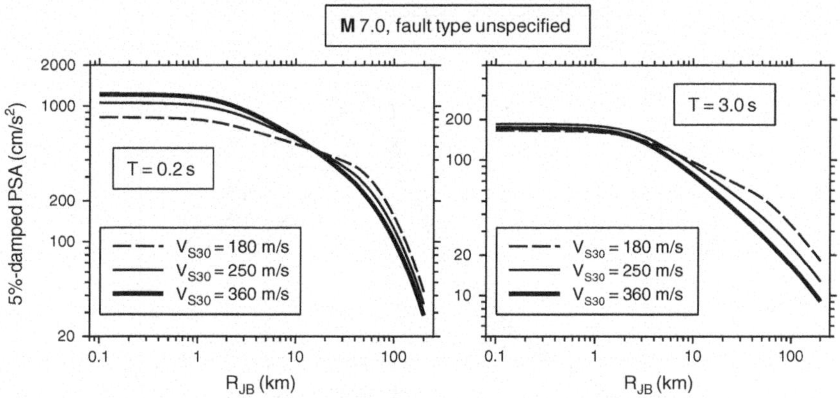

Fig. 1.8 *PSA* vs. distance, from BA GMPEs

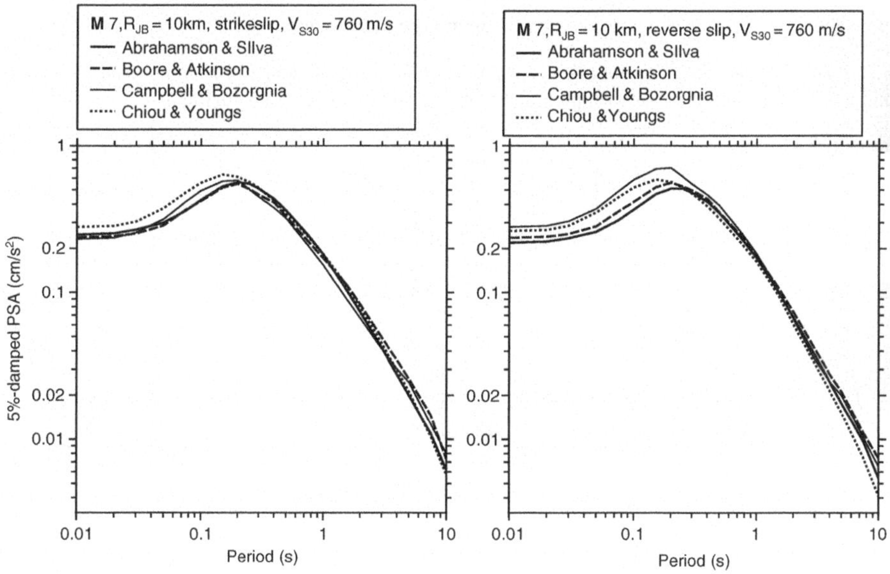

Fig. 1.9 *PSA* vs. period

Another comparison of the NGA GMPEs is shown in Fig. 1.11, which plots the event residuals as a function of magnitude (the residuals for the AS GMPEs were not available). Each symbol represents the average of the difference between observed and predicted motions for each event; the residuals for the 1999 Chi–Chi mainshock have been identified. These graphs show that the residuals have a similar scatter for the three GMPEs, but the values are not identical.

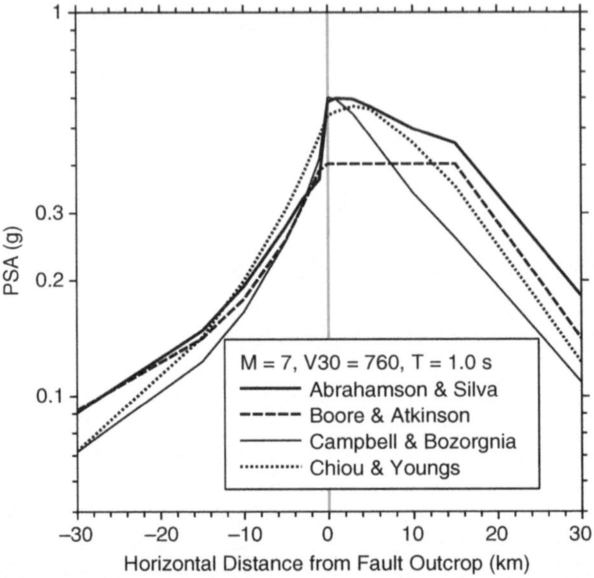

Fig. 1.10 *PSA* vs. distance, for sites along a transect perpendicular to the fault strike, over the midpoint of a reverse-slip fault dipping at 45° to the right, from 0 to 15 km depth

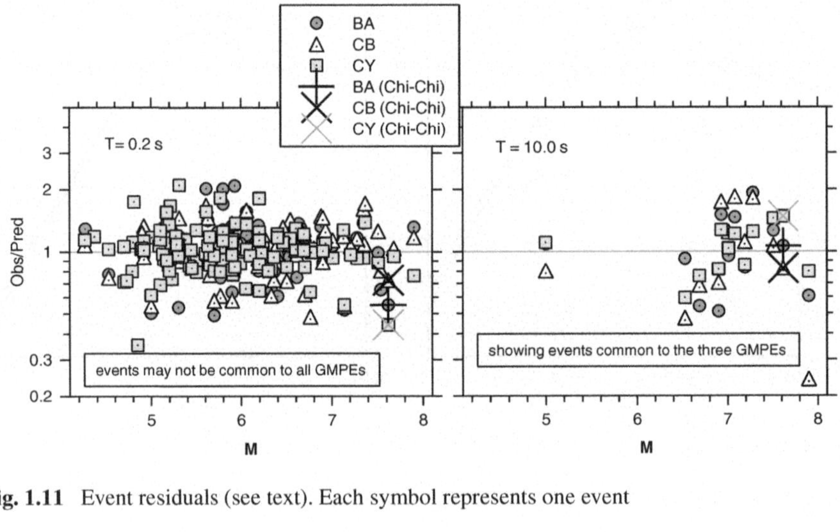

Fig. 1.11 Event residuals (see text). Each symbol represents one event

1.5 Comparisons with Other Studies

A number of studies have compared motion from the NGA GMPEs with data generally different than used in the derivation of the NGA GMPEs. The first such comparison was by Stafford et al. [13], who concluded "The analyses indicate

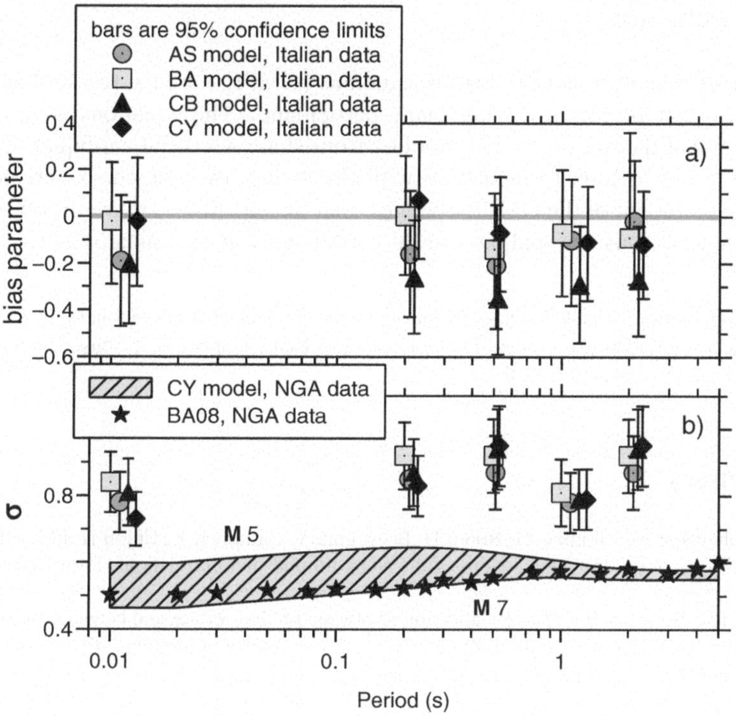

Fig. 1.12 **a** Bias in median ground motions between Italian strong-motion data and predictions from NGA GMPEs and **b** aleatory uncertainty for Italian data (relative to NGA GMPEs) and for the BA and CY GMPEs (the latter is magnitude dependent; shown are the values for magnitudes ranging from 5 to 7)

that for most engineering applications, and particularly for displacement-based approaches to seismic design, the NGA models may confidently be applied within Europe." Another study is that of Shoja-Taheri et al. [12], who compare the ground motions from the NGA GMPEs with data from Iran; they find that the mean residuals (log observed – log predicted) are close to unity, indicating that the Iranian strong-motion data are consistent with the NGA GMPEs. In a third comparison, Bindi et al. [3] showed that Italian strong-motion data are consistent with the NGA GMPEs. This is shown more quantitatively in Fig. 1.12, which is redrafted from Scasserra et al. [11]. The bias for median ground motions (Fig. 1.12a) is generally close to zero for most of the NGA GMPEs.

The standard deviation of Italian data is much higher than associated with the NGA GMPEs, as shown in Fig. 1.12b (that figure also shows that the magnitude-independent aleatory uncertainty associated with the BA GMPEs is similar to the **M** 7 aleatory uncertainty for the CY GMPEs). The uncertainties for the Italian data are relative to the NGA GMPEs because GMPEs for the Italian data were not available to the authors; the comparisons shown in Fig. 1.12a suggest that similar results would have been obtained if GMPEs fit to the Italian data had been used.

1.6 Conclusions

The NGA ground-motion prediction equations (GMPEs) are the result of an intensive multi-year project involving many participants. The equations were derived using a global dataset of ground motions from shallow crustal earthquakes in tectonically active regions. Comparisons with global data, most of which were not used in the development of the GMPEs, suggest that the equations are useful for predicting ground-motions for shallow crustal earthquakes in tectonically active regions worldwide.

Acknowledgments I thank Jon Stewart for providing the data used to construct Fig. 1.12, Sinan Akkar and Erol Kalkan for reviews of the paper, and Ken Campbell and Yousef Bozorgnia for material that helped in preparing this paper.

References

1. Abrahamson N, Atkinson G, Boore D, Bozorgnia Y, Campbell K, Chiou B, Idriss IM, Silva W, Youngs R (2008) Comparisons of the NGA ground-motion relations. Earthquake Spectra 24:45–66
2. Beyer K, Bommer JJ (2006) Relationships between median values and between aleatory variabilities for different definitions of the horizontal component of motion. Bull Seismol Soc Am 96:1512–1522
3. Bindi D, Luzi L, Massa M, Pacor F (2010) Horizontal and vertical ground motion prediction equations derived from the Italian accelerometric archive (ITACA). Bull Earthquake Eng 8: 1209–1230
4. Boore DM, Atkinson GM (2007). Boore-Atkinson NGA Ground Motion Relations for the Geometric Mean Horizontal Component of Peak and Spectral Ground Motion Parameters, *PEER Report 2007/01*, Pacific Earthquake Engineering Research Center, University of California, Berkeley, 234 pp.
5. Boore DM, Watson-Lamprey J, Abrahamson NA (2006) Orientation-independent measures of ground motion. Bull Seismol Soc Am 96:1502–1511
6. Campbell K, Abrahamson N, Power M, Chiou B, Bozorgnia Y, Shantz T, Roblee C (2009). Next Generation Attenuation (NGA) project: empirical ground motion prediction equations for active tectonic regions, Sixth International Conference on Urban Earthquake Engineering, March 3–4, 2009, Tokyo Institute of Technology, Tokyo, Japan (available from http://www.cuee.titech.ac.jp/Japanese/Publications/Doc/conference_6th.pdf)
7. Campbell KW, Bozorgnia Y (2007). Campbell-Bozorgnia NGA Ground Motion Relations for the Geometric Mean Horizontal Component of Peak and Spectral Ground Motion Parameters, *PEER Report 2007/02*, Pacific Earthquake Engineering Research Center, University of California, Berkeley, 238 pp.
8. Chiou B, Darragh R, Gregor N, Silva W (2008) NGA project strong-motion database. Earthquake Spectra 24:23–44
9. Huang Y-N, Whittaker AS, Luco N (2008) Maximum spectral demands in the near-fault region. Earthquake Spectra 24:319–341
10. Power M, Chiou B, Abrahamson N, Bozorgnia Y, Shantz T, Roblee C (2008) An overview of the NGA project. Earthquake Spectra 24:3–21
11. Scasserra G, Stewart JP, Bazzurro P, Lanzo G, Mollaioli F (2009) A comparison of NGA ground-motion prediction equations to Italian data. Bull Seismol Soc Am 99: 2961–2978

12. Shoja-Taheri J, Naserieh S, Ghofrani H (2010) A test of the applicability of NGA models to the strong ground-motion data in the Iranian plateau. J Earthquake Eng 14:278–292
13. Stafford PJ, Strasser FO, Bommer JJ (2008) An evaluation of the applicability of the NGA models to ground-motion prediction in the Euro-Mediterranean region. Bull Earthquake Eng 6:149–177
14. Watson-Lamprey JA, Boore DM (2007) Beyond SaGMRotI: conversion to Sa_{Arb}, Sa_{SN}, and Sa_{MaxRot}. Bull Seismol Soc Am 97:1511–1524

Chapter 2
Ground-Motion Models for Defining Seismic Actions in Eurocode 8

J.J. Bommer

Abstract One of the long-term aims of Eurocode 8 is to achieve uniform levels of seismic risk throughout the European Community. Although in its current implementation each country adopting the code develops its own seismic zonation map, the objective of harmonization of risk requires that there must be a harmonized seismic hazard map. A key component of such a harmonized hazard map is a coordinated approach to deriving, selecting or adjusting ground-motion prediction equations, whilst recognizing that there are compelling reasons for different combinations of equations to be used in different regions. Since such a unified hazard map is not likely to be the basis for defining seismic actions in Eurocode 8 at least until the first major revision of the code, the approach to defining ground-motion models for the preparation of seismic hazard maps should also contemplate the possibility that future editions of the code will go beyond the specification of elastic design spectra anchored only to the peak ground acceleration, PGA.

2.1 Harmonization of Seismic Hazard Assessment in Europe

Eurocode 8 is one of a series of European Standards for Civil Engineering design of buildings and civil works, known as the Structural Eurocodes, which are to be implemented throughout the European Community. Eurocode 8, or EC8, is entitled "Design of Structures for Earthquake Resistance", and it is intended to provide seismic safety in all structures. As each country adopts the Eurocodes, they are effectively transformed into national codes which must include the full Eurocode as published by CEN (Comité Européen de Normalisation) "which may be preceded by a National title page and National foreword, and may be followed by a National annex (informative)" [31]. The National annex "may only contain information on those parameters which are left open in the Eurocode for national choice, known

J.J. Bommer (✉)
Civil and Environmental Engineering Department, Imperial College London,
London SW7 2AZ, UK
e-mail: j.bommer@imperial.ac.uk

S. Akkar et al. (eds.), *Earthquake Data in Engineering Seismology*, Geotechnical, Geological, and Earthquake Engineering 14, DOI 10.1007/978-94-007-0152-6_2,

as National Determined Parameters, to be used for the design of buildings and civil engineering works to be constructed in the country concerned" [31]. In the case of EC8, the key information provided in the National annexes will be the seismic zonation map showing values of peak ground acceleration (PGA) which, in combination with classification of the site in terms of the near-surface geology, defines the basic seismic actions to be considered in structural design. This basic design acceleration, represented by the symbol a_g in the code, may be increased for the design of high-occupancy buildings and structures whose integrity is important for civil protection, through application of an importance factor γ_I. The code provides very little in the way of guidance or specifications for the preparation of these zonation maps other than to specify that it should be the peak ground acceleration in rock with a specified probability of exceedance, which for the no-collapse performance target (i.e., for life safety) is recommended as being a 10% probability of exceedance in 50 years, equivalent to a return period of 475 years. The code also implies that the definition of the horizontal component of the ground acceleration is the widely-used geometric mean of the two horizontal components from each accelerogram: "the horizontal seismic action is described by two orthogonal components assumed as being independent and represented by the same response spectrum" [31]. Therefore, as a bare minimum, all that is required to develop the zonation map for a National annex is a locally-applicable equation for the prediction of PGA in rock, defined as sites with an average shear-wave velocity over the uppermost 30 m of 800 m/s or greater. If the chosen equation predicts values of PGA that do not correspond to the geometric mean component, these can be easily converted using empirical relationships such as those of Beyer and Bommer [13].

The state of national seismic hazard mapping in Europe varies significantly from one country to another. A review conducted by García-Mayordomo et al. [36] of national seismic hazard maps for design codes in 16 European countries revealed that the approaches included some cases of deterministic rather than probabilistic seismic hazard analysis, the use of return periods other than 475 years, and that one quarter of the countries based their hazard maps on predictive equations for macroseismic intensity rather than PGA. Although the review by García-Mayordomo et al. [36] showed that the state of practice in national seismic hazard mapping varied greatly across Europe, some countries have derived new hazard maps in recent years based on fully probabilistic seismic hazard analyses (PSHA) including logic-tree formulations to capture the epistemic uncertainty in both the seismic source and ground-motion prediction models. In these cases, the hazard maps have been generated using several ground-motion prediction equations combined in a logic tree with weights assigned to reflect the relative confidence of the analysts' in their applicability to the target region (e.g. [22]). One example of such a hazard map generated specifically for a National annex of EC8 is the study by Musson and Sargeant [50] for the UK, a country of rather low seismic activity. The logic-tree for the UK PSHA included two equally-weighted ground-motion prediction equations (GMPEs), namely an early version of the NGA model of Campbell and Bozorgnia [30] and the European model of Bommer et al. [24]. For the seismic hazard map of Switzerland [64], a country of somewhat higher seismic activity, the adopted GMPE was the stochastic model of Bay et al. [11, 12] for Switzerland. A logic-tree formulation was also adopted but the branches were for alternative extrapolations of this stochastic model to magnitudes greater than those covered by the dataset from which the model was derived. For the new seismic hazard map of Italy [48], one of the most seismically active countries in Europe, a logic-tree was also adopted that included empirical GMPEs for Italy [54] and Europe [8], together with stochastic models for different regions within the country (e.g. [44, 45]). These recent studies are, on the one hand, very encouraging because they reflect the application of PSHA accounting for both aleatory variability and epistemic uncertainty in ground-motion predictions. On the other hand, they illustrate that approaches to adopting and combining GMPEs for such studies vary from one country to another, whence even if there were a pan-European seismic source characterization model, the resulting national hazard maps could show discontinuities and incompatibilities at their borders. The solution clearly does not have to lie in the adoption of a single logic-tree formulation for GMPEs to be used throughout Europe, but rather a coherent approach to the development of the ground-motion branches of the logic-tree for PSHA.

In the long term, it might be hoped that rather than each country adopting EC8 developing its own seismic zonation map in a National annex there will be a single seismic zonation map for all of Europe. Similar pan-European maps for other environmental loads, such as snow and wind (e.g. [32]), have been discussed in the European civil engineering community, with a view to their being included in Eurocode 0 "Basis for Design" or Eurocode 1 "Actions on Structures", but this has yet to be realized in practice. Nonetheless, pan-European seismic hazard maps have been produced, firstly within the Global Seismic Hazard Assessment Project (GSHAP). This mapping project was actually developed through the compilation and integration of a series of regional mapping projects, one being for Central, North and Northwest Europe [37]. The PSHA deployed logic-trees for ground-motion prediction equations, which varied according the tectonic framework of different areas: five equally-weighted models for the Fennoscnandian shield, two of which were originally developed for Eastern North America [10, 62]; a single model for the Vrancea region [43]; and three equally-weighted models in the remaining areas [8, 54, 59]. Reflecting the diversity of approaches across the broader European region, the mapping project for Northern Eurasia [63] was based on a single equation for the prediction of macroseismic intensity, subsequently transformed to PGA via an empirical relationship. In the mapping project for the Ibero-Maghreb region [39], it was revealed that each country was originally using a different ground-motion prediction equation (some in terms of intensity, others in terms of PGA), with those used for the major earthquakes affecting the Luso-Iberian peninsula, such as the 1755 Lisbon event, reflecting the low rate of attenuation expected for this predominantly oceanic source-to-site path. However, the final hazard map for the Ibero-Maghreb region was generated using a single predictive equation, namely that for California derived by Joyner and Boore [40]. The seismic hazard map for Italy and the countries on the eastern coast of the Adriatic Sea [58] used the pan-European equation of Ambraseys et al. [8]. The mapping of the North Balkan region [49] used the pan-European model of Ambraseys and Bommer [6] except for the Vrancea region where modified versions of azimuth-dependent equations by Lungu et al. [42] were adopted. Finally, the hazard map for Turkey [35] used three equally-weighted models from western North America, namely Boore et al. [26], Campbell [29] and Sadigh et al. [55]. After the completion of GSHAP, another project was launched, SESAME ("Seismotectonic and Seismic Hazard Assessment of the Mediterranean Basin"), which resulted in a new seismic hazard map, published by the European Seismological Commission, showing 475-year PGA values throughout Europe. The SESAME map was produced using primarily the pan-European equations of Ambraseys et al. [8], with other models deployed where seismicity occurs at sub-crustal depths, namely the Vrancea region [49] and the Hellenic subduction trench [52].

In 2009, a 3-year research project funded under the Seventh Framework Programme of the Council of European Communities, called SHARE ("Seismic Hazard Harmonization in Europe") was launched, aiming to address the inconsistencies that have existed to date in European seismic hazard assessments.

This paper provides a brief overview of the current state-of-the-art in ground-motion prediction in Europe, and discusses the issues to be resolved in developing a harmonized approach to the development and selection of GMPEs for seismic hazard analysis in the region. The paper also presents a critical review of the current specification of seismic actions in EC8; since revisions and updates of EC8 are expected to be implemented every 5 years or so, the paper looks to the next version of the code and proposes a new formulation for the presentation of seismic design loads that can overcome some of the identified shortcomings.

2.2 Current Status of Ground-Motion Prediction in Europe

Although most seismic hazard mapping for Europe to date has been performed only in terms of PGA, and despite the fact that the current formulation of EC8 only requires a zonation map in terms of this parameter, for ground-motion prediction models to be genuinely useful for engineering purposes they must also predict response spectral ordinates. The first pan-European equations for predicting spectral ordinates were those of Ambraseys et al. [8], which were modified by Bommer et al. [17] to include the influence of style-of-faulting. New European equations were presented by Ambraseys et al. [7] using an expanded database. Akkar and Bommer [2] re-processed the same accelerograms in order to extend the usable period range of the data, and then derived equations for spectral ordinates, PGA and PGV [3, 4]. Both the Ambraseys et al. [7] and Akkar and Bommer [3, 4] equations adopted a model with magnitude-dependent variability, which was subsequently shown to yield unstable predictions whence a new model for pseudo-spectral accelerations, PSA, PGA and PGV was derived by Akkar and Bommer [5] assuming magnitude-independent sigma values.

In addition to these pan-European equations, many GMPEs have been derived for application only within the borders of one country within the region, or in some cases only within a particular part of that country [23]. Space limitation in this paper precludes a comprehensive review of national and regional models for Europe, but it can be stated that they tend, necessarily, to be based on much smaller datasets than those used for the derivation of the pan-European equations. Consequently, whilst these models may faithfully capture ground motions in the target region for the magnitude-distance scenarios represented in the dataset, they will generally be less well-constrained for the full range of magnitudes and distances considered in PSHA calculations.

An interesting point to note is that Stafford et al. [61], using an updated version of the approach of Scherbaum et al. [56] to test the applicability of GMPEs using recordings from the target region, showed that the models derived primarily for application in coastal California within the Next Generation of Attenuation (NGA) project [1, 25, 30, 33] provide a good fit to the European strong-motion data and therefore can be considered to be applicable in the seismically active parts of Europe, the Middle East and North Africa.

For PSHA in Europe, one of the key issues is to select ground-motion prediction equations, combined within a logic-tree framework, that collectively capture the likely range of ground motions in future earthquakes in the region. Cotton et al. [34] proposed an approach to this challenge that begins by collating all available GMPEs and then excluding those that either are clearly not applicable in terms of tectonic regime (for example, using equations derived for shallow crustal earthquakes to estimate motions from subduction events) or judged to fail on criteria related to their quality and stability. Bommer et al. [16] have expanded the list of quality criteria, as well as adding more specific requirements to some of the original criteria. Many of these criteria are related to providing good coverage over a reasonably wide range of magnitudes and distances, and having a functional form amenable to extrapolation (i.e. non-linear magnitude scaling and magnitude-dependent distance decay). With this is mind, the question arises whether small national datasets should be used to derive national GMPEs through regression analyses, or whether such data should be used either for testing the applicability of more global models, using methods such those of Scherbaum et al. [56] or [57], or else to adjust such models in a way that retains the well-constrained scaling and attenuation functions (e.g. [9]). Clearly, the degree of rigor with which the exclusion criteria are applied to reduce the list of available equations may need to be significantly relaxed for those regions where less candidate models exist including the Hellenic subduction arc, the Vrancea region, offshore Portugal and Spain, and stable areas of north and northwestern Europe.

2.3 Current Definition of Seismic Actions in Eurocode 8

The basic representation of seismic actions in EC8, in common with the majority of seismic design codes around the world, is an elastic response spectrum of absolute acceleration with 5% of critical damping. The EC8 spectra do include some noteworthy features, including the definition of a separate spectral shape for the vertical component unlike many codes which use a constant scaling factor to transform the horizontal spectrum into the vertical component, an approach that does not reflect the nature of vertical response spectral shapes [27]. However, the EC8 spectra also have a number of serious shortcomings, as have been identified by Bommer and Pinho [20]. The most serious of these is that the spectral shape is fixed for each site class and scaled only with PGA, and hence provides a poor approximation to the uniform hazard spectrum (UHS) since the influence of magnitude on spectral shape is not accounted for, a problem originally identified more than 30 years ago by McGuire [46]. Eurocode 8 presents two spectral shapes, Type 2 being intended for lower seismicity regions where seismic hazard is associated with earthquakes of M_S 5.5 or smaller, but each national code needs to adopt either the Type 1 or Type 2 spectrum for each country or region within the country. The two spectral shapes allow a slightly better approximation to the UHS than if a single spectrum were specified throughout the European region, but the spectral shapes do not reflect the full influence of magnitude (Fig. 2.1). The two EC8 spectra are an inelegant and ineffective attempt to define UHS, which would have been much better addressed by mapping a second parameter in addition to PGA.

Another shortcoming is the control period, T_D, beyond which the acceleration ordinates decay in proportion to the reciprocal of the squared period, thus giving a constant displacement plateau. Even though this period has previously been underestimated in Europe because of the dominance of analog recordings in the strong-motion databank and the consequent application of excessively severe filter cut-offs, the value specified in EC8 of just 2 s is almost ridiculously low.

Another problem is related to the scaling factor applied to transform the default 5%-damped spectral ordinates to other damping ratios, which is a function only of the response period and the target damping level. Bommer and Mendis [19] showed that these scaling factors should also vary with the duration of the ground motion, which is reflected in a dependence on magnitude and distance. This led to a proposed modification of the scaling factors to be applied to the Type 2 spectrum to

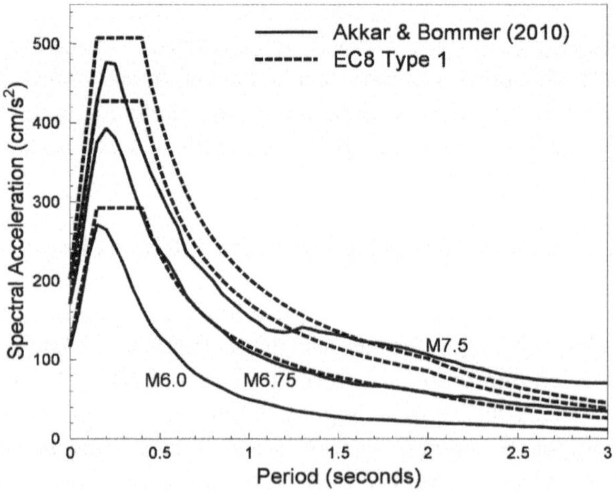

Fig. 2.1 Comparison of predicted median spectral ordinates from Akkar and Bommer [5] and the EC8 Type 1 spectra anchored to PGA for strike-slip earthquakes of three different magnitudes at 15 km from rock sites

account for the shorter duration of motions from the smaller magnitude events that correspond to this spectrum [47].

Eurocode 8 also provides guidance on the selection of acceleration time-histories for use in dynamic analyses of structures. The specifications share many of the shortcomings identified in other codes in this respect, the main limitation being that the code provides no information about the earthquake scenarios underlying the UHS [21]. Notwithstanding the inherent limitations, Iervolino et al. [38] present a procedure to obtain accelerograms from the European Strong-Motion Database that satisfy the criteria specified in EC8.

2.4 A New Formulation for Seismic Actions in Eurocode 8

Current practice, for example, in the United States seismic design codes is based on mapping three parameters to construct the elastic response spectrum, namely PGA and spectral accelerations at 0.2 and 1.0 s [41]. This approach could easily be adopted in future editions of EC8, but this would require GMPEs for spectral ordinates for all parts of Europe and the acceptance of two additional zonation maps. As a compromise that requires the introduction of only one more parameter, Bommer et al. [23] proposed adding zonation maps of PGV in addition to those for PGA. One of the advantages this offers is that the ratio of these two parameters is effectively a surrogate for the earthquake magnitude (Fig. 2.2). The curves in this figure show that the average PGV/PGA ratio is almost entirely insensitive to style-of-faulting

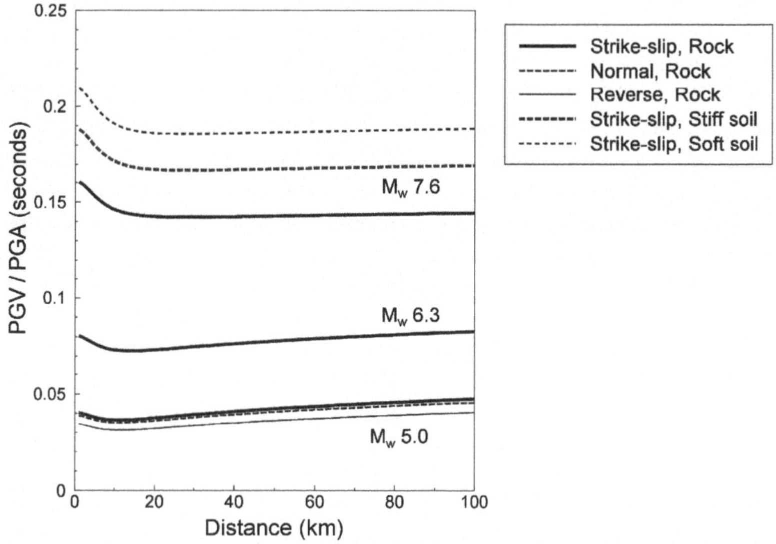

Fig. 2.2 Ratios of predicted median PGV to PGA from Akkar and Bommer [5] for different combinations of magnitude, style-of-faulting, distance and site classification

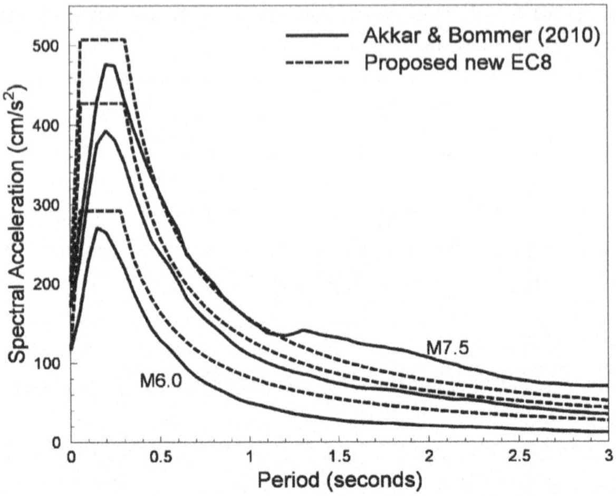

Fig. 2.3 Comparison of predicted median spectral ordinates from Akkar and Bommer [5] and the proposed new formulation for EC8 spectra anchored based on PGV/PGA ratios as proposed by Bommer et al. [23] for strike-slip earthquakes of three different magnitudes at 15 km from rock sites

and distance but strongly dependent on magnitude and site classification. The formulation of the response spectrum in the code can account directly for the influence of site class (e.g. [53], whence defining the control periods for a given site class as a function of the PGV/PGA ratio allows the influence of earthquake magnitude on the spectral shape to be incorporated into the definition of the seismic actions (Fig. 2.3). The spectra in Fig. 2.3 show that the formulation improves the approximation of the spectra (cf. Fig. 2.1) although it needs to be noted that the real test must be made comparing the proposed new EC8 spectra against UHS rather than median predicted ordinates. The formulation can always be refined to improve the approximation, which was originally derived using the models of Akkar and Bommer [3, 4] – and it can be seen that for M_W 7.5 the long-period ordinates are slightly underestimated – but the purpose here is to illustrate a possible approach rather than a definitive set of equations ready for implementation. Clearly at intermediate response periods (say 0.3–1.0 s) the agreement of the code and predicted spectra is greatly improved.

As well as the control periods that determine the spectral shape, including the constant-displacement plateau, the PGV/PGA ratio can also be used to refine the definition of the scaling factors for other damping levels [23] acting as a surrogate for such relationships defined in terms of magnitude [28] or duration [60].

The concept of using the PGV/PGA ratio is not new, and in many ways the formulation proposed by Bommer et al. [23] is nothing more than an update of the approach of Bommer et al. [18], which was in itself an adaptation of the spectral construction technique of Newmark and Hall [51]. Although on the one hand the proposed scheme does represent a tool of convenience, it also raises a challenge for application in the short-term since whilst there is an abundance of GMPEs for PGA and spectral ordinates, until now comparatively few models have been developed for the prediction of PGV despite the large number of applications that this parameter has in engineering seismology and earthquake engineering [15]. To implement the approach proposed herein, GMPEs for PGV in all regions of Europe (including both the active and stable regions, as well as special cases such as the Hellenic subduction arc, the Vrancea region, offshore Portugal and volcanic areas in Italy) will need to be developed.

Another issue for consideration is whether improvements could also be made in terms of the specifications for the selection and scaling of acceleration time-histories for dynamic analyses of structures. Although work, cited earlier in Section 2.3, has been published to develop suites of accelerograms compatible with the current version of the EC8 spectra, modifications could be made if the scheme for the construction of the spectrum presented in this paper were adopted. The use of the PGV/PGA ratio would, for example, allow the magnitude of the earthquake to be taken into account in the selection of real accelerograms, which is recommended rather than selecting records purely on the basis of conforming to the shape of the elastic design spectrum [14]. The recommendation of the author of this paper, however, would be to specify the elastic spectrum to be slightly conservative (as is the case in Fig. 2.3 for most periods) and then allow reduced ordinates through a site-specific PSHA, which would facilitate full definition of the required suite of acclerograms. The argument in support of this approach would be that any structure of sufficient importance to warrant dynamic analysis should also warrant a site-specific seismic hazard analysis.

References

1. Abrahamson N, Silva W (2008) Summary of the Abrahamson & Silva NGA ground-motion relations. Earthquake Spectra 24:67–97
2. Akkar S, Bommer JJ (2006) Influence of long-period filter cut-off on elastic spectral displacements. Earthquake Eng Struct Dyn 35:1145–1165
3. Akkar S, Bommer JJ (2007) Prediction of elastic displacement response spectra in Europe and the Middle East. Earthquake Eng Struct Dyn 36:1275–1301
4. Akkar S, Bommer JJ (2007) Empirical prediction equations for peak ground velocity derived from strong-motion records from Europe and the Middle East. Bull Seismol Soc Am 97: 511–530
5. Akkar S, Bommer JJ (2010) Empirical equations for the prediction of PGA, PGV and spectral accelerations in Europe, the Mediterranean region and the Middle East. Seismol Res Lett 81:195–206
6. Ambraseys NN, Bommer JJ (1991) The attenuation of ground accelerations in Europe. Earthquake Eng Struct Dyn 20:1179–1202
7. Ambraseys NN, Douglas J, Sarma SK, Smit PM (2005) Equations for the estimation of strong ground motions from shallow crustal earthquakes using data from Europe and the Middle East: horizontal peak ground acceleration and spectral acceleration. Bull Earthquake Eng 3:1–53
8. Ambraseys NN, Simpson KA, Bommer JJ (1996) Prediction of horizontal response spectra in Europe. Earthquake Eng Struct Dyn 25:371–400
9. Atkinson GM (2008) Ground-motion prediction equations for Eastern North America from a referenced empirical approach: implications for epistemic uncertainty. Bull Seismol Soc Am 98:1304–1318
10. Atkinson GM, Boore DM (1997) Some comparisons between recent ground-motion relations. Seismol Res Lett 68:24–40

11. Bay F, Wiemer S, Fäh D, Giardini D (2005) Predictive ground-motion relationships for Switzerland: best estimates and uncertainties. J Seismol 9:223–240
12. Bay F, Wiemer S, Malagini L, Giardini D (2003) Spectral shear wave ground-motion scaling in Switzerland. Bull Seismol Soc Am 93:414–429
13. Beyer K, Bommer JJ (2006) Relationships between median values and aleatory variabilities for different definitions of the horizontal component of motion. Bull Seismol Soc Am 96:1512–1522. Erratum 97:1769
14. Bommer JJ, Acevedo AB (2004) The use of real earthquake accelerograms as input as input to dynamic analysis. J Earthquake Eng 8 (special issue 1):43–91
15. Bommer JJ, Alarcón JE (2006) The prediction and use of peak ground velocity. J Earthquake Eng 10:1–31
16. Bommer JJ, Douglas J, Scherbaum F et al (2010) On the selection of empirical ground-motion prediction equations for seismic hazard analysis. Seismol Res Lett, 81:783–793
17. Bommer JJ, Douglas J, Strasser FO (2003) Style-of-faulting in ground motion prediction equations. Bull Earthquake Eng 1:171–203
18. Bommer JJ, Elnashai AS, Weir AG (2000) Compatible acceleration and displacement spectra for seismic design codes. Proc 12 World Conf Earthq Eng, paper no 207, Auckland
19. Bommer JJ, Mendis R (2005) Scaling of spectral displacement ordinates with damping ratios. Earthquake Eng Struct Dyn 34:145–165
20. Bommer JJ, Pinho R (2006) Adapting earthquake actions in Eurocode 8 for performance-based seismic design. Earthquake Eng Struct Dyn 35:39–55
21. Bommer JJ, Ruggeri C (2002) The specification of acceleration time-histories in seismic design codes. Eur Earthquake Eng 16:3–17
22. Bommer JJ, Scherbaum F, Bungum H et al (2005) On the use of logic trees for ground-motion prediction equations in seismic hazard analysis. Bull Seismol Soc Am 95:377–389
23. Bommer JJ, Stafford PJ, Akkar S (2010) Current empirical ground-motion prediction equations for Europe and their application to Eurocode 8. Bull Earthquake Eng 8:5–26
24. Bommer JJ, Stafford PJ, Alarcón JE et al (2007) The influence of magnitude range on empirical ground-motion prediction. Bull Seismol Soc Am 97:2152–2170
25. Boore DM, Atkinson GM (2008) Ground-motion prediction equations for the average horizontal component of PGA, PGV, and 5%-damped PSA at spectral periods between 0.01 s and 10.0 s. Earthquake Spectra 24:99–138
26. Boore DM, Joyner WB, Fumal TE (1997) Equations for estimating horizontal response spectra and peak acceleration from western North American earthquakes: a summary of recent work. Seismol Res Lett 68:128–153
27. Bozorgnia Y, Campbell KW (2004) The vertical-to-horizontal spectral ratio and tentative procedures for developing simplified V/H and vertical design spectra. J Earthquake Eng 4:539–561
28. Cameron WI, Green RU (2007) Damping correction factors for horizontal ground-motion response spectra. Bull Seismol Soc Am 97:934–960
29. Campbell KW (1997) Empirical near-source attenuation relationships for horizontal and vertical components of peak ground acceleration, peak ground velocity, and pseudo-absolute acceleration response spectra. Seismol Res Lett 68:154–179
30. Campbell KW, Bozorgnia Y (2008) NGA ground motion model for the geometric mean horizontal component of PGA, PGV, PGD and 5% damped linear elastic response spectra for periods ranging from 0.01s to 10 s. Earthquake Spectra 24:139–171
31. CEN (2004) Eurocode 8. Design of Structures for Earthquake Resistance – Part 1: general rules, seismic actions and rules for buildings. EN 1998-1: 2004, Comité Européen de Normalisation, Brussels
32. CEN (2005) Eurocode 1. Action on Structures. Part 1–4: wind actions. EN 1991-1-4: 2005, Comité Européen de Normalisation, Brussels
33. Chiou BS-J, Youngs RR (2008) An NGA model for the average horizontal component of peak ground motion and response spectra. Earthquake Spectra 24:173–215

34. Cotton F, Scherbaum F, Bommer JJ, et al (2006) Criteria for selecting and adjusting ground-motion models for specific target regions: applications to Central Europe and rock sites. J Seismol 10:137–156
35. Erdik M, Biro YA, Onur T et al (1999) Asssessment of earthquake hazard in Turkey and neighboring regions. Ann Geofis 42:1125–1138
36. García-Mayordomo J, Faccioli E, Paolucci R (2004) Comparative study of the seismic hazard assessments in European national seismic codes. Bull Earthquake Eng 2:51–73
37. Grünthal G, Boose C, Camelbeeck T et al (1999) Seismic hazard assessment for Central, North and Northwest Europe: GSHAP Region 3. Ann Geofis 42:999–1011
38. Iervolino I, Maddolini G, Cosenza E (2008) Eurocode 8 compliant real record sets for seismic analysis of structures. J Earthquake Eng 12:54–90
39. Jiménez M-J, García-Fernández M, Chadi M et al (1999) Seismic hazard assessment of the Ibero-Maghreb region. Ann Geofis 42:1057–1065
40. Joyner WB, Boore DM (1981) Peak acceleration and velocity from strong-motion records including records from the 1979 Imperial Valley, California, earthquake. Bull Seismol Soc Am 71:2011–2038
41. Leyendecker EV, Hunt RJ, Frankel AD et al (2000) Development of maximum considered earthquake ground motion maps. Earthquake Spectra 16:21–40
42. Lungu D, Cornea T, Aldea A et al (1997) Basic representation of seismic action. In: Lungu D et al (eds) Design of structures in seismic zones. Bridgeman Ltd, Timisoara, pp 9–60
43. Lungu D, Cornea T, Nedelcu C (1999) Hazard assessment and site-dependent response for Vrancea earthquakes. In: Wenzel F et al (eds) Vrancea earthquakes: tectonics, hazard and risk mitigation. Kluwer Academic Publishers, Dordrecht, pp 251–267
44. Malagnini L, Akinci A, Herrmann RB, et al (2002) Characteristics of the ground-motion in Northeastern Italy. Bull Seismol Soc Am 92:2186–2204
45. Malagnini L, Herrmann RB, Di Bona M (2000) Ground-motion scaling in the Apennines (Italy). Bull Seismol Soc Am 90:1062–1081
46. McGuire RK (1977) Seismic design spectra and mapping procedures using hazard analysis based directly on oscillator response. Earthquake Eng Struct Dyn 5:211–234
47. Mendis R, Bommer JJ (2006) Modification of the Eurocode 8 damping reduction factors for displacement spectra. Proc 13 Eur Conf Earthq Eng, paper no 1203, Geneva
48. Montaldo V, Faccioli E, Zonno G et al (2005) Treatment of ground-motion predictive relationships for the reference seismic hazard map of Italy. J Seismol 9:295–316
49. Musson RMW (1999) Probabilistic seismic hazard maps for the North Balkan region. Ann Geofis 42:1109–1124
50. Musson RMW, Sargeant SL (2007) Eurocode 8 seismic hazard zoning maps for the UK. Technical Report CR/07/125, British Geological Survey, Keyworth
51. Newmark NM, Hall WJ (1969) Seismic design criteria for nuclear reactor facilities. Proc 4 World Conf Earthq Eng 2:B5.1-B5.12, Santiago de Chile
52. Papaioannou Ch, Papazachos C (2000) Time-independent and time-dependent seismic hazard in Greece based on seismogenic sources. Bull Seismol Soc Am 90:22–33
53. Rey J, Faccioli E, Bommer JJ (2002) Derivation of design soil coefficients (S) and response spectral shapes for Eurocode 8 using the European strong-motion database. J Seismol 6: 547–555
54. Sabetta F, Pugliese A (1996) Estimation of response spectra and simulation of nonstationary earthquake ground-motions. Bull Seismol Soc Am 86:337–352
55. Sadigh K, Chang C-Y, Egan JA et al (1997) Attenuation relationships for shallow crustal earthquakes based on California strong motion data. Seismol Res Lett 68:180–189
56. Scherbaum F, Cotton F, Smit P (2004) On the use of response spectral-reference data for the selection and ranking of ground-motion models for seismic-hazard analysis in regions of moderate seismicity: the case of rock motion. Bull Seismol Soc Am 94:2164–2185
57. Scherbaum F, Delavaud L, Riggelsen C (2009) Model selection in seismic hazard analysis: an information-theoretic perspective. Bull Seismol Soc Am 99:3234–3247

58. Slejko D, Camassi R, Cecić I et al (1999) Seismic hazard assessment for Adria. Ann Geofis 42:1085–1107
59. Spudich P, Fletcher JB, Hellweg M et al (1997) SEA96 – a new predictive relation for earthquake ground motions in extensional tectonic regimes. Seismol Res Lett 68:190–198
60. Stafford PJ, Mendis R, Bommer JJ (2008b) Dependence of damping correction factors for response spectra on duration and number of cycles. ASCE J Struct Eng 134:1364–1373
61. Stafford PJ, Strasser FO, Bommer JJ (2008a) An evaluation of the applicability of the NGA models to ground-motion prediction in the Euro-Mediterranean region. Bull Earthquake Eng 6:149–177
62. Toro GR, Abrahamson NA, Schneider JF (1997) Model of strong ground motions from earthquakes in Central and Eastern United States: best estimates and uncertainties. Seismol Res Lett 68:41–57
63. Ulomov VI, Shumilina L, Trifonov V et al (1999) Seismic hazard of Northern Eurasia. Ann Geofis 42:1023–1038
64. Wiemer S, Giardini D, Fäh D et al (2009) Probabilistic seismic hazard assessment of Switzerland: best estimates and uncertainties. J Seismol 13:449–478

Chapter 3
Investigating Possible Regional Dependence in Strong Ground Motions

John Douglas

Abstract It is common practice to use ground-motion models, often developed by regression on recorded accelerograms, to predict the expected earthquake ground motions at sites of interest. An important consideration when selecting these models is the possible dependence of ground motions on geographical region, i.e., are median ground motions in the (target) region of interest for a given magnitude and distance the same as those in the (host) region where a ground-motion model is from, and are the aleatory variabilities of ground motions also similar? In this brief article, some of the recent literature with relevance to these questions is summarized. It is concluded that although some regions seem to show considerable differences in shaking it is currently more defensible to use well-constrained models, possibly based on data from other regions, rather than use local, often poorly-constrained, models. In addition, it is noted that the presence of "pseudo-regional dependency" due to differences in, for example, focal depths, average shear-wave velocity profiles or focal mechanisms can lead to apparent variations between areas when these variations could be captured in well-characterized ground-motion prediction equations.

3.1 Introduction

One of the main topics of debate in the recent engineering seismology literature is the question of whether strong ground motions show dependence on the region or country in which they occur or in other words "Have strong-motion data a nationality?" Whether median earthquake ground motions and their variabilities for the same magnitude and distance show a significant dependence on the area in which they were recorded is a fundamental but still open question. Almost all parts of the world and, in particular, the Euro-Mediterranean region do not have sufficient

J. Douglas (✉)
BRGM – RNSC/RIS, 3 avenue C. Guillemin, BP 36009, 45060 ORLEANS Cedex 2, France
e-mail: j.douglas@brgm.fr

John Douglas is also at BRGM, Orléans, France.

S. Akkar et al. (eds.), *Earthquake Data in Engineering Seismology*, Geotechnical, Geological, and Earthquake Engineering 14, DOI 10.1007/978-94-007-0152-6_3,

strong-motion data from which to derive robust estimates of median ground motions based solely on instrumental data from a small geographical area. Therefore, for many projects, including the Global Earthquake Model (GEM) and Seismic Hazard Harmonization in Europe (SHARE), whether ground-motion models derived for one region can be safely transferred for the prediction of shaking in another is a pressing issue. Recent articles on this topic include Douglas [19], on which this current paper is based, and Bommer et al. [9].

This article presents evidence taken from the available literature from both sides of the debate. No new results are shown but it seeks to provide a summary of recent studies on this topic. Drawing conclusions either for or against regional dependency based on visual inspection of median response spectra from published ground-motion prediction equations (GMPEs) (e.g. [16]). should be avoided since such comparisons can often be interpreted in support of either side of the argument due to the large epistemic uncertainties associated with GMPEs. Therefore, in the following other more objective methods are preferred.

3.2 Pseudo-Regional Differences

Before beginning the review of evidence for or against regional dependency, I will discuss what I am entitling "pseudo-regional dependency". This refers to an apparent dependence of ground motions on region that should disappear (or become negligible) if a GMPE that is sufficiently well characterized is used to estimate shaking. This idea is discussed in more detail in Douglas [19], where actual examples are given.

For example, in two regions the average focal depths (an important earthquake characteristic controlling shaking, particularly for small events) could differ thereby leading to a difference in median ground motions if a distance metric (such as Joyner-Boore distance, r_{jb}) that does not take into account the depth of the earthquake is used. If, however, a metric, such as rupture distance (r_{rup}), is employed the variation in ground motions due to differences in depths between the two regions could be modeled. Similarly, if in one region reverse-faulting earthquakes are prevalent whilst in another normal-faulting earthquakes are most common then this could lead to a difference of 10–40% in ground motions since shaking in reverse-faulting events are generally significantly higher than those in normal or strike-slip earthquakes (e.g. [8]). However, if a GMPE was used that had terms modeling style-of-faulting effects then this apparent regional dependence would disappear. As a final example, differences in the average soil profiles in one region could lead to differences in median ground motions. For example, if a GMPE used site classes based on broad ranges of average shear-wave velocities in the top 30 m, V_{s30} (e.g. the Eurocode 8 or NEHRP classifications) then if in one region rock sites were harder on average than those in another area then this could lead to overprediction of shaking but if V_{s30} was used directly this difference could be captured.

A difficulty that can complicate comparisons between ground motions, particularly of smaller events, in various regions is the requirement for a consistent magnitude scale. Douglas [17] notes that above roughly magnitude five earthquakes generally have moment magnitude (M_w) estimates reported by global or regional data centers (e.g. Global CMT or the National Earthquake Information Center) but for smaller events only local magnitudes (M_L) are available, which can be highly network dependent. For example, Scherbaum et al. [26] list various M_L estimates for the St Dié 2003 earthquake that occurred in France close to the Swiss and German borders. The M_L estimates range from 5.4 (LED, Germany) to 5.8 (LDG, France) (whereas the M_ws reported are 4.7–4.8). Therefore, comparisons between ground motions from French or German earthquakes associated with only M_Ls from the local networks should account for a possible difference of up to, for this case, 0.4 magnitude units. Such a large difference in magnitude scales would obviously have a significant effect on predicted ground motions.

3.3 Previous Studies

The following sections summarize evidence for and against regional dependency of ground motions. The next section deals with evidence based on physical differences between regions and those based on macroseismic intensities; the mapping of these differences to instrumental strong motion can be difficult. The subsequent section discusses evidence based on weak ground motion, which is becoming increasingly abundant with the installation of high-quality digital instruments and the consideration of combined accelerometric and broadband datasets. The final section presents evidence based on strong-motion data.

3.3.1 Evidence from Physical Reasoning or Macroseismic Intensities

Variations in certain physical properties of Earth's crust could be thought to lead to differences in strong ground motions, such as anelastic attenuation parameters (Q) measured by, for example, Mitchell and co-workers for many parts of the world (e.g. [7]) and crustal structural velocities (e.g. [24]). However, although variations in these physical properties will affect the shaking at long distances (> 50 km) they do not seem to significantly affect ground motions close to the source, where such estimates are vital for engineering purposes.

Similarly a number of authors have evidenced clear variations in the attenuation of macroseismic intensities in different parts of the world (e.g. [6]) (Fig. 3.1). Using macroseismic intensities to examine regional differences is attractive since for many parts of the globe these are the only observations of large earthquakes currently available. Differences in observed intensities generally become greater as source-to-site distance increases and they are clearest when comparing the felt areas (i.e. intensity III) of earthquakes in various regions. However, close to the source (< 100 km) observed macroseismic intensities appear to be similar in different areas (e.g [22]).

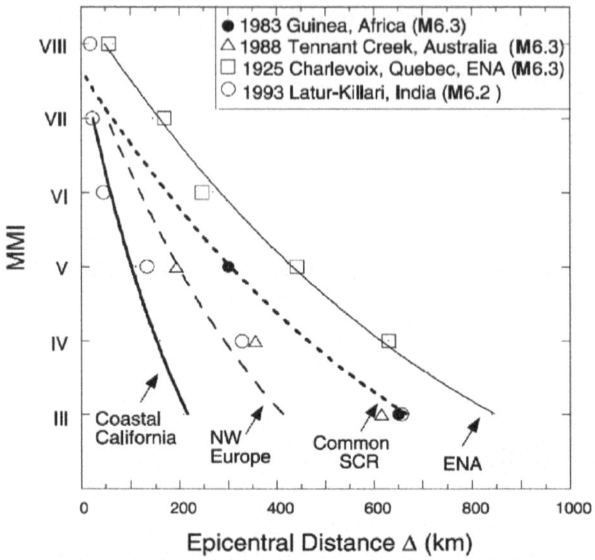

Fig. 3.1 Comparison between attenuation of Modified Mercalli intensity (MMI) in four different regions (from [6])

3.3.2 Evidence from Weak-Motion Data

One of the richest sources of evidence for regional dependence of ground motions are recent studies using weak-motion records from high-quality digital accelerometric and broadband networks that have been installed in the past couple of decades in many parts of the world. An early study is Campbell [13] who finds that near-source peak ground accelerations (PGAs) from small earthquakes in eastern North America (ENA) are consistent with a GMPE derived using data from small events in California, once differences in site effects and magnitude scales (see earlier comments) are accounted for. This study demonstrates two important points. Firstly, the importance of reducing "pseudo-regional differences", such as general site variations between two areas, and, secondly, comparing weak motions in one region to weak motions in another. As shown by, for example, Bommer et al. [10] and Cotton et al. [15] ground motions from small earthquakes scale differently with respect to magnitude and source-to-site distance in small and large events and this effect must be accounted for when examining variations between ground motions in one region and those in another.

Two studies that reach contrasting conclusions on the similarity of shaking in two different areas of Australia and ENA are those by Allen et al. [3] who find that on average ground motions in ENA are higher than those in *south-western Western* Australia whereas Allen and Atkinson [2] conclude that motions are similar between ENA and *south-eastern* Australia. They conclude, therefore, that it is valid to combine data from ENA and south-eastern Australia when deriving models for use in either area or stable continental regions in general.

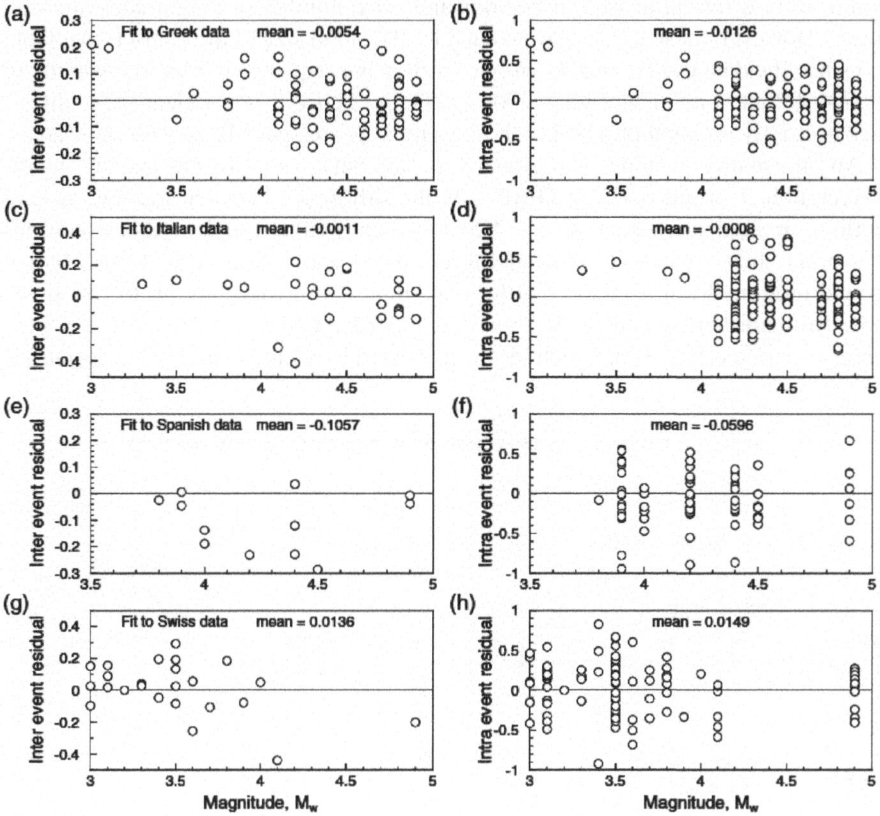

Fig. 3.2 Residuals with respect to the PGA GMPE of Bommer et al. [10] for four national subsets (from [10])

Bommer et al. [10] derive a set of GMPEs that are valid down to M_W 3 and examine inter- and intra-event residuals with respect to their model for four countries (Greece, Italy, Spain and Switzerland) in the Mw 3–5 range. They find that none of these sets of residuals shows a clear bias (Fig. 3.2). Consequently, apparent differences in GMPEs derived using data from these regions can be related to data coverage (particularly magnitude range), choices of functional form and regression techniques and, therefore, the strongest evidence for or against regional dependency comes from using the observations directly.

Douglas [18] developed such a technique based on analysis of variance (ANOVA) within small magnitude and distance bins that can be used to statistically test the null hypothesis of no difference between observed ground motions in two areas. This technique has been applied for different areas (see [19] for a summary) and some tests clearly demonstrated a difference in ground motions between regions. For example, observed shaking in Umbria-Marche and Molise (two areas of central Italy) was shown by Douglas [19] using the ANOVA technique to be significantly different at various periods and for a number of magnitudes and distances. However, these tests are weakened by the small size of the datasets available and since the data used comes from one earthquake sequence in Umbria-Marche in 1997–1998 and one in Molise in 2002–2003 and hence it is not clear if the difference is strictly regional or whether these sequences are special cases for their areas.

An informative example of regional variation between two areas that are invariably combined when deriving GMPEs is the difference between average ground motions in small earthquakes (M < 5.5) between southern and northern (central) California shown by Atkinson and Morrison [5] and Chiou et al. [14]. Median ground motions from southern Californian small earthquakes are up to two times those from northern events of the same size recorded at the same distance at a wide range of periods (Fig. 3.3), which has been related by Chiou et al. [14] to variations

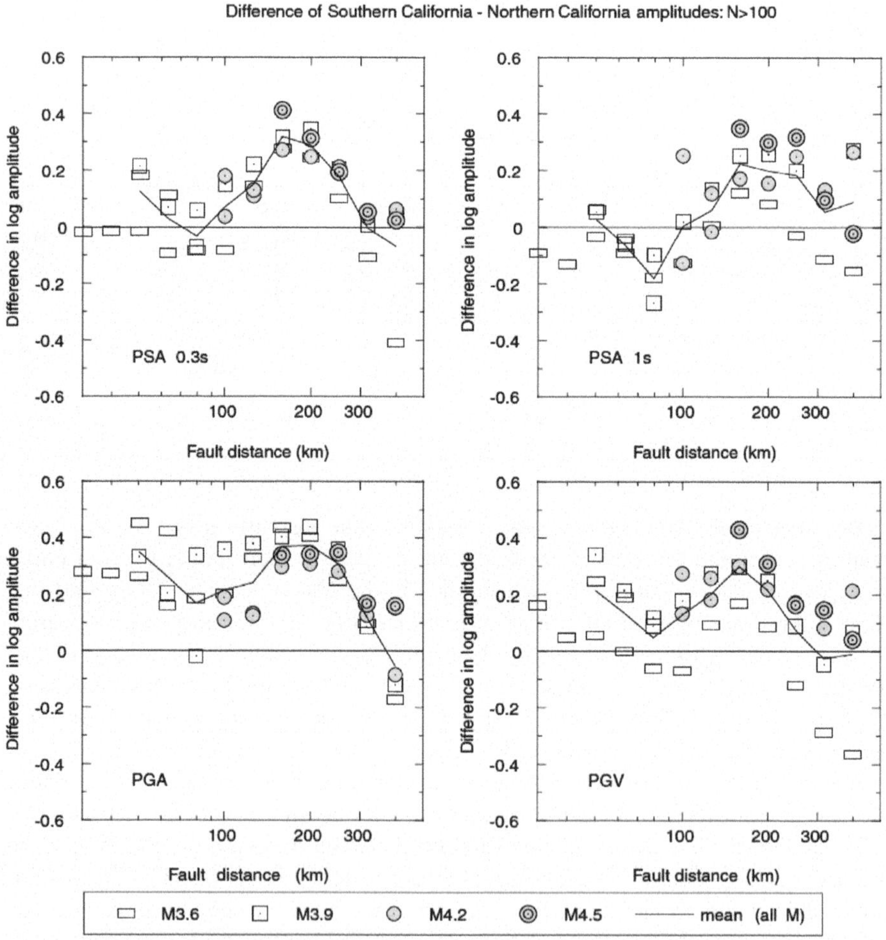

Fig. 3.3 Difference in log (base 10) amplitudes between southern and northern California ground motions for PGA, peak ground velocity (PGV) and pseudo-spectral accelerations (PSA) at two periods (from [5])

in stress drop between the two areas. However, for larger earthquakes, which are the main focus of seismic hazard assessments the clear differences in ground motions between the two regions become negligible. This leads Chiou et al. [14] to make this clear statement on the difficulty of using weak motions to examine regional dependency for strong motions (SMM is small-to-moderate magnitude range, NGA is Next Generation Attenuation, CCal is central California and SCal is southern California):

> Our results suggest that regional difference derived from small-to-moderate earthquake data is not an infallible indicator of regional difference expected in the moderate-to-large magnitude range of primary importance to the quantification of seismic hazard to civil structures. Differences observed in SMM range between the locally recorded data and California data should therefore be used with caution to infer NGA model applicability or to adjust NGA models for use in seismic hazard analysis. Finally, the insignificant difference in large-magnitude median motions between CCal and SCal is in agreement with the working hypothesis of NGA project, that is ground motions from moderate-to-large magnitude earthquakes in different active tectonic regions are similar.

3.3.3 Evidence from Strong-Motion Data

Two large geographical regions that have tended to develop roughly independent GMPEs are western North America (WNA) and Europe, Mediterranean and Middle East (EMME). The recent set of NGA GMPEs (e.g. [1]). do use some data from EMME but the models are not highly dependent on these data. Stafford et al. [27] quantitatively examine, using an extended Scherbaum et al. [26] technique, the ability of one of the NGA models [11] to predict observed ground motions in EMME. They conclude that this model provides a good match to the observed median shaking.

In a similar, but larger scale, study Allen and Wald [4] compute average residuals for a worldwide strong-motion dataset from shallow crustal earthquakes (containing data from most active regions with strong-motion networks) with respect to a number of recent robust GMPEs for WNA and EMME. They find that most models do a good job of predicting median ground motions and their variabilities within the magnitude-distance range of validity.

Douglas [19] shows that aleatory variabilities (standard deviations, sigmas) of empirical GMPEs derived using data from small geographical zones are not lower, and in many cases are higher, than those associated with GMPEs derived from combining data from many parts of the globe. If ground motions show a clear regional dependency then this variation between regions should show up as larger sigmas in GMPEs derived from global datasets. Bragato [12] calculates, using a large Italian weak-motion dataset (M_L 2.7–4.5), that regional variations in ground motions contributes only 4% to the total observed sigma.

The technique of Scherbaum et al. [26] for quantitatively comparing observed and predicted ground motions has been applied to numerous datasets. Some of these (e.g. [23]) have found that none of the models tested provide good predictions for some earthquakes outside their geographical zone of origin whereas others (e.g. [21]) have found some GMPEs closely predict observations from completely different geographical zones. The good or poor match between observations and predictions can often be related to the magnitude and distance ranges of validity of the considered GMPEs and different average site effects between regions.

3.4 Discussion and Conclusions

This brief article has sought to summarize some of the recent literature that has a bearing on the question of regional differences in earthquake ground motions. There are a number of recent studies showing strong evidence for differences in ground motions from small and moderate earthquakes occurring in different areas and also at long distances. However, these differences for weak motions rarely seem to carry over to shaking of engineering significance, i.e. close (<50 km) to earthquakes of magnitudes greater than roughly five. Evidence for regional dependency from weak motions does not imply regional dependency for strong motions. Due to the large and rapidly growing databanks of weak motions in many parts of the world, it is tempting to develop local GMPEs based on these data in the hope that they are more appropriate for that region than models derived for other locations. This temptation, however, should be resisted since a number of recent studies (e.g. [10, 15]) have demonstrated that the extrapolation of GMPEs derived from weak-motion data are likely to significantly over-estimate ground motions in future large earthquakes. Therefore, it is more defensible to make the assumption that ground motions from large earthquakes do not show a significant regional dependency and hence adopt GMPEs that are robust at high magnitudes even if local data was not used in their derivation. In order to match observations from small events in the local area adjustments for small magnitudes like those made by Chiou et al. [14] or Scasserra et al. [25] could be applied.

As shown by Douglas [20] ground-motion prediction is still affected by large, and only slowly decreasing, uncertainties even for well-instrumented areas with long histories of strong-motion observation (e.g. California). These large epistemic uncertainties caused by a lack of data and knowledge means that it is not currently possible to make firm conclusions on regional dependency of earthquake shaking. Consequently, the null hypothesis of no regional dependency cannot be rejected. In Scottish law a "Not proven" verdict would probably be returned.

Acknowledgments I thank Landsvirkjun and the University of Iceland for funding my 1-year visiting professorship at the Earthquake Engineering Research Centre, University of Iceland. Also I thank the organizers of the Second Euro-Mediterranean meeting on Accelerometric Data Exchange and Archiving for the invitation to present at this meeting. This article has benefited from participation in discussions during the PEGASOS Refinement Project and the EC-funded Seventh Framework Programme project SHARE and with Teraphan Ornthammarth.

References

1. Abrahamson N, Atkinson G, Boore D, Bozorgnia Y, Campbell K, Chiou B, Idriss IM, Silva W, Youngs R (2008) Comparisons of the NGA ground-motion relations. Earthquake Spectra 24(1):45–66. doi: 10.1193/1.2924363

2. Allen TI, Atkinson GM (2007) Comparison of earthquake source spectra and attenuation in eastern North America and Southeastern Australia. Bull Seismol Soc Am 97(4):1350–1354. doi: 10.1785/0120060206
3. Allen TI, Dhu T, Cummins PR, Schneider JF (2006) Empirical attenuation of ground-motion spectral amplitudes in Southwestern Western Australia. Bull Seismol Soc Am 96(2):572–585. doi: 10.1785/0120040238
4. Allen TI, Wald DJ (2009) Evaluation of ground-motion modeling techniques for use in Global ShakeMap – A critique of instrumental ground-motion prediction equations, peak ground motion to macroseismic intensity conversions, and macroseismic intensity predictions in different tectonic settings. US Geological Survey Open-File Report 2009-1047, 114p
5. Atkinson GM, Morrison M (2009) Observations on regional variability in ground-motion amplitudes for small-to-moderate earthquakes in North America. Bull Seismol Soc Am 99(4):2393–2409. doi: 10.1785/0120080223
6. Bakun WH, McGarr A (2002) Differences in attenuation among the stable continental regions. Geophys Res Lett 29(23):2121. doi:10.1029/2002GL015457
7. Baqer S, Mitchell BJ (1998) Regional variation of Lg coda Q in the continental United States and its relation to crustal structure and evolution. Pure Appl Geophys 153: 613–638
8. Bommer JJ, Douglas J, Strasser FO (2003) Style-of-faulting in ground-motion prediction equations. Bull Earthquake Eng 1(2):171–203
9. Bommer JJ, Stafford PJ, Akkar S (2010) Current empirical ground-motion prediction equations for Europe and their application to Eurocode 8. Bull Earthquake Eng 8(1):5–26. doi: 10.1007/s10518-009-9122-9. doi: 10.1785/0120080133
10. Bommer JJ, Stafford PJ, Alarcón JE, Akkar S (2007) The influence of magnitude range on empirical ground-motion prediction. Bull Seismol Soc Am 97(6):2152–2170. doi: 10.1785/0120070081
11. Boore DM, Atkinson GM (2008) Ground-motion prediction equations for the average horizontal component of PGA, PGV, and 5%-damped PSA at spectral periods between 0.01s and 10.0s. Earthquake Spectra 24(1):99–138. doi: 10.1193/1.2830434
12. Bragato PL (2009) Assessing regional and site-dependent variability of ground motions for ShakeMap implementation in Italy. Bull Seismol Soc Am 99(5):2950–2960. doi: 10.1785/0120090020
13. Campbell KW (1989) The dependence of peak horizontal acceleration on magnitude, distance, and site effects for small-magnitude earthquakes in California and eastern North America. Bull Seismol Soc Am 79(5):1311–1346
14. Chiou B, Youngs R, Abrahamson N, Addo K (2010) Ground-motion attenuation model for small-to-moderate shallow crustal earthquakes in California and its implications on regionalization of ground-motion prediction models. Earthquake Spectra 26(4):907–926. doi: 10.1193/1.3479930
15. Cotton F, Pousse G, Bonilla F, Scherbaum F (2008) On the discrepancy of recent European ground-motion observations and predictions from empirical models: analysis of KiK-net accelerometric data and point-sources stochastic simulations. Bull Seismol Soc Am 98(5):2244–2261. doi: 10.1785/0120060084
16. Douglas J (2003a) Earthquake ground motion estimation using strong-motion records: a review of equations for the estimation of peak ground acceleration and response spectral ordinates. Earth Sci Rev 61(1):43–104
17. Douglas J (2003b) A note on the use of strong-motion data from small magnitude earthquakes for empirical ground motion estimation. Skopje Earthquake 40 Years of European Earthquake Engineering (SE-40EEE)
18. Douglas J (2004) An investigation of analysis of variance as a tool for exploring regional differences in strong ground motion. J Seismol 8(4):485–496

19. Douglas J (2007) On the regional dependence of earthquake response spectra. ISET J Earthquake Tech 44(1):77–99
20. Douglas J (2010) Consistency of ground-motion predictions from the past four decades. Bulletin of Earthquake Engineering 8(6):1515–1526. DOI 10.1007/s10518-010-9195-5
21. Drouet S, Scherbaum F, Cotton F, Souriau A (2007) Selection and ranking of ground motion models for seismic hazard analysis in the Pyrenees. J Seismol 11(1):87–100. doi: 10.1007/s10950-006-9039–6
22. Hanks TC, Johnston AC (1992) Common features of the excitation and propagation of strong ground motion for north American earthquakes. Bull Seismol Soc Am 82(1):1–23
23. Hintersberger E, Scherbaum F, Hainzl S (2007) Update of likelihood-based ground-motion model selection for seismic hazard analysis in western central Europe. Bull Earthquake Eng 5(1):1–16. doi: 10.1007/s10518-006-9018–x
24. Mooney WD, Laske G, Masters TG (1998) CRUST 5.1: a global crustal model at 5 degrees × 5 degrees. J Geophys Res 103:727–747
25. Scasserra G, Stewart JP, Bazzurro P, Lanzo G, Mollaioli F (2009) A comparison of NGA ground-motion prediction equations to Italian data. Bull Seismol Soc Am 99(5):2961–2978
26. Scherbaum F, Cotton F, Smit P (2004) On the use of response spectral-reference data for the selection and ranking of ground-motion models for seismic-hazard analysis in regions of moderate seismicity: the case of rock motion. Bull Seismol Soc Am 94(6):2164–2185
27. Stafford PJ, Strasser FO, Bommer JJ (2008) An evaluation of the applicability of the NGA models to ground-motion prediction in the Euro-Mediterranean region. Bull Earthquake Eng 6(2):149–177

Chapter 4
A Predictive Ground-Motion Model for Turkey and Its Comparison with Recent Local and Global GMPEs

Z. Çağnan, Sinan Akkar, and Polat Gülkan

Abstract We present a local ground motion prediction equation (GMPE) for estimating the peak ground-motion values that was derived using the recently compiled Turkish strong motion database. The new GMPE is comparable with the recent global GMPEs in terms of model sophistication and quality of underlying database. Using this equation, we explore the inter-event, inter-station ground motion variability of the recent Turkish strong motion database as well as suitability of some local and global GMPEs for regional seismic hazard assessment analyses. The inter-event error underlines the distinguishing characteristics of few earthquakes, suggesting that the use of local GMPEs can be important especially when specific scenario studies are to be carried out. The inter-station variability allowed us to detect stations with outlier site response and to investigate the goodness of the employed site effects model. Results also indicate suitability of some global and local GMPEs for use in regional seismic hazard assessments together with the model presented in this study. These should be combined through a logic tree scheme to reduce both aleatory and epistemic uncertainty in local hazard assessments.

4.1 Introduction

In Turkey, strong motion recordings have been available since the second half of 1970s. As the accelerometric data accumulated, various researchers compiled the seismic catalog information of the corresponding events as well as the relevant strong-motion data with the aim of improving seismic hazard and risk mitigation studies in Turkey (e.g. [9, 10, 20, 25–28, 35, 42]). These studies, however, are mostly out-dated lacking the recent seismic catalog information and focusing only on parts

Z. Çağnan (✉)
Department of Civil Engineering, Middle East Technical University Northern Cyprus Campus, Kalkanlı, Güzelyurt, KKTC, Mersin 10, Turkey
e-mail: cagnan@metu.edu.tr

S. Akkar et al. (eds.), *Earthquake Data in Engineering Seismology*, Geotechnical, Geological, and Earthquake Engineering 14, DOI 10.1007/978-94-007-0152-6_4,

of the data available. In 2005, ERD-GDDA[1] and EERC-METU[2] jointly launched the project "Compilation of Turkish strong-motion network according to the international standards" with the financial support of TUBITAK.[3] The aim of the project was the revision of the entire strong-motion database acquired by ERD-GDDA over the period 1976–2007 in Turkey and conducting in-situ tests for strong-motion site characterization. The seismic event, station, and instrument metadata have been revised, and the raw records have been individually processed. The main result of the project is the online accelerometric archive (http://www.daphne.deprem.gov.tr [7, 36]) where strong motion recordings of earthquakes that occurred in Turkey can be downloaded and metadata about stations and earthquakes can be obtained.

One of the important input parameters of a seismic hazard analysis is reliable empirical ground motion prediction equations (GMPEs). These equations are typically determined by fitting the assumed ground motion model to a set of observed strong motion parameters, such as the peak ground acceleration (PGA), velocity (PGV) or different spectral ordinates at several periods. Recent studies [19, 39] have shown that GMPEs, which are based on carefully-compiled global strong-motion databases estimate comparable ground motions provide a similar level of complexity in their functional forms. Despite these observed overall agreements between global models, other evaluations have emphasized that there might be considerable differences for some particular earthquake scenarios [1, 14]. These complementary studies highlight the continuing need of GMPEs to include additional parameters to trace regional differences [24]. A more important conclusion of these studies however is the significance of epistemic uncertainty requiring the consideration of a set of predictive equations for a better quantification of seismic hazard in a region. One way of accounting for epistemic uncertainty while determining the regional seismic hazard levels can be the utilization of both region-specific and global GMPEs. This task would require more elaborate regional predictive equations that are compatible with the general features of global GMPEs.

In this study, we present a recently developed prediction equation [6] for PGA that is based on the aforementioned new Turkish strong-motion database. The equation is valid for a distance (R_{jb}) range of 0–200 km and is derived for moment magnitudes between $5 \leq \mathbf{M} \leq 7.6$. This equation considers the effects of nonlinear soil behavior and style-of-faulting. We explore the suitability of some recent local and global prediction equations (i.e. recent NGA and pan-European equations) for Turkey by quantitatively comparing them with the developed GMPE.

---

[1] Earthquake Research Department, General Directorate of Disaster Affairs, Ministry of Settlement and Public Works, Republic of Turkey

[2] Earthquake Engineering Research Center, Middle East Technical University

[3] Scientific and Technological Research Council of Turkey

4.2 Database

The database used in this study comprises of 1,259 records from 573 earthquakes with moment magnitude (**M**) and source-to-site distance (Joyner-Boore distance, R_{jb}) ranges of $3.5 \leq \mathbf{M} \leq 7.6$ and $0\ km \leq R_{jb} \leq 200\ km$, respectively (Fig. 4.1a, b). 65% of this data belongs to recordings of small events with $\mathbf{M} < 5$ as can also be inferred from the magnitude histograms plotted on the right hand side of Fig. 4.1a, b. The two largest magnitude events are the 1999 Düzce (**M**7.2) and Kocaeli (**M**7.6) earthquakes whose ground motions exhibit relatively low amplitude peak motions when compared with other similar size earthquakes. The low amplitude waveforms were attributed to the observed surface rupture in these events [30]. The majority of the ground motions are from stations with NEHRP C ($360\ m/s \leq V_{S30} < 760\ m/s$) and D ($180\ m/s \leq V_{S30} < 360\ m/s$) site classes. Only 8% of the waveforms have corresponding V_{S30} values exceeding 760 m/s (NEHRP B).

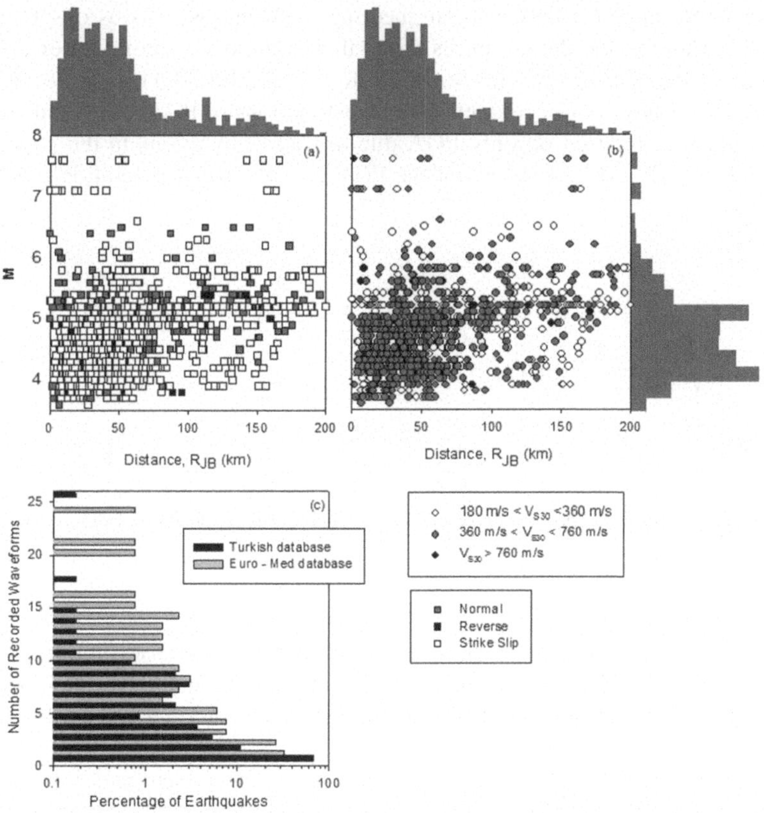

Fig. 4.1 Magnitude vs. Distance distribution of the data in the Turkish strong motion database according to (**a**) faulting style and (**b**) site conditions. (**c**) comparison of number of recorded waveforms obtained for each event in the Turkish and Euro-Med. databases. (Modified from [6])

As illustrated in Fig. 4.1a, the database is dominated by the strike-slip events (70% of the entire dataset). Records corresponding to normal events make up the 28% of the database and records corresponding to reverse/thrust events constitute only the 2% of the database. Another feature of our database is the higher percentage of singly recorded events that is inferred by the variation of earthquakes as a function of recorded waveforms (Fig. 4.1c). For comparative purposes the same distribution for the Euro-Mediterranean database (used in [5]) is also presented in Fig. 4.1c. The number of singly recorded events in the Turkish strong-motion database can be seen to exceed that of Euro-Mediterranean database by a factor of 2.

4.3 Developed Local GMPE

Details of the GMPE presented here can be found in [6]. Several functional forms were explored during its derivation with the aim of keeping a balance between a rigorous model (for meaningful and reliable estimations) and a robust expression (for wider implementation in engineering applications). It was observed that the general trend in the data requires terms that account for saturation effects and magnitude-dependent decay. Hence the form given in Eq. (4.1) was chosen to be the basic expression for the geometric mean estimations of PGA, PGV and spectral acceleration (We focused on only PGA due to space limitations in the paper. The reader is referred to [6] for details of the PGV and spectral acceleration prediction equations).

For $\mathbf{M} \le c_1$:

$$\ln(Y) = a_1 + a_2(M - c_1) + a_4(8.5 - M)^2 + [a_5 + a_6(M - c_1)] \ln\sqrt{R_{jb}^2 + a_7^2} + [a_8 F_N + a_9 F_R] \tag{4.1.a}$$

For $\mathbf{M} > c_1$:

$$\ln(Y) = a_1 + a_3(M - c_1) + a_4(8.5 - M)^2 + [a_5 + a_6(M - c_1)] \ln\sqrt{R_{jb}^2 + a_7^2} + [a_8 F_N + a_9 F_R] \tag{4.1.b}$$

In Eq. (4.1), the constant c_1 defines the limiting magnitude for saturation effects and is considered as 6.5 in this study. The parameters F_N and F_R are dummy variables for the influence of faulting, taking values of 1 for normal and reverse faults, respectively, and are 0 otherwise.

Given the fully defined V_{S30} values at each station, the linear and nonlinear site effects were described by the site response function used in [15]. The original form of this model is proposed in [21]. This site response model among various alternatives was chosen due to its simplicity and its fairly good performance with the strong-motion database of this study; as normalized intra-event residuals of [15] yield no trend with changes in V_{S30} and their distributions follow closely the standard normal distribution [6]. The explicit forms of the linear and nonlinear site response function will not be given here; the details are provided in [15].

To derive the PGA GMPE and its associated uncertainty, one-stage maximum likelihood regression that was first introduced in [17, 18], then improved in [2], and later re-examined in [29] was used. References [17, 18] use an Expectation Maximization algorithm while [2, 29] use a Certain Search algorithm to obtain the maximum likelihood estimates of the coefficients and the intra-event (σ) (combined ground-motion variability from record specific factors) and inter-event (τ) standard deviation (combined ground-motion variability resulting from event specific factors) components. The Expectation Maximization algorithm does not necessarily work in the absence of good initial estimates. Unreasonable initial estimates can lead to biased results. Methods proposed in [2, 29], on the other hand, are numerically more stable. We used the maximum likelihood methodology of [29] in this study. The two-stage regression was not employed as it underestimates aleatory variability in case of datasets consisting heavily of singly-recorded events [38]. This is the case in our database as shown in Fig. 4.1c.

GMPEs derived from small and large magnitude recordings differ due to rapid decay of small events with distance and due to different ground motion amplitude scaling between small and large magnitudes [22]; so $\mathbf{M} \geq 5$ events were considered only for the predictive model. A total of 433 waveforms from 137 events (including aftershocks and mainshocks) were considered in our regression analysis. Table 4.1 lists the regression coefficients and the inter-event, intra-event standard deviations of the GMPEs derived for PGA.

Note that the style-of-faulting coefficients given in Table 4.1 indicate smaller horizontal ground-motion amplitudes for normal events when compared to strike-slip and reverse faulting cases (reverse events attaining, on average, 2% larger values with respect to strike-slip cases). This ratio is within the ranges reported by previous studies (see [13] for a comprehensive literature review of the influence of faulting style on ground-motion components). However, given the relatively small number of reverse faulting records in the database, the style-of-faulting dependency of the GMPE can be subject to further refinements with the accumulation of reverse fault events in the future. As can be seen from Table 4.1, the sigma values obtained in this study are slightly larger in comparison with the recently derived global GMPEs (e.g., NGA or pan-European global models; see relevant discussions in [40]).

Table 4.1 Regression coefficients obtained in this study for the GMPE for PGA

	a_1	a_2	a_3	a_4	a_5	a_6	a_7	a_8	a_9
PGA	8.92418	–0.5130	–0.6950	–0.18555	–1.25594	0.18105	7.33617	–0.02125	0.01815
	σ^a	τ^a	$\sigma_{Tot}{}^a$						
PGA	0.6527	0.5163	0.8322						

[a]Symbols σ and τ refer to intra- and inter-event standard deviations, respectively; σ_{Tot} is the square root of sum of σ^2 and τ^2.

This can be attributed to the database features or the selected explanatory variables in the functional form. However, when the overall sigma variation is of concern, our sigma values are within the expected limits of previously published values for the last 4 decades [40]. Further discussion on possible sources of this obtained sigma is given in the following section.

4.4 Inter-event and Inter-station Distribution of Residuals

Inter- and intra-event variations of the considered model for different estimator parameters were reported in [6]. However, to quantify further the error associated with each event and each station of the database we analyzed the estimated inter-event and inter-station distributions (Fig. 4.2a, b).

Figure 4.2a indicates that all the earthquakes show errors within the ± 0.6 range, with the exception of three aftershocks of the 1999 Düzce earthquake (**M** = 5.0, **M** = 5.3, **M** = 5.1), 2007 Tutak (Ağrı) earthquake (**M** = 5.1), 2007 Muğla earthquake (**M** = 5.1) and 1998 Adana-Ceyhan earthquake (**M** = 6.2) that are all overestimated by the predictions except for the latter.

The underestimation of the 1999 Düzce aftershocks could be partially attributed to the uncertainties associated with the estimated magnitude and faulting style of these events. The first of these parameters was obtained by employing magnitude conversion relations and the latter by epicentral proximity to known faults (see [7] for details on the magnitude conversion relations and the epicentral location based faulting style assignment method employed). It should be underlined that as the main 1999 Düzce earthquake had an important dip-slip component in addition to the main faulting style of strike-slip, its aftershocks are likely to be associated with a similar component as well (Ulusay, R., 2010, personal communication). For the 2007 Tutak and Muğla earthquakes, the behavior however was found to be closely linked to the stations that recorded these earthquakes as will be discussed below.

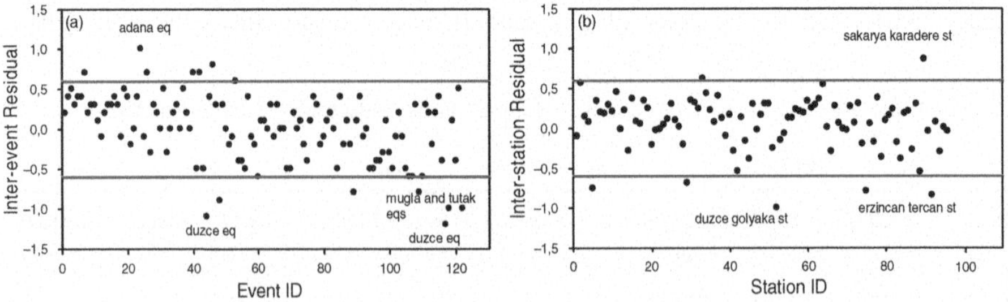

Fig. 4.2 **a** Distribution of inter-event residuals and **b** distribution of inter-station residuals for each event and station, respectively, considered in the regression analysis

Most of the inter-station residuals of this study have absolute values smaller than 0.6 as well (Fig. 4.2b). In particular, the largest overestimations are observed at Erzincan Tercan and Düzce Gölkaya stations; whereas the largest underestimation is observed at Sakarya Karadere station. It is interesting to note that the latter two are the only stations that recorded the 1999 Düzce $M = 5.1$ aftershock which led to the lowest inter-event residual value mentioned earlier. Similarly, Ağrı-Merkez station, which corresponds to one of the high overestimations, is the only station that recorded the 2007 Tutak (Ağrı) earthquake. Muğla Merkez station is another location that yields both extreme inter-event and inter-station values (2007 Muğla earthquake). Figure 4.2a, b indicate that the inter-station and inter-event residuals are comparable to each other for this study; this contradicts findings based on the Italian data [11, 12].

To investigate these observations further, the residual vs. residual analysis proposed in [32] was carried out. In this method, only stations that recorded at least two events are considered. The residuals are corrected for the inter-event component of error and random pairs of records belonging to the same station are combined and plotted. A good correlation between pairs indicates that site effect is the dominant factor in controlling the residuals.

As can be seen in Fig. 4.3a, the residuals display scatter along the diagonal with a correlation coefficient of 0.62. Because the number of available pairs is 101 (i.e., 99 degrees of freedom), the null hypothesis of no correlation among pairs was rejected at 5% level of significance.

This same approach was also applied to inter-event residuals. For each earthquake recorded by at least two stations, after correcting the residuals for inter-station error, pairs of records for each earthquake are randomly combined and plotted.

Figure 4.3b shows that the residuals have a less clear trend in this case. The correlation coefficient was calculated to be 0.47, indicating that uncertainties on the

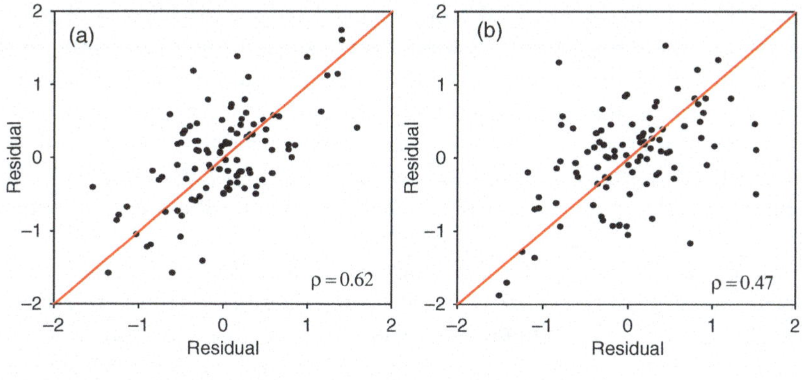

Fig. 4.3 Residual-residual plots after correction for (**a**) inter-event errors and (**b**) inter-station errors

event parameters are not the main source of the ground-motion variability for this dataset but site effects are.

4.5 Quantitative Comparison of Developed GMPE with Other Local and Global GMPEs

Given the revised Turkish strong motion dataset, the question we answer in this section is: how applicable are other recently developed local and global GMPEs (Table 4.2) for Turkey in comparison with the ground motion model developed in this study? In this section, the comparisons are made using analyses of model residuals and the likelihood approach of [37]. The modeling assumption made in regression analysis is that if the observed data distribution agrees well with the distribution of predictions then normalized differences of these two sets of values (normalized residuals) should yield the standard normal distribution. Both the analysis of model residuals and the likelihood approach is based on this assumption. In the case of likelihood approach, if the observed data coincides with the mean predicted value, the corresponding likelihood value becomes 1. The likelihood value reduces with decreasing quality of the fit; if the normalized residuals follow the standard normal distribution, corresponding likelihood values are distributed uniformly between 0 and 1.

From the Next Generation Attenuation (NGA) relationships, only the GMPE of [15] could be employed due to the absence of reliable estimates of some of the predictor variables (such as depth to top of rupture, sediment depth) for other NGA GMPEs. Also the recent local GMPE of [41] could not be included in these quantitative comparisons because the model has been published without its variance. It should be underlined that all the GMPEs were employed within their limits of applicability (both magnitude and distance) and care was taken to achieve parameter compatibility.

The performance of the models in Fig. 4.4 can be assessed by considering both the distributions of the normalized total residuals and the distribution of the likelihood values. In the former case, a model is considered to be performing well if the distribution of observed normalized residuals (the solid black line) agrees well with the standard normal distribution (the solid red line). An agreement between these two curves indicates that the model is not biased and that the standard deviation of the model appropriately captures the variability in the observed PGA values. As mentioned earlier a uniform distribution of likelihood values also supports this result.

As can be seen from Fig. 4.4, the global predictive model in [5] and the local predictive model of [11], which is based on Italian data, perform well with the Turkish strong motion data. On the other hand, considerable overestimation should be expected with the model in [15] as well as other NGA models [39]. Previously developed local Turkish GMPEs can be seen to have the tendency of overestimating the ground motions as well. This can be attributed to the assumptions made by

Table 4.2 List of compared GMPEs

References	Abbr.[a]	Region	N_R[b], N_E[b]	Comp.[c]	M[d], M_{min}[d], M_{max}[d]	R[e], R_{min}[e], R_{max}[e]	Fault[f]	Site[g]
[5]	AB10	Euro-Med	532, 131	PGA_{GM}	**M**, 5.0, 7.6	R_{jb}, 0, 100	SS, N, R	R, ST, SF
[8]	Aetal05	Euro-Med	595, 135	PGA_{Max}	**M**, 5.0, 7.6	R_{jb}, 0, 100	SS, N, T, O	R, ST, SF
[11]	Betal09	Italy	561, 107	PGA_{Max}	**M**, 4.0, 6.9	R_{jb}, 0, 100	SS, N, R	R, SA, DA
[15]	BA08	Worldwide	1574, 58	PGA in GMRotI50	**M**, 5.0, 8.0	R_{jb}, 0, 200	SS, N, R	V_{S30}: 180–1300 m/s
[23]	DT07	Greece	335, 151	PGA_{AM}	**M**, 4.5, 6.9	R_{epi}, 0, 136	SS+R, N	R, ST, SF
[31]	KG04	Turkey	112, 57	PGA_{Max}	**M**, 4.0, 7.4	R_{jb}, 1.2, 250	U	R, ST, SF
[34]	Oetal04	NW Turkey	195, 17	PGA_{GM}	**M**, 5.0, 7.4	R_{jb}, 5, 300	U	R, ST, SF
NA	This study	Turkey	433, 137	PGA_{GM}	**M**, 5.0, 7.6	R_{jb}, 0, 200	SS, N, R	V_{S30}: 180–900 m/s

[a] Abbreviations of GMPEs used in this study.
[b] N_R is number of records and N_E is number of earthquakes in the dataset.
[c] GMRotI50 is rotation-independent average horizontal component [16], subscripts Max, GM, AM refer to maximum of horizontal components, geometric mean of horizontal components, and arithmetic mean of horizontal components, respectively.
[d] M is magnitude scaling, M_{min} and M_{max} are minimum and maximum magnitude range of GMPE, **M** is moment magnitude.
[e] R is distance metric, R_{min} and R_{max} are minimum and maximum distance range of GMPE, R_{rup} is closest distance to rupture surface, R_{jb} is closest distance to horizontal projection of rupture surface, R_{epi} is epicentral distance.
[f] SS, N, R and T refer to strike-slip, normal, reverse and thrust faulting, O is odd faulting, SS+R indicates strike-slip and reverse are considered together, U means faulting style is either unknown or not considered in GMPE.
[g] V_{S30} is average shear-wave velocity in the upper 30 m soil profile, R is rock, SA is shallow alluvium, DA is deep alluvium, ST is stiff soil, SF is soft soil.

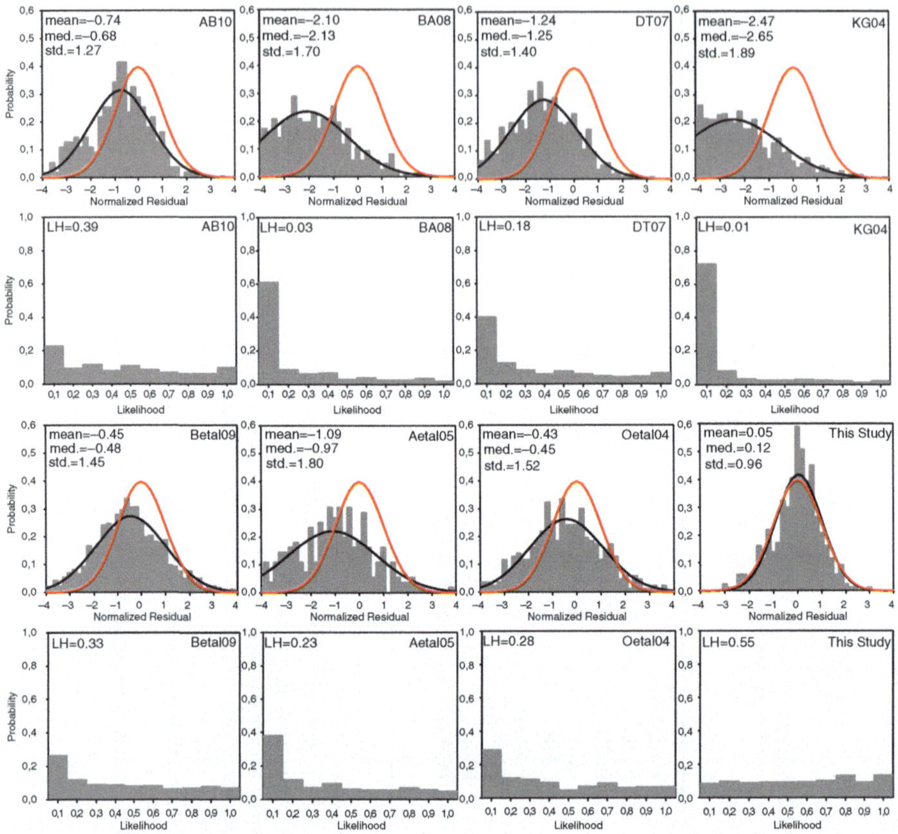

Fig. 4.4 Histograms of the normalized total model residuals and likelihood values for [5] (AB10), [15] (BA08), [23] (DT07), [31] (KG04), [11] (Betal09), [8] (Aetal05) and [34] (Oetal04). The plots of the normalized total model residuals also include the standard normal distribution (*solid red line*) and the normal distribution fitted to the residuals (*solid black line*). Ranking parameters of the median, mean and standard deviation of the total normalized residuals (med., mean, std.) and median likelihood values (LH) are also given as part of the plots

their developers due to insufficient level of database knowledge at the time they were derived. Also the fact that [31] and [15] are based on only mainshock data contributes to the observed performance of these models with the Turkish strong motion database as the database includes aftershocks as well. According to the classification scheme of [37], the GMPE of [5] and [11] belong to class B, the GMPE of [34] belongs to class C and the model of this study is not unexpectedly the most suitable for Turkey with class A. Another important finding is the consistently high standard deviation values obtained for total residuals (Fig. 4.4). This is because the considered models have lower standard deviation values than the variability of ground motions in the Turkish strong motion dataset. This observation is consistent with the findings of previous sections.

As underlined by [39], for databases where certain events provide large numbers of records to the overall database, results obtained by total residuals only can be biased. Similar analyses as described above were repeated for inter-event and intra-event residuals, however results were found not to improve drastically with this change. Hence only results obtained by total model residuals are presented.

4.6 Conclusions

An updated database of strong motion data recorded in Turkey was recently made available that includes state of the art metadata relevant to both stations and earthquakes [7, 36]. We present a new empirical ground motion equation for the ranges $5 \leq \mathbf{M} \leq 7.6$ and $0 \leq R_{jb} \leq 200$ km that was derived using data from 137 earthquakes recorded by 86 stations from this database. The adopted functional form takes into account magnitude saturation, magnitude dependent attenuation, a quadratic term in magnitude, style of faulting as well as nonlinear site effects.

As in [12, 33] estimates of the intra-station and inter-station variability were investigated in detail. Both inter-station and inter-event residuals were found to be comparable, varying in the range ± 0.6 mostly. The residual distribution analysis allowed identification of earthquakes and stations with significant deviation from the average, caused by the strong under or overestimation of the predictions. Overlap was found between outlier stations and outlier events (i.e. some outlier events were observed to be recorded by outlier stations only). Hence carrying out further detailed geological and geophysical studies at these special locations would be beneficial for explaining the physical bases behind these outlier situations.

The inter-station and inter-event residual analyses further indicate that site effect is the dominant factor governing the developed ground motion model's variability. In this study, linear and nonlinear site effects model of [21] is used. Our findings show that developing local linear and nonlinear site effects model in the future, possibly including parameters such as depth of soil deposit, resonance period as well as V_{s30}, would reduce the total variability of the local GMPE presented here.

Based on quantitative comparison of Turkish strong motion database with recently developed other local and global GMPEs, a number of models were identified suitable for ground motion prediction in Turkey. However, it should be underlined that tested models on average overestimate PGA values, and underestimate corresponding variability. As both the median estimations as well as model variability have significant impact on the results of probabilistic seismic hazard analyses, our findings can be considered useful by the analysts when forming a logic tree. Further discussions on modeling uncertainty between local and global GMPEs are addressed in [6].

Finally, we would like to stress here that although considerable effort was given to the improvement of the Turkish strong motion database within the scope of this study and to the development of GMPEs with up to date and robust functional forms, with accumulation of additional strong motion data over time, which would improve the current event-record ratio of the database, still more representative GPMEs for Turkey would be possible.

Acknowledgments This study is financially supported by the Scientific and Technological Council of Turkey under the award no. 105G016. Mr. E. Yenier and Mr. A. Sandıkkaya processed many records and compiled the entire dataset. We sincerely appreciate their efforts.

References

1. Abrahamson NA, Atkinson G, Boore DM, Bozorgnia Y, Campbell K, Chiou B, Idriss IM, Silva W, Youngs R (2008) Comparison of the NGA ground-motion relations. Earthquake Spectra 24:45–66
2. Abrahamson NA, Youngs RR (1992) A stable algorithm for regression analysis using the random effects model. Bull Seismol Soc Am 82:505–510
3. Akkar S, Bommer JJ (2007a) Empirical prediction equations for peak ground velocity derived from strong-motion records from Europe and the Middle East. Bull Seismol Soc Am 97: 1275–1301
4. Akkar S, Bommer JJ (2007b) Prediction of elastic displacement response spectra in Europe and the Middle East. Earthquake Eng Struct Dynam 36:1275–1301
5. Akkar S, Bommer JJ (2010) Empirical equations for the prediction of PGA, PGV and spectral accelerations in Europe, the Mediterranean region and the Middle East. Seismol Res Lett 81: 195–206
6. Akkar S, Çağnan Z (2010) A local ground-motion predictive model for Turkey and its comparison with other regional and global ground-motion models. Bull Seismol Soc Am Doi: 10.1785/0120090367
7. Akkar S, Çağnan Z, Yenier E, Erdoğan O, Sandıkkaya A, Gülkan P (2010) The recently compiled Turkish strong motion database: preliminary investigation for seismological parameters. J Seismol 14:457–479
8. Ambraseys N, Douglas J, Sarma SK, Smit PM (2005) Equations for the estimation of strong ground motions from shallow crustal earthquakes using data from Europe and the Middle East: horizontal peak ground acceleration and spectral acceleration. Bull Earthquake Eng 3:1–53
9. Ambraseys N, Durukal E, Free M (1993) Re-evaluation of strong-motion data in Tukey. ESEE research report, No. 93/2. Imperial College of Science, Technology and Medicine, Civil Engineering Department, London
10. Aydan O, Hasgür Z (1997) The characteristics of acceleration waves of Turkish earthquakes, 4. Ulusal Deprem Muhendisligi Kongresi, Ankara, Deprem Muhendisligi Turk Milli Komitesi, pp 30–37, (in Turkish)
11. Bindi D, Luzi L, Pacor F (2009) Interevent and interstation variability computed for the Italian accelerometric archieve (ITACA). Bull Seismol Soc Am 99:2471–2488
12. Bindi D, Luzi L, Pacor F, Franceschina G, Castro RR (2006) Ground-Motion prediction from empirical attenuation relationships versus recorded data: the case of the 1997–1998 Umbria-Marches, Central Italy, strong motion data set. Bull Seismol Soc Am 96(3): 984–1002
13. Bommer JJ, Douglas J, Strasser FO (2003) Style-of-faulting in ground motion prediction equations. Bull Earthquake Eng 1:171–203
14. Bommer JJ, Stafford P, Akkar S (2010) Current empirical ground-motion prediction equations for Europe and their application to Eurocode 8. Bull Earthquake Eng 8: 5–26
15. Boore DM, Atkinson GM (2008) Ground-motion prediction equations for the average horizontal component of PGA, PGV, and 5%-damped PSA at spectral periods between 0.01s and 10.0s. Earthquake Spectra 24:99–138
16. Boore DM, Watson-Lamprey J, Abrahamson NA (2006) GMRotD and GMRotI: orientation independent measures of ground motion. Bull Seismol Soc Am 96:1502–1511

17. Brillinger DR, Preisler HK (1984) An exploratory analysis of the Joyner-Boore attenuation data. Bull Seismol Soc Am 74:1441–1450
18. Brillinger DR, Preisler HK (1985) Further analysis of the Joyner-Boore attenuation data. Bull Seismol Soc Am 75:611–614
19. Campbell KW, Bozorgnia Y (2006) Next Generation Attenuation (NGA) empirical ground motion models: can they be used in Europe? Proceedings of the First European Conference on Earthquake Engineering and Seismology, Geneva, Switzerland, Paper No. 458
20. Çelebi M, Akkar S, Gülerce U, Şanlı A, Bundock H, Salkın A (2001) Main shock and aftershock records of the 1999 İzmit and Düzce, Turkey earthquakes, USGS/OFDA Project [USGS Project No.: 1-7460-63170], USGS Open-File Report 01–163
21. Choi Y, Stewart JP (2005) Nonlinear site amplification as function of 30m shear wave velocity. Earthquake Spectra 21:1–30
22. Cotton F, Pousse G, Bonilla F (2008) On the discrepancy of recent European ground motion observations and predictions from empirical models: analysis of KiK-Net accelerometric data and point-source stochastic simulation. Bull Seismol Soc Am 98:2244–2261
23. Danciu L, Tselentis G (2007) Engineering ground motion parameters attenuation relationships for Greece. Bull Seismol Soc Am 97:162–183
24. Douglas J (2007) On the regional dependence of earthquake response spectra. ISET J Earthquake Tech 44:71–99
25. Durukal E, Alpay Y, Biro T, Mert A, Erdik M (1998) Analysis of strong motion data of the 1995 Dinar, Turkey Earthquake, Second Japan-Turkey Workshop: Earthquake Disaster Prevention Research in Turkey, 23–25 February 1998, Technical University of İstanbul
26. Erdik M (1984) Report on the Turkish earthquake of October 30, 1983. Earthquake Spectra 1:151–172
27. Erdik M, Durukal E (2001) 1999 Kocaeli and Düzce, Turkey earthquakes: strong ground motion, XV ICSMGE TC4 'Lessons learned from recent strong earthquakes', 25 August 2001, İstanbul, Turkey
28. İnan E, Çolakoğlu Z, Koç N, Bayülke N, Çoruh E (1996) Earthquake catalogs with acceleration records from 1976 to 1996, General Directorate of Disaster Affairs, Earthquake Research Department, Ankara, Turkey (98 pp, in Turkish)
29. Joyner WB, Boore BM (1993) Methods for regression analysis of strong-motion data. Bull Seismol Soc Am 83:469–487
30. Kagawa T, Irikua K, Somerville PG (2004) Differences in ground motion and fault rupture between the surface and buried rupture earthquakes. Earth Planets Space 56:3–14
31. Kalkan E, Gülkan P (2004) Site-dependent spectra derived from ground motion records in Turkey. Earthquake Spectra 20:1111–1138
32. Lee Y, Zeng Y, Anderson JG (1998) A simple strategy to examine the sources of errors in attenuation relations. Bull Seismol Soc Am 88:291–296
33. Niazi M, Bozorgnia Y (1991) Behavior of near-source peak horizontal and vertical ground motions over SMART-1 array, Taiwan. Bull Seismol Soc Am 81:715–732
34. Özbey C, Sarı A, Manuel L, Erdik M, Fahjan Y (2004) An empirical attenuation relationship for Northerwestern Turkey ground motion using a random effects approach. Soil Dynam Earthquake Eng 24:115–125
35. Rathje EM, Stokoe KH, Rosenblad BL (2003) Strong-motion station characterization and site effects during the 1999 earthquakes in Turkey. Earthquake Spectra 19:653–676
36. Sandıkkaya MA, Yılmaz MT, Bakır BB, Yılmaz O (2010) Site classification of Turkish national strong-motion stations. J Seismol 14: 543–563
37. Scherbaum F, Cotton F, Smit P (2004) On the use of response spectral-reference data for the selection and ranking of ground-motion models for seismic-hazard analysis in regions of moderate seismicity: the case of rock motion. Bull Seismol Soc Am 94:2164–2185
38. Spudich P, Joyner WB, Lindh AG, Boore DM, Margaris BM, Fletcher JB (1999) SEA99: a revised ground motion prediction relation for use in extensional tectonic regimes. Bull Seismol Soc Am 89:1156–1170

39. Stafford PJ, Strasser FO, Bommer JJ (2008) An evaluation of the applicability of the NGA models to ground motion prediction in the Euro-Mediterranean region. Bull Earthquake Eng 6:149–177
40. Strasser FO, Abrahamson NA, Bommer JJ (2009) Sigma: issues, insights, and challenges. Seismol Res Lett 80:40–56
41. Ulusay R, Tuncay E, Sönmez H, Gökçeoğlu C (2004) An attenuation relationship based on Turkish strong motion data and iso-acceleration map of Turkey. Eng Geol 74:265–291
42. Zaré M, Bard PY (2002) Strong motion dataset of Turkey: data processing and site classification. Soil Dynam Earthquake Eng 22:703–718

Chapter 5
Strong-Motion Networks in Italy and Their Efficient Use in the Derivation of Regional and Global Predictive Models

L. Luzi, M. Massa, D. Bindi, and F. Pacor

Abstract Italy is a country characterized by high seismic hazard so strong-motion monitoring represents a relevant issue. Several strong-motion networks have been installed in the Italian territory during the last decades, with the aim of recording the ground motion generated by moderate to strong events or to monitor single regions. The collection of the strong-motion recordings of the Italian earthquakes was recently fulfilled and data are distributed through the ITACA database (http://itaca.mi.ingv.it). The new data set was used to develop a set of ground motion prediction equations (hereinafter GMPEs) for the Italian territory (Bindi et al., 2009a,c), in order to update the well known GMPEs developed by Sabetta and Pugliese [22]. The recent M_w 6.3 earthquake that occurred in central Italy on April 2009 and the upgrades of the ITACA database gave us the possibility to validate the predictive capability of the newly developed GMPEs and to explore the regional variability inside the Italian territory.

5.1 Strong Motion Networks in Italy and the Italian Strong Motion Data Set

Strong-motion monitoring in Italy started in the early 1970s, when a national strong motion network was designed and installed by ENEA (Italian Energy and Environment Organization) and ENEL (Italian Electricity Company), to evaluate the seismic risk for the construction of nuclear power plants. The first analog strong motion record is dated 1972 and refers to the 6th February Ancona M_l 3.0 earthquake. Since 1997 the Italian strong motion network (*Rete Accelerometrica Nazionale, RAN*) is owned by the Italian Civil Protection. At the end of 2008 the network was composed by 265 digital and 119 analogue permanent stations [17], with an expected final set up of 500 free field digital installations on a grid of about 25 × 25 km, covering the areas with high seismic risk. Since a decade the National

L. Luzi (✉)
Istituto Nazionale di Geofisica e Vulcanologia, Milano, Italy
e-mail: luzi@mi.ingv.it

S. Akkar et al. (eds.), *Earthquake Data in Engineering Seismology*, Geotechnical, Geological, and Earthquake Engineering 14, DOI 10.1007/978-94-007-0152-6_5,

Seismometric Network (*CNT, Rete Sismometrica Nazionale*, http://cnt.rm.ingv.it), handled by the *Istituto Nazionale di Geofisica e Vulcanologia*, has been improved by co-locating about 60 velocimetric and accelerometric sensors, in order to guarantee the ground motion recording in case of strong events.

Besides the two national networks, many local networks have been installed in areas of moderate to high seismic hazard. The Irpinia Seismic Network (ISNet) has been deployed in the Southern Apennines along the active fault system which caused the 1980, November 23, M_W 6.9, Campania–Lucania earthquake. The ISNet is mainly devoted to estimate real-time earthquake magnitude and location. It is composed of 28 stations and covers an area of about 100 × 70 km, as described by Iannaccone et al. [18].

The Basilicata region (southern Italy), affected as well by the intense seismic activity of the Apennine chain, is monitored by a local strong-motion network, handled by the University of Basilicata, the Institute of Methodologies for Environmental Analysis of the Italian research council (IMAA–CNR) and the local government of the Province of Potenza [3].

In north-east Italy the Friuli Venezia Giulia Accelerometric Network (*RAF Rete Accelerometrica del Friuli Venezia Giulia*), handled by the University of Trieste, covers the area struck by the 1976 M_W 6.4 earthquake. It was installed in the early nineties in the framework of international scientific projects and today it is joined to other network stations operating in Italy as well as in Austria and Slovenia [12].

Finally, in northern Italy, a further strong-motion network has been set up by the Milano-Pavia department of the *Istitituto Nazionale di Geofisica e Vulcanologia*, in the area struck by the 2004 Salò earthquake, composed of 24 stations [4]. The geographic distribution of the above mentioned networks is displayed in Fig. 5.1.

The data most widely used for engineering-seismology purposes come from the RAN network, since they have been continuously collected since 1972 by several institutions and at the moment archived and distributed through a web database (ITACA, http://itaca.mi.ingv.it, [19]).

The ITACA database includes records of 1,017 earthquakes: 1,002 from the period range 1972–2004 ($1.1 < M_l <= 6.9$), 2 from the 2008 Parma sequence (M_W 4.9 and $M_W = 5.4$) and 13 from the 2009 L'Aquila sequence ($4 \leq M_W \leq 6.3$).

There are 2,550 3-component waveforms in the uncorrected version and 2,401 of them have been processed. Acceleration, velocity and displacement time series and the acceleration response spectra at 121 periods up to 4 s (5% damping) are distributed. About 650 temporary or permanent stations are included, but only 15% of them have been characterized by geotechnical or geophysical parameters.

The ITACA data set was recently exploited to derive a set of ground motion prediction equations for Italy (hereinafter referred to as ITA08) for the prediction of maximum horizontal and vertical peak ground acceleration, peak ground velocity and 5% damped acceleration response spectra (Bindi et al., 2009a). The data set is composed by 561 3-component waveforms from 107 earthquakes with moment magnitude in the range 4.0–6.9, that occurred in Italy from 1972 to 2004 and recorded by 206 stations at distances up to 100 km.

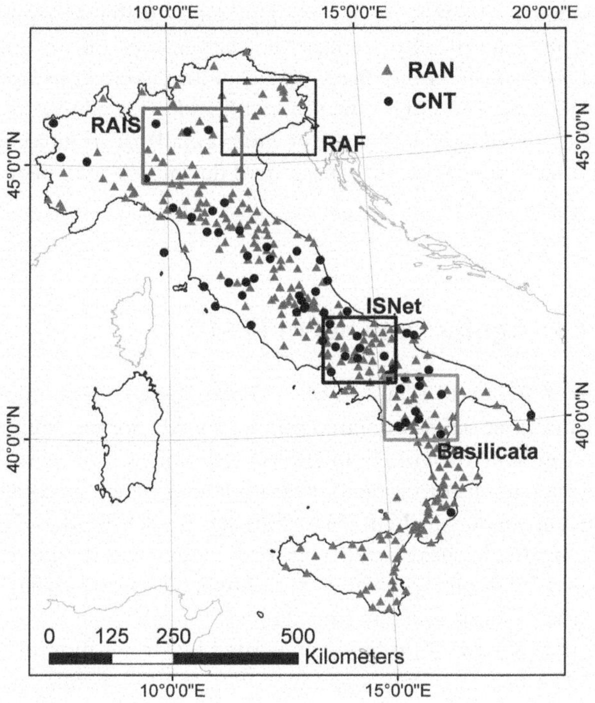

Fig. 5.1 Strong-motion networks in Italy: *gray triangles* represent the stations of the Italian strong-motion network (RAN); *black dots* represent the stations of the National Seismometric Network (CNT); *squares* indicate the areas covered by regional networks: RAIS (northern Italy strong-motion network); RAF (Friuli Venezia Giulia accelerometric network); ISNet (Irpinia Seismic Network); Basilicata (strong motion network of the Basilicata Region)

The adopted functional form includes a linear and a quadratic term for magnitude, a magnitude-dependent geometrical spreading, style of faulting and site terms as:

$$\log_{10} Y = a + b_1(M_W - M_{ref}) + b_2(M_W - M_{ref})^2 + \left[c_1 + c_2(M_W - M_{ref})\right] \log_{10}\sqrt{(R_{JB}^2 + h^2)} + e_i S_i + f_j F_j \tag{5.1}$$

where Y is the response variable; M_{ref} is a reference magnitude; R_{jb} is the Joyner-Boore distance when $M_w \geq 5.5$, or the epicentral distance (km); h is the pseudo-depth (km); S_i with i=1,2,3 are dummy variables that assume either the value 0 or 1 depending on soil type (rock, class C_0: $S_1=1$ and $S_2=S_3=0$; shallow alluvium, class C_1: $S_2=1$ and $S_1=S_3=0$; deep alluvium, class C_2: $S_3=1$ and $S_1=S_2=0$); F_j are dummy variables that take either the value 0 or 1 depending on the style of faulting (normal fault: $F_1=1$ and $F_2=F_3=0$; strike-slip: $F_2=1$ and $F_1=F_3=0$; reverse fault: $F_3=1$ and $F_1=F_2=0$); e_i and f_j are the site and the style-of-faulting coefficients, respectively. Since the regression accounting for the style of faulting provided coefficients f_i is not significantly different from zero, and its introduction did not significantly reduce the variance of the residuals, only the results obtained without including the style of faulting were considered. A regression scheme based on the random effect model [1, 11] was adopted to describe the errors which are assumed to be independent and normally distributed. Both the inter-event (σ_{inter}) and the intra-event (σ_{intra}) standard deviations have been evaluated, as described by Bindi et al. (2009b).

5.2 Fit of the L'Aquila Data Set to ITA08

On April 6th, 2009, 01:32:40 UTC, an M_w 6.3 earthquake occurred in the Abruzzo region (Central Italy), at 9.5 km depth along a NW-SE normal fault with SW dip, very close to L'Aquila, a town of about 70,000 inhabitants. The mainshock was followed by seven aftershocks of moment magnitude larger than or equal to 5, the two strongest ones occurred on April 7th (M_W=5.6) and April 9th (M_W=5.4). The mainshock and its aftershocks have been recorded by several digital stations of the RAN, the National Seismometric Network and by a temporary strong-motion array installed by the INGV (http://rais.mi.ingv.it).

This event represents the third largest recorded by strong-motion instruments in Italy, after the 1980, M_w 6.9, Irpinia and the 1976, M_w 6.4, Friuli earthquakes. The data set relative to this sequence includes 954 strong-motion waveforms with M_w larger than 4.0 (http://itaca.mi.ingv.it). The peak ground acceleration in the near-fault region ranges from 347 to 647 cm/s^2, the latter representing one of the highest values ever recorded in Italy.

This strong-motion data set is unique in Italy, since it is entirely digital and includes observations from near-fault distances to some hundred kilometres, so that it can be used to verify the prediction performance of the recently developed GMPEs for Italy (Bindi et al., 2009a). The goodness of fit of the main peak ground motion parameters (PGA, PGV and acceleration response spectra ordinates at 5% damping) is evaluated applying the method proposed by Spudich et al. [23], where the authors estimate the goodness of fit of their model by analyzing the residuals between observed and expected ground motion. In particular, they used the maximum likelihood formalism to calculate the mean value of the residuals (bias), and the dependence of the residuals on the magnitude and logarithm of distance.

In this work, the residuals were partitioned into the inter-event (η) and intra-event (ε) components, which are assumed to be independent, normally distributed with variances $\sigma^2{}_{eve}$ (inter-event component of variance) and $\sigma^2{}_{intra}$ (intra-event component of variance), respectively. The analysis of the inter-event error allows us to quantify the error associated to each event in the dataset, hence to investigate the dependence of errors on magnitude. On the other hand, the analysis of the intra-event distribution allows us to investigate the dependence of errors on attenuation by analyzing its dependence on distance. Figure 5.2a–d show the results of the goodness of fit obtained for the maximum horizontal PGA.

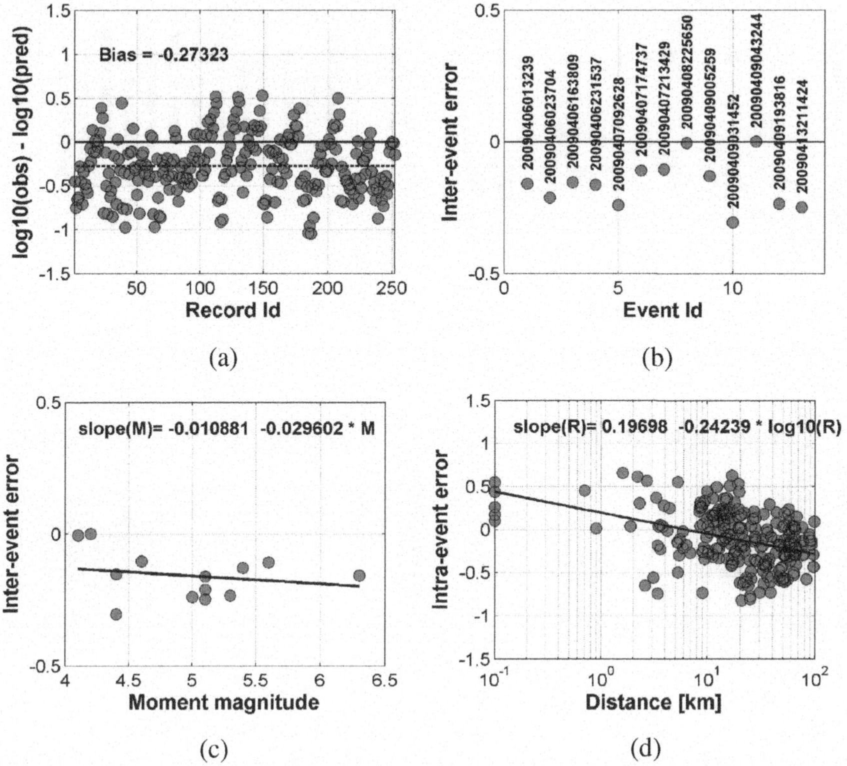

Fig. 5.2 Goodness of fit of PGA recorded during L'Aquila sequence to ITA08: (**a**) bias; (**b**) inter-event error; (**c**) dependence of the inter-event errors on magnitude; (**d**) dependence on Joyner Boore distance of the intra-event errors

The overall bias for PGA is negative (–0.27323), denoting a general overestimation of the prediction. Almost the totality of the events has a negative inter-event error, while the dependence of inter-event on magnitude is small. Finally a strong dependence is observed with distance, which means that PGA attenuates faster than the Italian average. The same analysis was performed on PGV, which is a ground motion parameter representative of intermediate frequencies (Fig. 5.3a–d). The fit of PGV to ITA08 results in a negative bias (–0.25933), confirming the overestimation of the prediction also obtained for PGA. The inter-event error shows no dependence on magnitude, while the intra-event error has a weaker dependence on distance than PGA.

Finally, when the acceleration response spectra ordinates at 2 s are taken into account (Fig. 5.4a–d), the overall bias reduces substantially to –0.12932, denoting only a slight overestimation of ground motion at large periods. The dependence of the inter-event error on magnitude is positive, while the dependence of the intra-event error on distance is negligible, denoting the same trend as the Italian average.

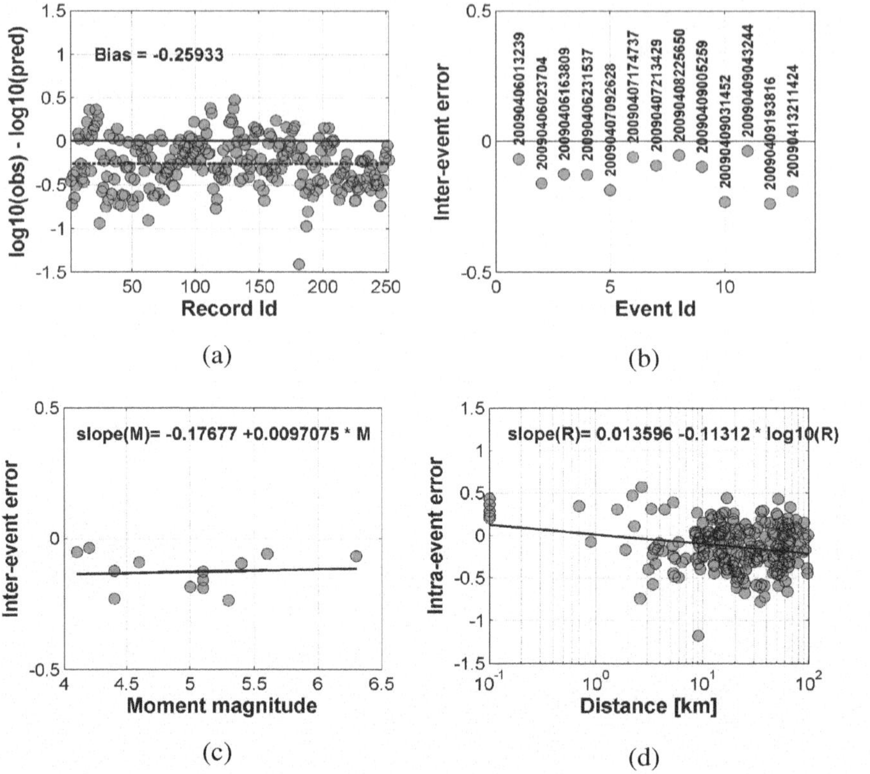

Fig. 5.3 Goodness of fit of PGV recorded during L'Aquila sequence to ITA08: (**a**) bias; (**b**) inter-event error; (**c**) dependence of the inter-event errors on magnitude; (**d**) dependence on Joyner Boore distance of the intra-event errors

We can conclude that the Italian GMPEs, developed by Bindi et al. (2009a), in general over predict the ground motions observed in the Abruzzo region, especially at high and intermediate frequencies. In particular, the lack of the anelastic coefficient in the functional form can be the cause the large overestimation of the ground motion at high frequencies and large distances.

The observations made for the Abruzzo region can be interpreted as a peculiarity of ground motion attenuation of this region, therefore an attempt to evaluate whether regional differences exist in Italy is made in the following paragraph, exploiting the updated ITACA dataset.

5.3 *Derivation of Regional GMPEs*

In the past decade, the improvement and expansion of strong motion networks in the world, led to an increase in the number of regional GMPEs, as reported by Douglas [13], number that has been increased certainly in the recent years. The idea

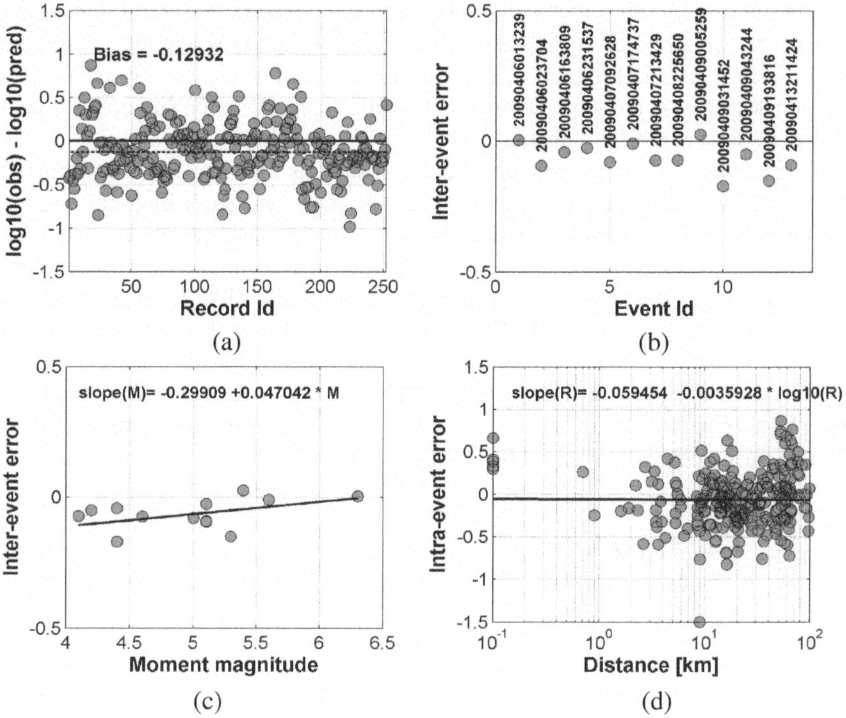

Fig. 5.4 Goodness of fit of SA (2 s) recorded during L'Aquila sequence to ITA08: (**a**) bias; (**b**) inter-event error; (**c**) dependence of the inter-event errors on magnitude; (**d**) dependence on Joyner Boore distance of the intra-event errors

behind the derivation of regional GMPEs is that the aleatoric uncertainty can be reduced, as the examined areas are expected to be homogeneous in terms of source, attenuation and site effects. Opposite to the expectation, the aleatoric uncertainty increases when compared to that obtained by global GMPEs, and one reason can be that small magnitude events, which are generally taken into account when deriving regional GMPE, are characterized by larger variability in the observed ground motion than large events [14]. An extreme example of regionalization occurs in Italy, where regionalized predictive relationships for ground motion are used for the development of ShakeMaps [20] and many GMPEs have been derived using regional datasets composed of weak and strong events ([10]; NE Italy), weak events ([16]; NW Italy) and datasets limited to single seismic sequences, such as the 1997 Umbria-Marche [7, 24] or the 2002 Molise [21]. Nevertheless, the examination of the parameters used to derive these GMPEs (i.e. magnitude and distance ranges, soil classification and functional forms) makes a direct comparison extremely difficult. Different regions can be characterized by different stress drops, focal depths, style-of-faulting or site conditions. Some of these effects, such as focal depth, can be modelled by the use of an appropriate distance metrics. The fault type causing the earthquake, is another factor that has been introduced in the functional form of the recently developed GMPE, as it has been demonstrated [9] that the motion generated from reverse faulting are in average higher (from 10 to 30% for PGA) than those generated by strike-slip and normal faulting events. Finally, soil conditions for different regions can differ substantially.

The advantage of a homogeneous database, such as ITACA, is the possibility of deriving regional datasets, unbiased in terms of event and recording station metadata, since it obtained from homogeneous and well documented sources. Therefore, in this study we did not chose to compare the existing regional GMPEs developed for Italy, but to explore the regional variability using the ITACA dataset. First we selected a set of well characterized seismic events (in terms of magnitude, style-of-faulting and focal depths), in the magnitude range from 4.0 to 6.9 and distance up to 300 km, then we evaluate a set of GMPEs for PGA and PGV for the Italian territory (hereinafter ITA-test) to estimate the average coefficients related to the style-of-faulting. We adopted the following functional form:

$$\begin{aligned}\log_{10} Y = a + b_1(M_w - M_{ref1}) + \left[c_1 + c_2(M_w - M_{ref2})\right] \\ \log_{10}\left(\sqrt{(R_{JB}^2 + h^2)}/R_{ref}\right) + k\left(\sqrt{(R_{JB}^2 + h^2)} - R_{ref}\right) + e_i S_i + f_j F_j\end{aligned} \quad (5.2)$$

where Y is the predicted ground motion (geometric mean of horizontal components of PGA, in cm/s^2, and PGV, in cm), M_w is the moment magnitude, M_{ref1} is the reference magnitude for the source term, M_{ref2} is a reference magnitude for the distance term, R_{JB} is the Joyner-Boore distance, or the epicentral distance (km) when $M_w < 5.5$, h is the pseudo-depth (km); k is the anelastic attenuation term. The adopted soil classification is the same as in Eurocode 8 [15], and S_i with $i = 1, 2, 3, 4, 5$ are dummy variables that assume either the value 0 or 1 depending on soil type (class A: $S_1 = 1$ and $S_2 = S_3 = S_4 = S_5 = 0$; class B $S_2 = 1$ and $S_1 = S_3 = S_4 = S_5 = 0$; class C: $S_3 = 1$ and $S_1 = S_2 = S_4 = S_5 = 0$, class D: $S_4 = 1$ and $S_1 = S_2 = S_3 = S_5 = 0$; class E: $S_5 = 1$ and $S_1 = S_2 = S_3 = S_4 = 0$); finally, F is the style-of-faulting term, which assumes the value 0 or 1 depending on the fault type (normal fault: $F_1 = 1$ and $F_2 = F_3 = 0$; strike-slip: $F_2 = 1$ and $F_1 = F_3 = 0$; reverse fault: $F_3 = 1$ and $F_1 = F_2 = 0$). M_{ref1} and M_{ref2} are set to 6.0 and 5.0, respectively and R_{ref} is set to unity. In the regression scheme we constrained the normal fault and the rock site coefficients to zero. A modified functional form was used to account for the hypocentral distance:

$$\begin{aligned}\log_{10} Y = a + b_1(M_w - M_{ref1}) + \left[c_1 + c_2(M_w - M_{ref2})\right] \\ \log_{10}\left(R_{hypo}/R_{ref}\right) + k\left(R_{hypo}/R_{ref}\right) + e_i S_i + f_j F_j\end{aligned} \quad (5.3)$$

where the variables are the same as in Eq. (5.2), except R_{hypo} which represents the hypocentral distance, in km. The same regression scheme adopted for ITA08 was used to derive the coefficients which are shown in Table 5.1. The style-of-faulting

Table 5.1 Coefficients obtained for PGA and PGV (*Area* = examined region; *R* = distance (km): R_{jb} Joyner-Boore distance, R_{hy} = hypocentral distance; a, b_1, c_1, c_2, h, k, f_1, f_2, f_3, s_1, s_2, s_3, s_4, s_5 = coefficients of Eqs. (5.2) and (5.3); σ_{inter}, σ_{intra}, σ_{tot}: inter-event, intra-event and total standard deviation of Eqs. (5.2) and (5.3))

Area	R	GM	a	b_1	c_1	c_2	H	k	f_1	f_2	f_3	s_1	s_2	s_3	s_4	s_5	σ_{nter}	σ_{intra}	σ_{tot}
Italy	Rjb	PGA	3.847	0.131	−1.831	0.263	10.034	−0.0003	0	0.117	0.018	0	0.172	0.225	0.104	0.442	0.193	0.295	0.353
Italy	Rhy	PGA	3.670	0.353	−1.485	0.154		−0.0027	0	0.191	0.073	0	0.211	0.218	0.044	0.397	0.220	0.300	0.371
Italy	Rjb	PGV	2.327	0.235	−1.670	0.311	9.350	0	0	0.065	0.032	0	0.200	0.264	0.300	0.307	0.199	0.270	0.336
Italy	Rhy	PGV	2.159	0.400	−1.368	0.229		−0.0014	0	0.129	0.077	0	0.235	0.258	0.255	0.269	0.226	0.277	0.358
Zone2	Rjb	PGA	3.984	0.221	−1.857	0.209	9.528	−0.0007				0	0.118	0.245	0.070	0.539	0.180	0.275	0.329
Zone2	Rhy	PGA	3.850	0.524	−1.501	0.050		−0.0026				0	0.175	0.232	0.017	0.512	0.188	0.287	0.343
Zone2	Rjb	PGV	2.330	0.3998	−1.577	0.202	7.656	0				0	0.158	0.273	0.302	0.388	0.175	0.267	0.319
Zone2	Rhy	PGV	2.391	0.589	−1.441	0.116		−0.00022				0	0.212	0.263	0.269	0.366	0.203	0.277	0.344

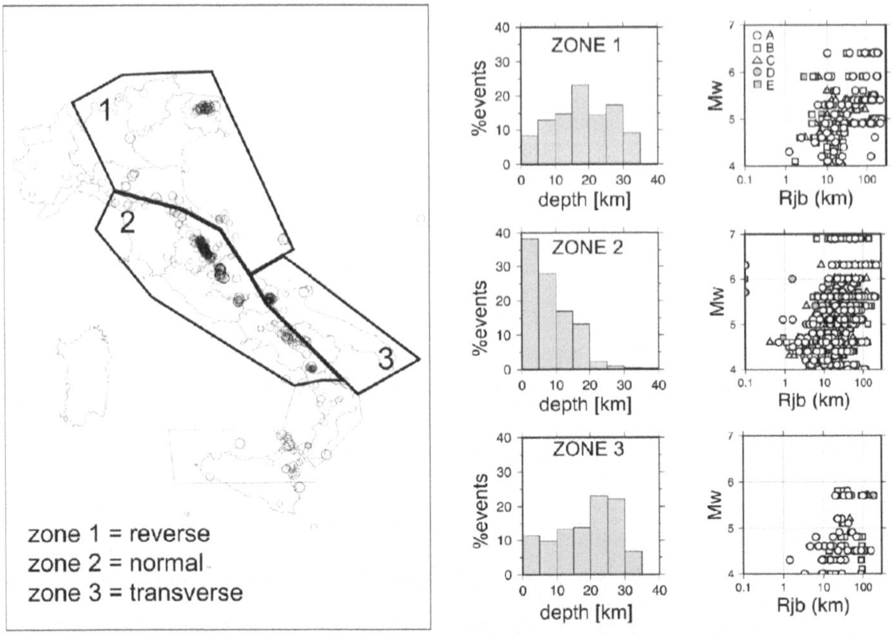

Fig. 5.5 Selected zones for the evaluation of regional effects (*left*), focal depth distribution and magnitude versus distance distribution of the recordings for each zone (*right*)

coefficients are significant and, in particular, the coefficients obtained for reverse focal mechanisms are larger than the coefficients of strike-slip events, especially in the case of PGA. In order to explore the possible regional differences in terms of ground motion attenuation, we identified three zones characterized by homogeneous tectonic regimes and focal depths, after an examination of the seismotectonic setting of the Italian peninsula (Fig. 5.5):

1. Zone 1 (NE Italy and northern Apennines), mainly characterized by a compressional tectonic regime, which includes the area struck by the 1976 Friuli (M_W 6.4) earthquake. The earthquake focal depth in Zone 1 is generally between 15 and 25 km.
2. Zone 2 (central-southern Apennines) characterized by an extensional regime, including the area struck by the strongest earthquake ever recorded by strong motion instruments in Italy, that is the 1980 Irpinia (M_W 6.9), the 1997 Umbria-Marche (M_W 6.0) and the recent L'Aquila event of April 6th 2009 (M_W 6.3); in Zone 2 earthquakes occurs in the upper crust at depths lower than 10 km.
3. Zone 3 (Apulian foreland) characterized by strike-slip regime, which includes the area struck by the 2002 Molise earthquake (M_W 5.7) and the 1995 Gargano (M_W 5.2); the earthquake focal depth in Zone 3 is generally between 20 and 30 km.

We derived a GM predictive equation for Zone 2, characterized by the largest data set, and then evaluated the goodness of fit of the datasets relative to Zone 1 and Zone 3, after a correction for the style-of-faulting coefficients obtained from ITA-test (Table 5.1). The used functional forms to derive the GMPE for Zone 2 (hereinafter referred to as ITA2) are the same as Eqs. (5.2) and (5.3), except that the style-of-faulting term has been neglected, being all FMs normal. The coefficients resulting from the regression are listed in Table 5.1 (ITA2 column). In order to explore the goodness of fit of the datasets relative to Zone 1 and 3 to ITA2 we used the method proposed by Spudich et al. [23], applied to the L'Aquila data set in paragraph 2. The results of the PGA indicate that, if the same site classification is adopted and the residuals are corrected for the style-of-faulting term, the bias evaluated for Zone 1 is negligible (Fig. 5.6a, b). The dependence of the inter-event on magnitude error is positive and mainly influenced by the error associated to the mainshock of the 1976 Friuli sequence, especially when the Joyner – Boore distance is considered (Fig. 5.6c, d). The dependence of the intra-event error on the logarithm of distance shows a positive trend for Zone 1, evidencing that, in general, ground motion attenuates slower than in Zone 2 (Fig. 5.6e, f). As the average focal depths of Zone 1 are larger than Zone 2, the adoption of the Joyner-Boore distance causes an underestimation of the true source-to-site distance; therefore the distance dependence is weaker when the hypocentral distance is used.

Similar considerations can be drawn for Zone3. The bias for PGA is very small and has almost the same value irrespective of whether hypocentral or Joyner- Boore distance is considered (Fig. 5.7a, b). The dependence on magnitude of the inter-event error is positive, but the paucity of the events does not allow drawing robust conclusions (Fig. 5.7c, d). There is no dependence of the intra-event error on the logarithm of distance when the hypocentral distance is adopted, where a positive trend is visible when Joyner-Boore distance is adopted. As Zone 3 is characterized by deep events, the use of Joyner-Boore distance instead of the hypocentral metrics results in a strong difference of the intra-event dependence on distance (Fig. 5.7e, f).

Smaller differences among zones are found for a parameter correlated to intermediate frequencies, such as PGV. The bias evaluated for Zone 1 is irrelevant, as shown in Fig. 5.8a, b, and the inter-event dependence on magnitude as well as the intra-event dependence on distance are also very small (Fig. 5.8b–f).

When the goodness of fit is evaluated for PGVs relative to Zone 3, similar conclusions can be drawn as for Zone 1 (Fig. 5.9b–f).

5.4 Conclusions

The recent release of the Italian strong-motion database (ITACA http://itaca.mi.ingv.it, [19]) created the opportunity to evaluate new ground motion prediction equations for Italy (ITA08, Bindi et al., 2009a) and update the well known GMPEs by Sabetta and Pugliese [22], which was based on a limited data set.

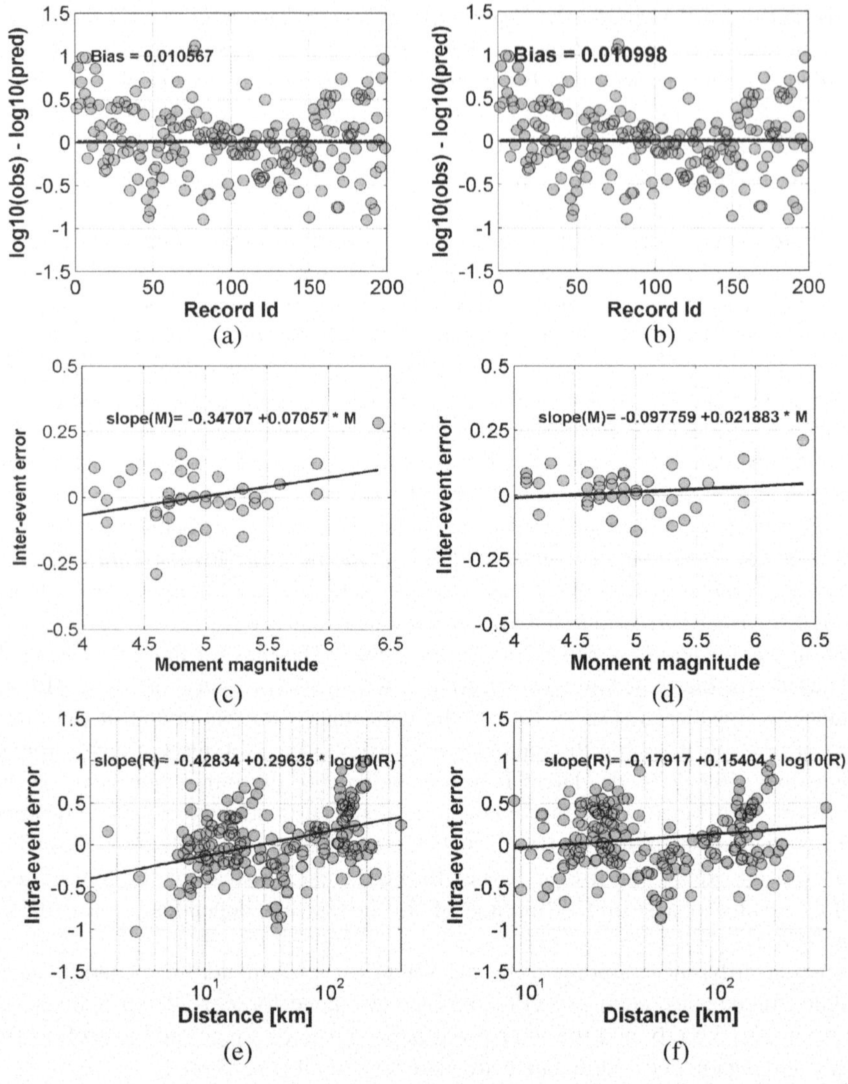

Fig. 5.6 Goodness of fit of PGA of Zone1 to ITA2: (**a**) bias obtained with Eq. (5.2); (**b**) bias obtained with Eq. (5.3); (**c**) dependence of the inter-event on moment magnitude for Eq. (5.2); (**d**) dependence of the inter-event on magnitude for Eq. (5.3); (**e**) dependence of the intra-event on Joyner-Boore distance; (**f**) dependence of the intra-event on hypocentral distance

The M_W 6.3 earthquake that occurred in central Italy on April 2009, and the upgrades of the ITACA database gave us the possibility to validate the predictive capability of the newly developed GMPEs and to explore the regional variability inside the Italian territory. The peak ground motions of L'Aquila sequence, correlated with high and intermediate frequencies (PGA and PGV), are generally

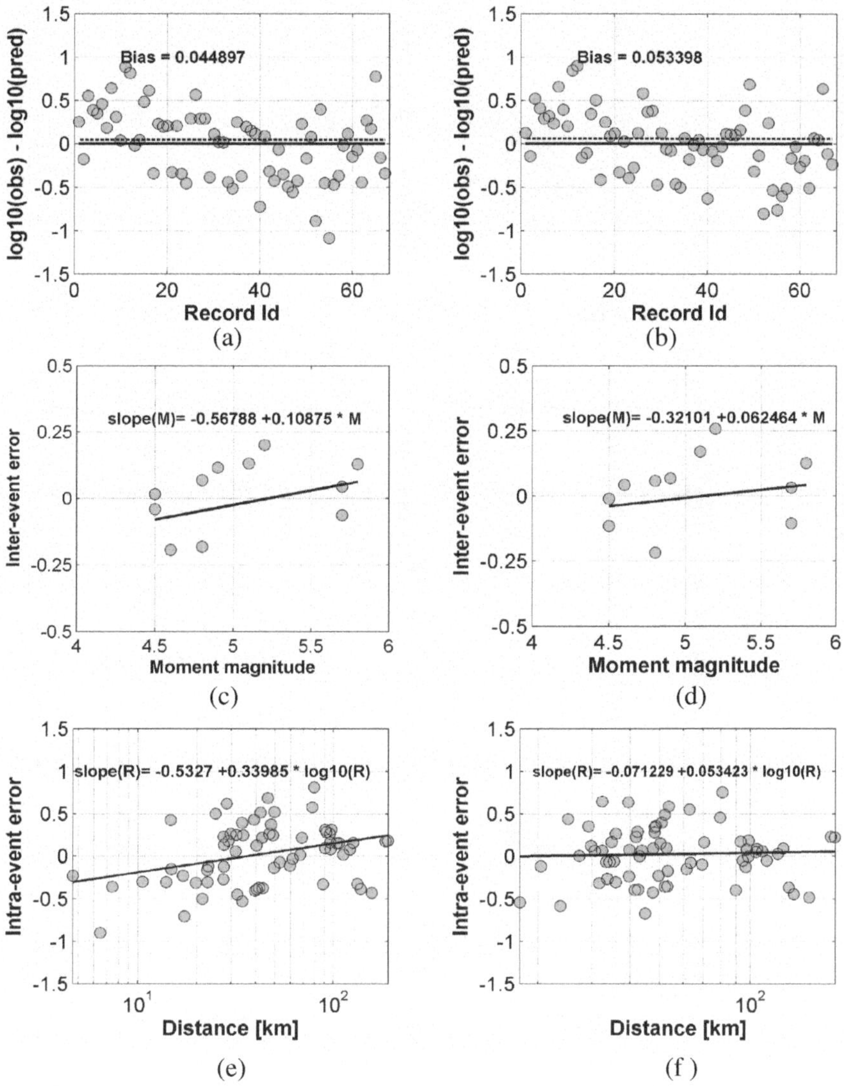

Fig. 5.7 Goodness of fit for PGA of Zone3 to ITA2: (**a**) bias obtained with Eq. (5.2); (**b**) bias obtained with Eq. (5.3); (**c**) dependence of the inter-event on moment magnitude for Eq. (5.2); (**d**) dependence of the inter-event on magnitude for Eq. (5.3); (**e**) dependence of the intra-event on Joyner-Boore distance; (**f**) dependence of the intra-event on hypocentral distance

over-predicted by the ITA08. The over-prediction is considerable for intermediate and large distances, while ground motions are underestimated in the near fault. While this misfit can be ascribed to the lack of the near fault recordings in the data set used to calibrate ITA08, the low PGA and PGV values at intermediate and large distances may be attributable to a strong attenuation of the crust occurring in

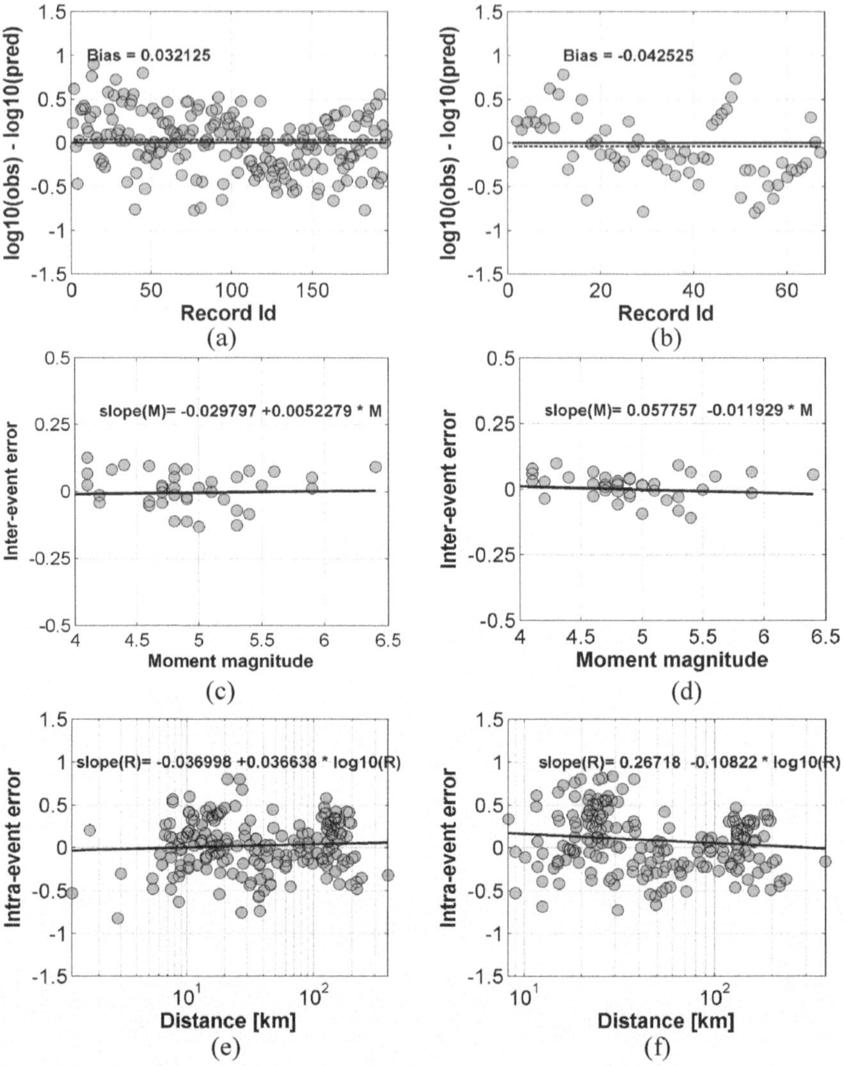

Fig. 5.8 Goodness of fit of PGV of Zone 1 to ITA2: (**a**) bias obtained with Eq. (5.2); (**b**) bias obtained with Eq. (5.3); (**c**) dependence of the inter-event on moment magnitude for Eq. (5.2); (**d**) dependence of the inter-event on magnitude for Eq. (5.3); (**e**) dependence of the intra-event on Joyner-Boore distance; (**f**) dependence of the intra-event on hypocentral distance

the western sector of the Apennines, as suggested by Ameri et al. [2]. Although no dependence of the inter-event error on magnitude is observed, all errors are negative, indicating a possible bias in the magnitude estimation of the L'Aquila seismic sequence.

The L'Aquila data set might suggest an evidence of a regional difference in the attenuation, and, in order to verify whether these observations are limited to the

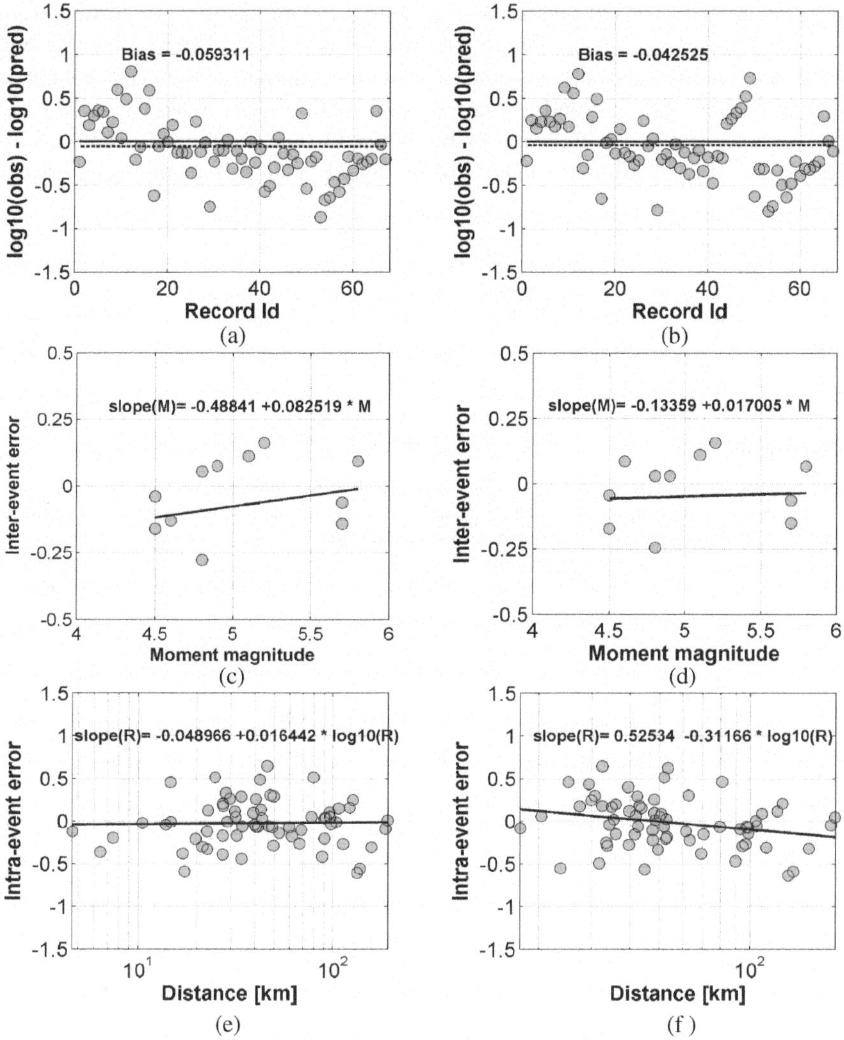

Fig. 5.9 Goodness of fit of PGV of Zone3 to ITA2: (**a**) bias obtained with Eq. (5.2); (**b**) bias obtained with Eq. (5.3); (**c**) dependence of the inter-event on moment magnitude for Eq. (5.2); (**d**) dependence of the inter-event on magnitude for Eq. (5.3); (**e**) dependence of the intra-event on JB distance; (**f**) dependence of the intra-event on hypocentral distance

examined sequence, we used the ITACA database to explore the different behaviour of three zones in Italy, characterized by different style of faulting and focal depths: north-east Italy, central southern Apennines and the Apulian foreland.

The results show that in the three regions the regional differences in the attenuation of PGA and PGV are negligible, when a homogeneous data set is used in terms of metadata, the opportune corrections are made for the style-of-faulting and the appropriate metrics is used to describe the source-to-site distance. In particular, the hypocentral distance allows reducing the error introduced by different focal depth.

After the conclusion of the ongoing project on the update of the ITACA database new data will be gathered and a final revision of the Italian GMPEs will be performed. Other regions in Italy might be explored (i.e. southern Tyrrhenian) and different soil classification schemes will be tested in order to reduce the error of the prediction.

Acknowledgments We are grateful to the organizing committee of the 2nd Euro-Mediterranean meeting on Accelerometric Data Exchange and Archiving (S. Akkar, P. Gülkan, A. Askan Gündoğan, T. van Eck, R. Bossu and F. Cotton) for inviting us to attend and contribute to the workshop.

References

1. Abrahamson NA, Youngs RR (1992) A stable algorithm for regression analyses using the random effects model. Bull Seismol Soc Am 82:505–510
2. Ameri G, Massa M, Bindi D, D'Alema E, Gorini A, Luzi L, Marzorati S, Pacor F, Paolucci R, Puglia R, Smerzini C (2009) The 6 April 2009, Mw 6.3, L'Aquila (Central Italy) earthquake: strong-motion observations. Seismol Res Lett 80:951–966
3. Attolico A, Gallipoli MR, Harabaglia P, Lapenna V, Mucciarelli M, Rosa AB (2009) A review of the activity of two accelerometric networks in Basilicata (Italy). Bull Earthquake Eng. doi: 10.1007/s10518-009-9129–2
4. Augliera P, D'Alema E, Marzorati S, Massa M (2009) A strong motion network in northern Italy: detection capabilities and first analysis. Bull Earthquake Eng. doi: 10.1007/s10518-009-9165–y
5. Bindi D, Luzi L, Massa M, Pacor F (2009a) Horizontal and vertical ground motion prediction equations derived from the Italian accelerometric archive (ITACA). Bull Earthquake Eng. doi: 10.1007/s10518-009-9130–9
6. Bindi D, Luzi L, Pacor F (2009b) Interevent and interstation variability computer for the Italian accelerometric archivi (ITACA). Bull Seismol Soc Am 99:2471–2488
7. Bindi D, Luzi L, Pacor F, Franceschina G, Castro RR (2006) Ground-motion predictions from empirical attenuation relationships versus recorded data: the case of the 1997–1998 Umbria-Marche, Central Italy, strong-motion data set. Bull Seismol Soc Am 96:984–1002
8. Bindi D, Luzi L, Pacor F, Sabetta F, Massa M (2009c) Towards a new reference ground motion prediction equation for Italy: update of the Sabetta-Pugliese (1996). Bull Earthquake Eng 7:591–608
9. Bommer JJ, Douglas J, Strasser FO (2003) Style-of-faulting in ground-motion prediction equations. Bull Earthquake Eng 1:171–203
10. Bragato PL, Slejko D (2005) Empirical ground-motion attenuation relations for the Eastern Alps in the magnitude range 2.5–6.3. Bull Seismol Soc Am 95:252–276
11. Brillinger DR, Preisler HK (1985) Further analysis of the Joyner–Boore attenuation data. Bull Seismol Soc Am 75:611–614
12. Costa G, Moratto L, Suhadolc P (2009) The Friuli Venezia Giulia accelerometric network: RAF. Bull Earthquake Eng. doi: 10.1007/s10518-009-9157–y
13. Douglas J (2003) Earthquake ground motion estimation using strong-motion records: a review of equations for the estimation of peak ground acceleration and response spectral ordinates. Earthquake Sci Rev 61(1–2):43–104
14. Douglas J (2007) On the regional dependence of earthquake response spectra. ISET J Earthquake Tech 44:71–99

15. ENV 1998-1-1, EUROCODE 8 (2002) Design Provisions for Earthquake Resistance of Structures. Seismic Actions and General Requirements of Structures. CEN/TC 250, Draft, May 2002
16. Frisenda M, Massa M, Spallarossa D, Ferretti G, Eva C (2005) Attenuation relationships for low magnitude earthquakes using standard seismometric records. J Earthquake Eng 9:23–40
17. Gorini A, Nicoletti M, Marsan P, Bianconi R, De Nardis R, Filippi L, Marcucci S, Palma F, Zambonelli E (2009) The Italian strong motion network. Bull Earthquake Eng. doi: 10.1007/s10518-009-9141–6
18. Iannaccone G, Zollo A, Elia L, Convertito V, Satriano C, Martino C, Festa G, Lancieri M, Bobbio A, Stabile TA, Vassallo M, Emolo A (2009) A prototype system for earthquake early-warning and alert management in southern Italy. Bull Earthquake Eng. doi: 10.1007/s10518-009-9131–8
19. Luzi L, Hailemikael S, Bindi D, Pacor F, Mele F (2008) ITACA (Italian accelerometric archive): a web portal for the dissemination of Italian strong motion data. Seismol Res Lett. doi:10.1785/gssrl.79.5
20. Michelini A, Faenza L, Lauciani V, Malagnini L (2008) ShakeMap implementation in Italy. Seismol Res Lett 79:688–697
21. Morasca P, Zolezzi F, Spallarossa D, Luzi L (2008) Ground motion models for the Molise region (Southern Italy). Soil Dyn Earth Eng 28:198–211
22. Sabetta F, Pugliese A (1996) Estimation of response spectra and simulation of nonstationary earthquake ground motions. Bull Seismol Soc Am 86:337–352
23. Spudich P, Joyner WB, Lindh AG, Boore DM, Margaris BM, Fletcher JB (1999) SEA99: a revised ground motion prediction relation for use in extensional tectonic regimes. Bull Seismol Soc Am 89:1156–1170
24. Zonno G, Montaldo V (2002) Analysis of strong ground motions to evaluate regional attenuation relationships. Ann Geoph 45:439–454

Chapter 6
Strong-Motion Networks in Greece and Their Efficient Use in the Derivation of Regional Ground-Motion Prediction Models

B. Margaris, A. Skarlatoudis, A. Savvaidis, N. Theodoulidis, I. Kalogeras, and S. Koutrakis

Abstract The destructive earthquakes that occurred over the last years in the broader Greek region urged the need for acquiring high quality strong ground motion recordings. This necessity led to the enhancement of accelerographic networks by deploying a significant number of new sensors all around Greece, improving their spatial coverage. Within the framework of this work, a new, more efficient, strong motion data processing technique is presented and the properties of the updated Greek strong motion database are presented. The resulted high quality dataset will be used in various applications of engineering seismology, soil dynamics and earthquake engineering with the main goal being the derivation of new ground motion prediction equations (GMPEs) for the broader Aegean area. Updated seismological and geotechnical information such as site classification of ITSAK accelerographic stations based on $\mathbf{V}_{S30}$, is adopted for the derivation of new GMPEs.

6.1 Introduction

The development of the Eurasia, Anatolia, African and Apulian plates define the seismotectonic setting of the broader Aegean region (Fig. 6.1). The convergence between the Eurasian and the African plate is taking place in the southern part of the Aegean in a N–S direction, at a rate of less than 1 cm/yr. However, the convergence of the Aegean plate to the African plate is taking place in a SW direction with a rate of ~3–3.5 cm/yr. This latter movement results into a well-defined Wadati – Beniof zone with intermediate-depth events due to the subduction of the East-Mediterranean lithosphere below the Aegean plate, dipping at a low angle (~30°) at shallow depths and a high angle (~45°) on the deeper branch [6]. The Anatolian Plate is moving westwards at a high rate (20–25 cm/yr) resulting in a strike-slip

B. Margaris (✉)
ITSAK – Institute of Engineering Seismology and Earthquake Engineering, 55102 Thessaloniki, Greece
e-mail: margaris@itsak.gr

S. Akkar et al. (eds.), *Earthquake Data in Engineering Seismology*, Geotechnical, Geological, and Earthquake Engineering 14, DOI 10.1007/978-94-007-0152-6_6,

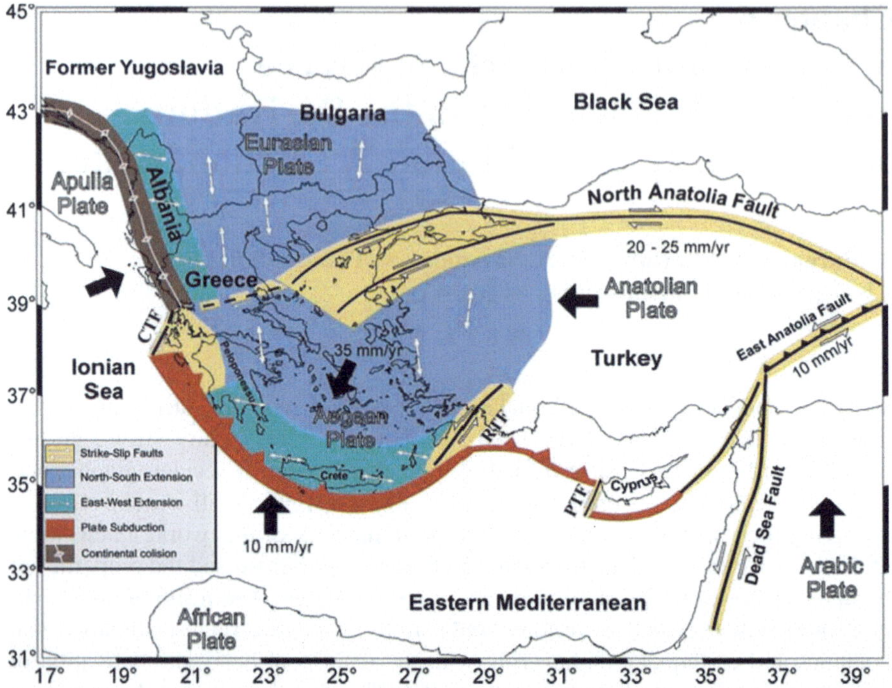

Fig. 6.1 Plate motions that affect the active tectonics in the broader Aegean area and its corresponding stress-field (modified from [7])

faulting in the North Anatolia-North Aegean and Cephalonia area. This highly complex seismotectonic setting in the broader Aegean area results in the highest seismicity release in Europe. A more complete description of these movements is provided in Papazachos et al. [6].

In the present work a newly developed data processing technique is applied to Greek strong motion data for the derivation of next-generation GMPEs. In addition, some preliminary results of the derivations of the new ground motion prediction equations in Greece are presented.

6.2 Greek Strong-Motion Network and Data Processing

The basic aim of the Greek strong motion network is to obtain the necessary information required for ground motion prediction, dynamic response of structures and improvement of seismic codes. Gradual installation of the first analog strong motion instruments began in the early 1970s by the Geodynamic Institute of National Observatory of Athens (NOA-GI), followed by the Institute of Engineering Seismology and Earthquake Engineering (ITSAK) in the 1980s, areas with the highest seismicity in Greece. The development and the spatial distribution of the accelerographic network in Greece were based mainly on the study of Theodoulidis

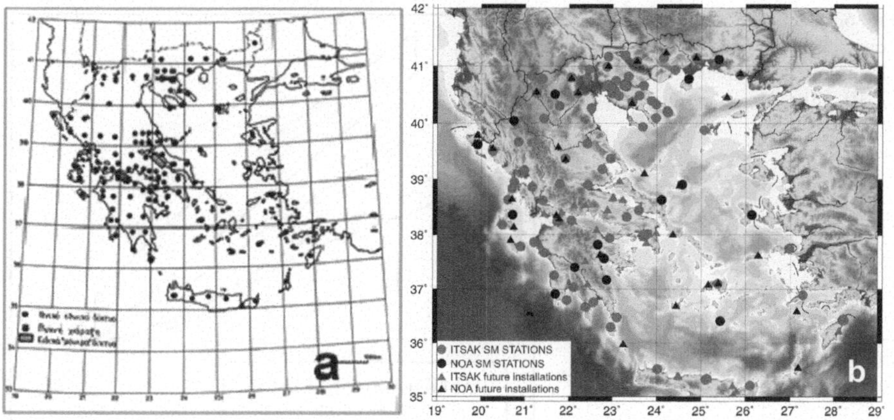

Fig. 6.2 **a** Proposed strong motion network in Greece [16]. **b** Current status and future installations of the Greek (ITSAK and NOA-GI) strong motion networks. *Circles* depict existing stations; *triangles* present stations being installed

et al. [16], regarding the seismotectonic setting, the seismicity and seismic hazard of the region. The original planning for the deployment of strong motion networks in Greece is depicted in Fig. 6.2a. After almost 25 years and two major upgrade projects (one is still ongoing), the current status of the national accelerographic network presented in Fig. 6.2b, is in good agreement with the one proposed in the mid 1980s by Theodoulidis et al. [16].

Both analog and digital instruments have been employed for recording strong ground motions in Greece. From the early 1980s until the late 1990s ITSAK and NOA-GI strong motion networks (both referred to as national strong motion network hereinafter) consisted mainly of, analog, SMA-1-type, instruments. Analog record processing was based on the technique suggested by Skarlatoudis et al. [10], aiming at reducing the noise level introduced during recording and digitization. The analog accelerograms were digitized using a high-resolution scanner and an appropriate software tool [5], thus reducing the digital noise introduced, compared to previous (manual or semi-automatic) digitization techniques. To further improve the quality of the strong motion dataset, a new correction procedure was applied in the uncorrected accelerograms. This procedure involves the estimation of the characteristic frequencies of the digital band-pass filter, based on comparisons of the Fourier Amplitude Spectra (FAS) of the digitized horizontal components with the FAS of the corresponding fixed traces. Attempts to further reduce noise levels were made by estimating characteristic frequencies based only on the energy window from 5 to 95% of the total energy of the accelerogram [9]. For this reason the FAS for both horizontal components and fixed traces, were recalculated for the equivalent window ($T_{5\text{–}95}$) and comparisons with results of Skarlatoudis et al. [10] and Trifunac and Todorovska [18] are shown in Fig. 6.3. These efforts led to the compilation of the unified Hellenic Accelerograms Database (HEAD) by Theodoulidis et al. [15].

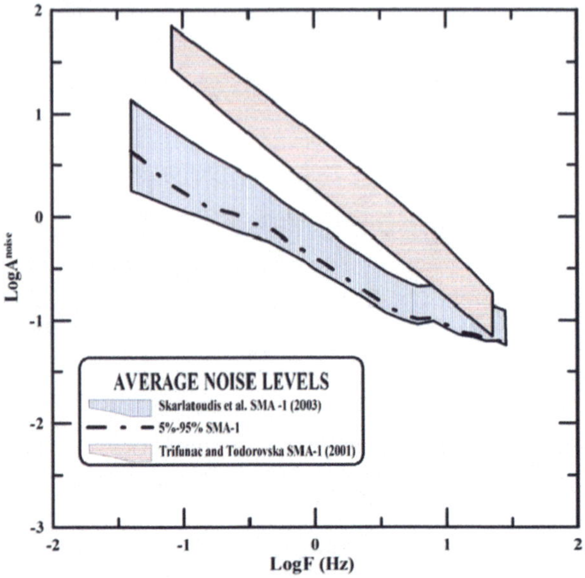

Fig. 6.3 Average spectral noise levels from various processing techniques [9, 10, 18]

Recently, Boore [1, 2] proposed a more efficient technique, based on both time and frequency domain analysis, which was used to reprocess the complete strong motion dataset. An example of this analysis is presented in Fig. 6.4. By visual inspection of the FAS of all components of ground motion we define a preliminary frequency window, in order to estimate the cut-off frequency, fc, of the high-pass filter applied to each record. For this filtered frequency window, the displacement

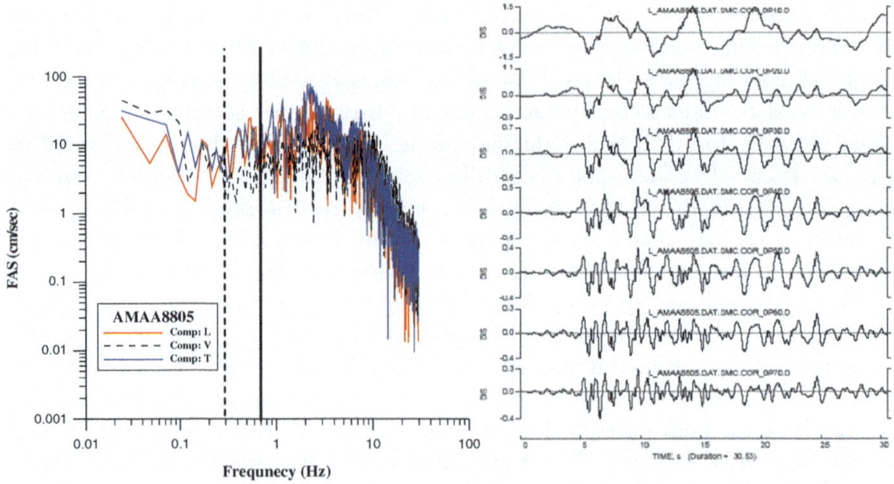

Fig. 6.4 (*Left*) FAS of a 3-component accelerogram (AMAA8805). The *vertical lines* correspond to cut-off frequencies based on Skarlatoudis et al. [10] (*continuous line*) and present study (*dashed line*) results (*Right*). Horizontal component (L-longitudinal) of displacement time histories corresponding to the various cut-off frequencies applied (0.10–0.70 Hz). The selected displacement time-history corresponds to a cut-off frequency of 0.30 Hz (third from *top*)

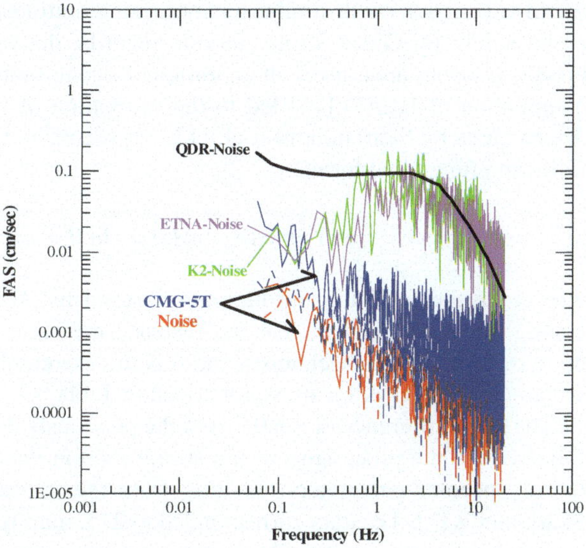

Fig. 6.5 Spectral noise comparison for the various digital sensors used by the Greek national strong motion network

time history is calculated for ten specific fc (logarithmically equally spaced) and after visual inspection the appropriate fc is selected in which the displacement values are stable without including long period effects or transients. The filtering procedure involves zero-padding in the beginning and at the end of each record based on the order of the applied digital filter and the use of a causal band-pass filter for digital noise removal [1]. The two vertical lines in FAS (Fig. 6.4) depict the characteristic cut-off frequencies determined based on Skarlatoudis et al. [10] (continuous line) and on Boore [1; 2] (dashed line) approaches, respectively. From this example it is evident that the adopted procedure improves significantly the usable frequency content as compared to the earlier results by Skarlatoudis et al. [10].

Since the early 2000s the national strong motion network was upgraded using various types of accelerographs (Kinemetrics QDR-11bit, A800/A900-18bit, ETNA-18bit, K2-19bit and Güralp CMG-5TD-24bit).

In order to shed some light into the average noise properties of various sensors and define an effective frequency window from which information for the strong motion characteristics could reliably be extracted, additional noise analysis was performed [9]. The spectral noise levels among the various accelerographs are shown in Fig. 6.5. As expected, higher resolution sensors exhibit lower levels of digital spectral noise.

6.3 Greek Ground Motion Prediction Equations

The first efforts for estimating empirical prediction relations (GMPEs) for the broader Aegean area were made by Theodoulidis [14] and Theodoulidis and Papazachos [17] (TP92), based on strong motion data recorded in Greece and other worldwide regions with similar seismotectonic settings. Margaris et al. [4] (Mea02), based solely on Greek strong motion records, derived updated relations for the broader Aegean area, used on an enriched strong motion dataset at that time. The compilation of HEAD [15] led to the estimation of new GMPEs for the broader Aegean area by Skarlatoudis et al. [12, 13] (Sea03). All previous GMPEs have the following functional form:

$$\ln Y = c_0 + c_1 M + c_2 \ln R + c_3 F + c_4 S \tag{6.1}$$

where, Y is the predicting strong motion parameter, M is the magnitude (type varies among the three studies), R is the distance measure, F is a factor accounting for the type of the focal mechanism and S is the site condition. The coefficients of the aforementioned GMPEs are summarized in Table 6.1.

The main difference of Sea03 from the previously derived relations is the method followed for the processing of the strong motion dataset [10] and the more accurate hypocenter parameters resulting from the relocation technique described in Skarlatoudis [8]. The data corresponded to 225, mainly normal and strike-slip faulting, shallow earthquakes in Greece spanning the time period 1973–1999. In the map of Fig. 6.6, the spatial distribution of the corresponding earthquake locations (black circles) is shown.

Table 6.1 Coefficients of TP92, Mea02 and Sea03 GMPEs for peak ground acceleration

Source	c_0	c_1	c_2	c_3	c_4
TP92	3.88	1.12	−1.65	–	0.41
Mea02	4.16	0.69	−1.24	–	0.12
Sea03	0.86	0.45	−1.27	0.1	0.06

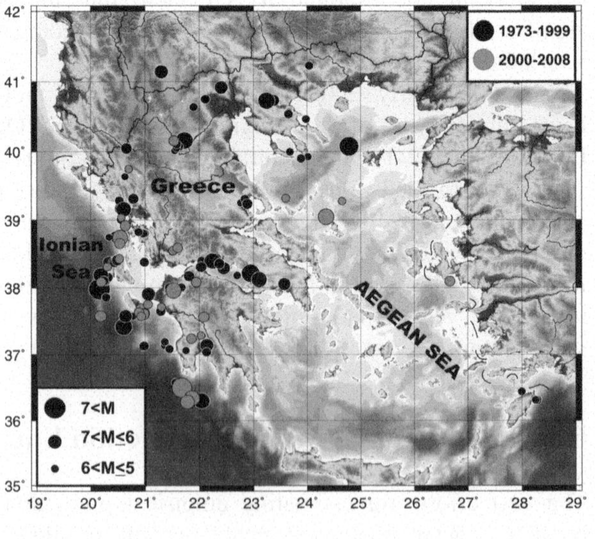

Fig. 6.6 Spatial distribution of the epicenters of earthquakes with magnitude $M > 5$. Epicenters from HEAD are denoted with *black circles* while epicenters for the time period 2000–2008, for earthquakes with processed strong motion data, are shown with *grey circles*

The dataset used for the derivation of Sea03 relations comprised of about 700 strong motion recordings with magnitudes 4.5⇐*M*⇐7.0 and hypocentral distances 1⇐R⇐170 km. The selected records satisfied at least one of the following criteria: (a) the earthquake had moment magnitude $M \geq 4.5$, (b) The strong motion record had peak ground acceleration $PGA \geq 0.05$ g regardless of earthquake magnitude or, (c) Records with $PGA < 0.05$ g should be accompanied by records with $PGA \geq 0.05$ g from the same earthquake.

The upgrade of the national network started in the late 1990s resulted in a high-quality strong motion network with a dense spatial coverage in the broader Aegean area. During the period of 2000–2008, a large number of events with magnitude $M \geq 5$ were recorded by the national strong motion network. In the following sections, the term "magnitude" will be used without any distinguish between different *M* scales. In Fig. 6.6 the spatial distribution of earthquakes (grey circles), with strong motion records included in the present study, is shown. The strong motion dataset for the period 2000–2008 will be further enhanced with data from earthquakes with magnitude $M > 5$ that produced at least two strong motion recordings. The final enhanced and updated dataset will be employed to derive new GMPEs for the broader Aegean area.

The distribution of peak ground acceleration (PGA) of the currently processed strong motion records as a function of magnitude *M* is shown in the left plot of Fig. 6.7. The ordinates of the plot correspond to the gmroti50 of the horizontal values of PGA as described in the work of Boore et al. [3]. The right plot depicts the magnitude, *M*, distribution as a function of hypocentral distance. The black circles correspond to data for the time period 1980–1999 and the grey squares correspond

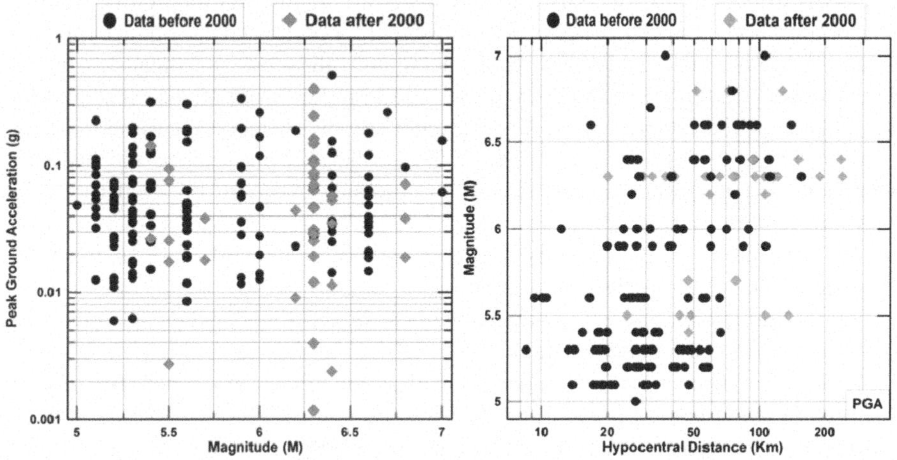

Fig. 6.7 (*Left*) Distribution of Peak Ground Acceleration (in g units) as a function of magnitude *M* and (*Right*) Distribution of magnitude *M* as a function of hypocentral distance. Data from HEAD (before 2000) are denoted with the *black circles* while data after 2000 are shown with the *grey squares*

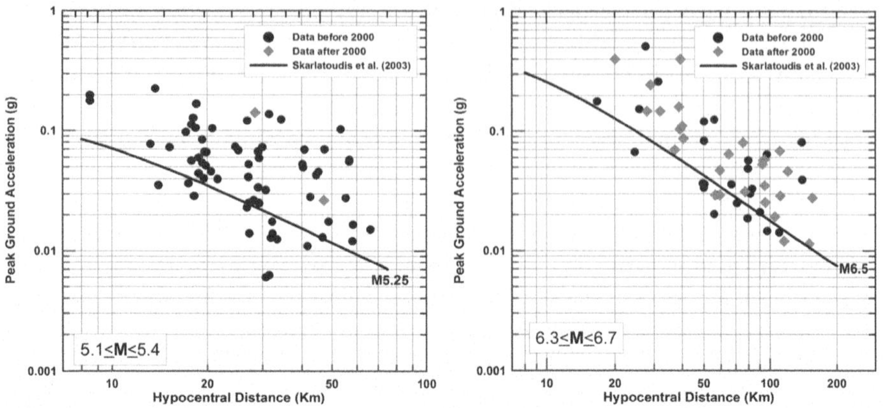

Fig. 6.8 Comparison of the Sea03 empirical prediction relation plotted for *M*5.25 (*left*) and *M*6.5 (*right*), normal fault and rock soil conditions, together with available strong motion data. Data within the magnitude range of *M*5.1–*M*5.4 (*left*) and *M*6.3–*M*6.7 (*right*) are used, respectively. Symbols and coloring are described in the figure legend

to data from 2000 to present. As it can be seen in Fig. 6.7, data after 2000 (processed so far) improve the completeness, in terms of magnitude and hypocentral distances, of the existing strong motion data set.

In Fig. 6.8 the Sea03 empirical prediction relation, plotted for magnitudes *M*5.25 (left) and *M*6.5 (right), for normal faults and rock soil condition is shown together with PGA values of the horizontal components. Only data within the magnitude range of *M*5.1–*M*5.4 are used in the left plot and *M*6.3–*M*6.7 in the right plot, respectively.

Underestimation of observed values by the Sea03 prediction relation is evident. This preliminary comparison shows that new GMPEs for crustal earthquakes should be computed in order to describe more accurately the observed data. The functional formulation of new GMPEs will be based on the formulations presented by the authors of the NGA project. Thus, except for the distance measures used in Sea03 (epicentral and hypocental), R_{jb} will also be used to describe the distance dependence of the data. Various strong motion measures will be studied (PGA, PGV, gmrotd50, gmroti50, etc) in order to define more accurate prediction relations. Furthermore, for the majority of the strong motion network stations, the $\mathbf{V}_{S30}$ measure will be used for site-classification.

Due to the subduction process taking place in the broader Aegean area (see Fig. 6.1) intermediate depth earthquakes are also contributing to the high seismicity of the area. The source properties as well as the effects of deeper structure of this type of earthquakes are significantly different compared to the shallow crustal earthquakes. Therefore different realizations of GMPEs should be used in order to accurately predict the induced strong ground motions. In Fig. 6.9 a comparison of the Sea03 GMPE with data from the 2006 Kythera (*M*6.7) intermediate depth (D=67 km) event is shown, suggesting the need for derivation of new GMPEs for intermediate depth earthquakes.

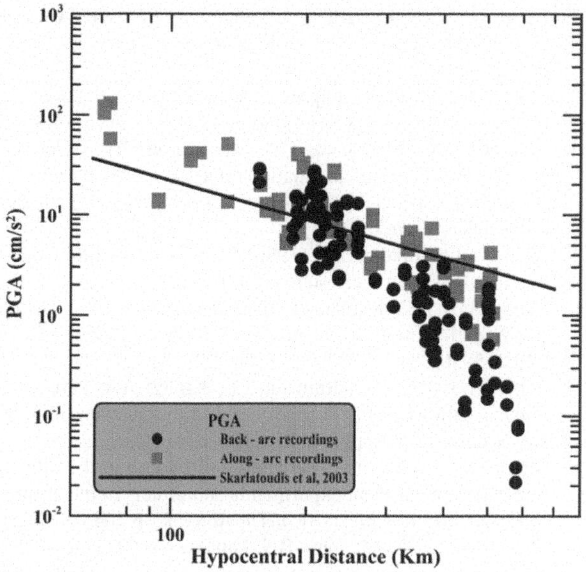

Fig. 6.9 Comparison of the Sea03 GMPE for shallow earthquakes with observed data from the 2006 Kythera (*M*6.7) intermediate depth earthquake (D=67 km) (Figure modified from [11])

6.4 Conclusions

An updated strong motion record database is currently being prepared for the broader Aegean area based on data from the national strong motion network (ITSAK – NOA). This database is compiled using high-quality and efficiently processed strong motion recordings as well as updated and more accurate metadata information such as, source parameters (magnitude, fault plane solutions etc) and site characterization ($\mathbf{V}_{S30}$ for most of the sites of the network). These data will form the basis for the derivation of new ground motion prediction equations (GMPEs) for the area of Greece encompassing the knowledge gained from the NGA project. The uncorrected and processed strong motion data will be disseminated by ITSAK and NOA-GI Website databases for engineering purposes.

Acknowledgements We are profoundly indebted to Dr. David M. Boore for his significant contribution in the strong motion data processing and his comments of the present paper. We would like to thank, Prof. J. Stewart (UCLA), Pr. N. Klimis (DUTH-Greece), Profs. G Mylonakis and G. Athanasopoulos (TUPatras-Greece), Dr. R. Kayen (USGS) and Pr. S. Foti (Pol. Torino – Italy) for their valuable help in the site classification in Greek accelerographic sites. Special thanks are due to Dr Toril Van Eck for his critical reviews/corrections and, Pr. George Mylonakis for his careful reading of the manuscript as well. This work is supported by European Community research project, Transfer of Knowledge, ITSAK-GR: MTKD-CT-2005-029627.

References

1. Boore DM (2005) On pads and filters: processing strong-motion data. Bull Seismol Soc Am 95:745–750

2. Boore DM (2009) TSPP – A Collection of FORTRAN programs for processing and manipulating time series, U.S. geological survey open file rept 2008-1111, V2.0, 10 Dec 2009, p. 52.
3. Boore DM, Watson-Lamprey J, Abrahamson NA (2006) GMRotD and GMRotI: orientation-independent measures of ground motion. Bull Seismol Soc Am 96:1502—1511
4. Margaris BN, Papazachos CB, Papaioanou CH, Theodoulidis N, Kalogeras I, Skarlatoudis AA (2002) Empirical attenuation relations for the horizontal strong ground motion parameters of shallow earthquakes in Greece, *in the Proc. of 12th European conference on earthquake engineering*, London.
5. Nigbor RL, Kodama. DA (1990) Scanner-based film accelerogram digitization system user's manual, Kinemetrics/Systems, 1.1–6.2.
6. Papazachos BC, Karakostas VG, Papazachos CB, Scordilis EM (2000) The geometry of the Wadati-Benioff zone and the lithospheric kinematics in the Hellenic arc. Tectonophysics 319:275–300
7. Papazachos BC, Papadimitriou EE, Kiratzi AA, Papazachos CB, Louvari EK (1998) Fault plane solutions in the Aegean sea and the surrounding area and their tectonic implications. Boll Geofis Teor Appl 39:199–218
8. Skarlatoudis AA (2002) Relocation of hypocentral earthquake parameters of Greece with the use of data from local experiments and effects in the strong motion attenuation relations, *Msc Thesis, Aristotle University of Thessaloniki*, p 204 (*in Greek*).
9. Skarlatoudis AA, Margaris BN, Papazachos CB (2004) Recent advances in Greece on strong motion networking and data processing, Proc. COSMOS meeting, Berkeley, CA, May 2004.
10. Skarlatoudis AA, Papazachos CB, Margaris BN (2003a) Determination of noise spectra from strong motion data recorded in Greece. J Seismol 7:533–540
11. Skarlatoudis AA, Papazachos CB, Margaris BN, Papaioannou Ch, Ventouzi Ch, Vamvakaris D, Bruestle A, Meier T, Friederich W, Stavrakakis G, Taymaz T, Kind R, Vafidis A, Dahm T EGELADOS group (2009) Combination of strong-and weak-motion recordings for attenuation studies: the case of the January 8, 2006 Kythera intermediate-depth earthquake. Bull Seismol Soc Am 99:694–704
12. Skarlatoudis AA, Papazachos CB, Margaris BN, Theodoulidis N, Papaioannou CH, Kalogeras I, Scordilis EM, Karakostas VG (2003b) Empirical peak ground motion predictive relations for shallow earthquakes in Greece. Bull Seismol Soc Am 93:2591–2603
13. Skarlatoudis AA, Papazachos CB, Margaris BN, Theodoulidis N, Papaioannou CH, Kalogeras I, Scordilis EM, Karakostas VG (2007) Erratum in paper "Empirical peak ground motion predictive relations for shallow earthquakes in Greece". Bull Seismol Soc Am 97:2219–2221
14. Theodoulidis NP (1991) Contribution to strong ground motion study in Greece, *PhD Thesis* (in Greek), 500 pp.
15. Theodoulidis N, Kalogeras I, Papazachos CB, Karastathis V, Margaris BN, Papaioannou Ch, Skarlatoudis AA (2004) HEAD v1.0: a unified HEllenic Accellerogram Database. Seismol Res Lett 75:36–45
16. Theodoulidis N, Margaris B, Papastamatiou D (1986) Planning of strong motion Network, Report ITSAK, Rept: 86-04, 148 pp.
17. Theodoulidis N, Papazachos B (1992) Dependence of strong ground motion on magnitude-distance, site geology and macroseismic intensity for shallow earthquakes in Greece: i, peak horizontal acceleration, velocity and displacement. Soil Dyn Earthq Eng 11:387–402
18. Trifunac MD, Todorovska. MI (2001) A note on the useable dynamic range of accelerographs recording translation. Soil Dyn Earthq Eng 21:275–286

Chapter 7
Ground Motion Simulation Using the Hybrid Empirical Method: Issues and Insights

K.W. Campbell

Abstract The widespread application of the hybrid empirical method (HEM) has made it a viable approach for developing ground motion prediction equations (GMPEs) in regions where there are few strong motion recordings but ample weak motion data from small-to-moderate magnitude earthquakes. The HEM uses empirical estimates of ground motion in a host region to provide estimates of ground motion in a target region by taking into account differences in source, path, and site effects between the two regions. Empirical ground motion estimates in the host region are transferred to the target region using adjustment factors that are calculated from regionally constrained seismological models using stochastic simulation. In this paper, I discuss the issues and demonstrate the epistemic uncertainty involved in applying the HEM using an example application to eastern North America (ENA) based on the Campbell-Bozorgnia NGA GMPE for western North America (WNA) and updated seismological models for ENA.

7.1 Introduction

The number and use of ground motion prediction equations (GMPEs) for seismic hazard studies in eastern North America (ENA) and other stable continental regions (SCRs) throughout the world have progressed rapidly over the last decade. As a result of the limited number of strong motion recordings in ENA and other SCRs, these models have been developed using a variety of theoretical and semi-theoretical methods (see history and summaries in [21, 23]). One method that has gained increasing popularity during the last decade is the hybrid empirical method (HEM), first introduced in the early 1980s by the author and later formalized by Campbell [20, 21]. Of the seven ENA ground motion models that have been selected for use in the 2008 update of the US national seismic hazard maps [38], two were developed using the HEM [21, 22, 46]. Other recent applications of the HEM have

---

K.W. Campbell (✉)
EQECAT, Inc., Beaverton, OR, USA
e-mail: kcampbell@eqecat.com

S. Akkar et al. (eds.), *Earthquake Data in Engineering Seismology*, Geotechnical, Geological, and Earthquake Engineering 14, DOI 10.1007/978-94-007-0152-6_7,

been conducted by Atkinson [5] in ENA, Scherbaum et al. [43] in central Europe, Atkinson [3] in the US Pacific Northwest, and Douglas et al. [31] in southern Spain and southern Norway. Atkinson [5] refers to her method as the referenced empirical approach, but considers it similar in concept to the HEM in that it makes use of an empirical GMPE (GMPE) from an other region [17]. The difference is that it uses empirical rather than theoretical regional adjustment factors.

In this paper, I demonstrate some of the issues involved in applying the HEM using as an example application to the development of a set of hybrid empirical hard-rock ground motion estimates for ENA (HE-GMPE) that was initially intended to serve as an update to the HE-GMPE developed by Campbell [21, 22]. This is the same application attempted by Campbell [23], except that the ground motions are for hard-rock sites ($V_{S30} = 2{,}000$ m/s) rather than firm-rock sites ($V_{S30} = 760$ m/s). It updates and extends a similar study by Campbell [24]. The updated HE-GMPE incorporates a revised ENA seismological model that was developed from an expanded set of weak motion data [2, 8] and a revised E-GMPE that was developed from an expanded set of strong motion data from western North America (WNA) and other active tectonic regimes [27].

The tentatively revised HE-GMPE is shown to provide ground motion estimates at moderate-to-large magnitudes that are similar to those predicted from the point-source stochastic model at large distances and the finite-source stochastic model at moderate-to-large distances when the same ENA seismological model is used to perform the ground motion simulations. However, as discussed later and confirmed by Atkinson et al. [6] and Boore [15], I find that this agreement can only be achieved if a larger stress drop then that used by Atkinson and Boore [8] is used with the point-source stochastic model. Because of its reliance on a well-constrained E-GMPE, the HE-GMPE offers an alternative, more empirically based, method for predicting near-source ground motions from large-magnitude earthquakes in ENA and other SCRs as long as the issues raised by Campbell [23, 24, 25] and augmented and reinforced in this paper are adequately addressed.

7.2 Example Application to ENA

The five steps that are required to implement the HEM are (1) the selection of a host and target region, (2) the calculation of empirical ground motion estimates in the host region, (3) the calculation of regional adjustment factors between the target and host regions, (4) the calculation of hybrid empirical ground motion estimates in the target region, and (5) the development of a HE-GMPE for the target region. I refer the reader to Campbell [21] for a detailed explanation of the mathematical framework involved in applying these steps. For the current example application, I selected the target region to be that area of ENA bounded on the west by the Rocky Mountains and on the south by the Gulf Coast region of the US. I selected the host region to be that area of WNA located west of the eastern front of the Rocky Mountains. Because of the preliminary nature of this study, I did not execute step 5 to formally develop a GMPE. Instead, I directly used the ground motion estimates from step 4. Nonetheless, I refer to these hybrid empirical estimates as if they represent a HE-GMPE in the remainder of the paper as a matter of convenience.

7.2.1 WNA Empirical Ground Motion Estimates

The E-GMPE of Campbell and Bozorgnia [27] is used to derive the empirical ground motion estimates in WNA. This model was developed as part of the Pacific Earthquake Engineering Research Center (PEER) Next Generation Attenuation (NGA) project [39]. Although most of the strong motion recordings used to develop this E-GMPE are from California, there are many other recordings from other parts of the western US and the world with tectonic characteristics similar to California. The validity of using strong motion data from crustal earthquakes in these other geographical regions has been verified for southern Europe and the Middle East [26, 37, 45], Taiwan [34], Italy [42], and Iran [44]. The E-GMPE is used to predict the horizontal components of peak ground acceleration (PGA), peak ground velocity (PGV), and acceleration response spectra (PSA) for periods ranging from 0.01 to 10 s. It is evaluated for moment magnitudes (**M**) ranging from 5.5 to 8.0 and earthquake rupture distances (R_{RUP}) ranging from 1 to 70 km. The reason for restricting the calculations to moderate-to-large magnitudes and near-source distances is explained later in the paper. The remaining dependent variables in the E-GMPE are assigned values consistent with the WNA seismological model [23].

7.2.2 WNA-to-ENA Regional Adjustment Factors

Based on its success in modeling a wide range of ground motion parameters [13], I use the stochastic method and a Brune omega-square single-corner point-source (SCPS) source spectrum to calculate seismological estimates of ground motion. A general discussion of the application of the stochastic method within the mathematical framework of the HEM is given by Campbell [21]. Campbell [23] found that the stochastic model parameters for WNA did not require modification from those used to develop the previous HE-GMPE [21, 22], after comparing stochastic simulations with the Campbell-Bozorgnia NGA E-GMPE for moderate magnitudes and short distances where finite-source and attenuation effects are negligible.

Seismological model parameters for ENA are updated based on the seismological studies of Atkinson [2] and Atkinson and Boore [8]. These latter investigators performed stochastic finite-source ground motion simulations for two different generic site profiles: (1) a traditional ENA hard-rock profile ($V_{S30} = 2{,}000$ m/s) and (2) a softer National Earthquake Hazard Reduction Program (NEHRP) BC ($V_{S30} = 760$ m/s) site profile [33]. Since I restricted the example application to hard-rock site conditions, I use the hard-rock site profile for the ENA stochastic ground motion simulations. There are three notable differences between the new ENA seismological parameters used in the current example application and those used by Campbell [21]: (1) the median stress drop is increased from 150 to 280 bars (see

Table 7.1 Seismological parameters used in the stochastic simulations

Parameter	WNA	ENA
Source spectrum	Brune ω-square SCPS	Brune ω-square SCPS
S-velocity at source (km/s)	3.5	3.7
Density at source (gm/cc)	2.8	2.8
Stress drop (bars)	100	140, 280
Source duration (s)	$1/f_0$ (f_0 = corner frequency)	$1/f_0$ (f_0 = corner frequency)
Geometrical spreading	$r^{-1.0}$ ($r < 40$ km) $r^{-0.5}$ ($r \geq 40$ km)	$r^{-1.3}$ ($r < 70$ km) $r^{+0.2}$ ($70 \leq r < 140$ km) $r^{-0.5}$ ($r \geq 140$ km)
Path duration (s)	$0.05r$	0.0 ($r < 10$ km) $+0.16r$ ($10 \leq r < 70$ km) $-0.03r$ ($70 \leq r < 130$ km) $+0.04r$ ($r \geq 130$ km)
Path attenuation (Q)	$180\,f^{0.45}$	$893\,f^{0.32}$ (1,000 minimum)
Site profile	WNA generic rock	ENA hard rock
Site amplification	¼-wavelength method	¼-wavelength method
Site attenuation, κ_0 (s)	0.04	0.005

explanation below), (2) the exponent n in the near-source geometrical spreading term r^n is decreased from –1.0 to –1.3 for hypocentral distances less than 70 km, and (3) the hard-rock crustal amplification at short periods is increased from a value slightly over unity to 1.41 consistent with a reduction in surface shear-wave velocity from 2,800 to 2,000 m/s. The ENA and WNA seismological parameters used in the stochastic simulations are summarized in Table 7.1. In this table, the site profile and amplification factors for WNA generic rock ($V_{S30} = 620$ m/s) were taken from Boore and Joyner [19] and those for ENA hard rock ($V_{S30} = 2{,}000$ m/s) were taken from Atkinson and Boore [8]. The amplification factors are calculated using the quarter-wavelength method described by Boore [13].

7.2.3 ENA Hybrid Empirical Ground Motion Estimates

I calculated the hybrid empirical ground motion estimates using the formulation of Campbell [21], assuming, as recommended by Boore [13], that the hypocentral distance used in the stochastic model (r) could be equated to the fault distance measure (R_{RUP}) used in the E-GMPE for purposes of applying the regional adjustment factors. An important limitation of the empirical and, therefore, the hybrid empirical ground motion estimates is their reduced reliability beyond 100 km. Consistent with the approach taken by Campbell [21], I avoided this limitation by substituting the hybrid empirical estimates with the ENA stochastic simulations for distances beyond 70 km after adjusting the stochastic simulations to have the same amplitude as the hybrid empirical estimates at a distance of 70 km. This is particularly an issue for those E-GMPEs that do not have a realistic anelastic attenuation term.

7.3 Application Issues

There are five important issues that were identified during the example application to ENA that need to be addressed or resolved before an updated HE-GMPE can be developed. I consider these issues to be generic to the application of the HEM in any region. These issues are: (1) whether a Brune omega-square SCPS source spectrum is appropriate for estimating regional adjustment factors for large-magnitude earthquakes, (2) what value of stress drop should be used for ENA earthquakes and whether this stress drop is model dependent, (3) what rate of near-source spreading is appropriate for ENA and WNA ground motions, (4) whether the E-GMPE is valid at small magnitudes, and (5) whether the magnitude-saturation characteristics of ground motion predicted by the E-GMPE at large magnitudes is applicable to ENA. Until these issues are addressed satisfactorily, they represent a significant source of epistemic uncertainty that will need to be accounted for in the development of the HE-GMPE.

7.3.1 Source Spectrum

Tavakoli and Pezeshk [46] proposed that the use of a Brune SCPS source spectrum by Campbell [21, 22] caused his HE-GMPE to underestimate ground motion amplitudes from near-source large-magnitude earthquakes in ENA. This conclusion was based on published studies that found that the use of a double-corner point-source (DCPS) source spectrum, or alternatively a SCPS source spectrum with a stress drop that decreases with magnitude, together with a focal depth that increases with magnitude was required to match a dataset of strong motion recordings from moderate-to-large earthquakes in California [11]. These investigators also noted that Atkinson and Boore [7] had proposed the use of a DCPS source spectrum to model the source spectra of large earthquakes in ENA. This led these investigators to suggest that using a DCPS source spectrum with the Campbell [21] HEM constituted an improvement in the method.

Campbell [25] thoroughly reviewed this issue and concluded that it appears that Tavakoli and Pezeshk [46] incorrectly used a magnitude-dependent stress drop with the DCPS source spectrum in the parameterization of their WNA seismological model. He also noted that, although their use of a DCPS source spectrum in both ENA and WNA, aside from the stress drop issue, did not necessarily constitute an error, there was considerable evidence in the literature to suggest that the use of such a spectrum, or alternatively a SCPS source spectrum with magnitude dependent stress drop in WNA, is only necessary when the SCPS stochastic method is used to estimate *absolute* amplitudes of ground motion from large-magnitude earthquakes. The HEM avoids this issue by using *relative* rather than *absolute* amplitudes, which requires only that the source spectral shape be the same in the host and target regions. Campbell [25] found that there is sufficient evidence to suggest that source spectral shapes are the same (within observational uncertainty) in ENA and WNA once differences in stress drop are taken into account, but this still remains a potential source of uncertainty. In response to Campbell's comments, Tavakoli and Pezeshk [47] defended their model without specifically addressing his comments. I suggest that these investigators have not improved the HEM but instead have actually overly complicated it, thus making it less transparent to the user. Additional research will be needed to confirm this.

7.3.2 Stress Drop

Atkinson and Boore [8] used their updated set of ENA seismological parameters along with the finite-source stochastic simulation program EXSIM [36] to infer a median stress drop of 140 bars for eight instrumentally recorded events in ENA. However, Campbell [23] found that using these same seismological parameters (Table 7.1) with the SCPS stochastic simulation program SMSIM [14] required a stress drop approximately double that in order to closely match the response spectral results of Atkinson and Boore [8] (Fig. 7.1). Figure 7.1 shows that the same conclusion can be drawn for the HE-GMPE at large magnitudes. The discrepancy between the HE-GMPE and the stochastic model results at small magnitudes is discussed in a subsequent section. The apparent need for a larger stress drop in SMSIM is critical, since I propose that the SCPS stochastic model can be used with the HEM. A similar discrepancy has been noted by Motazedian and Atkinson [36] and Assatourians [1].

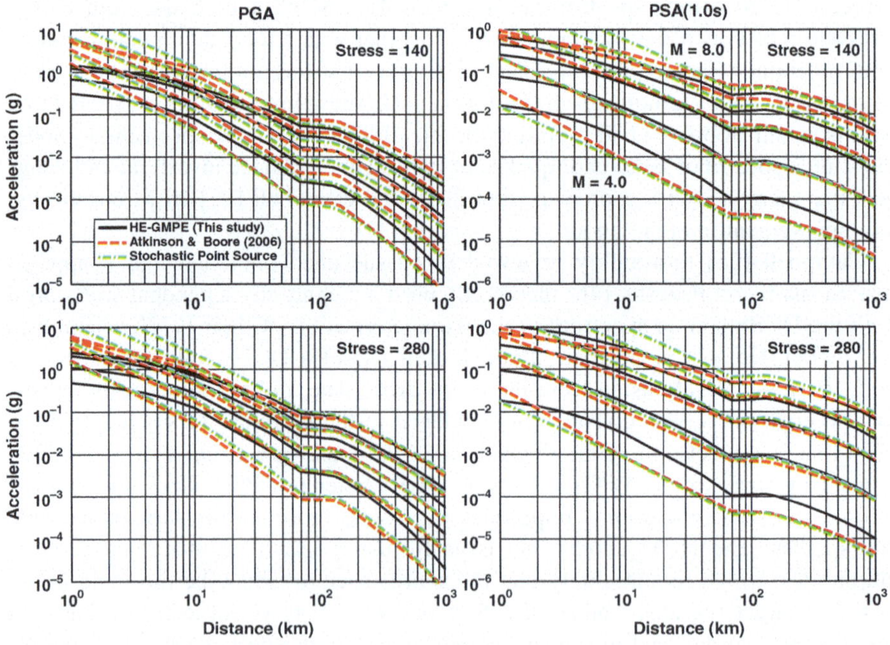

Fig. 7.1 Ground motion predictions for ENA hard rock for stress drops of 140 and 280 bars

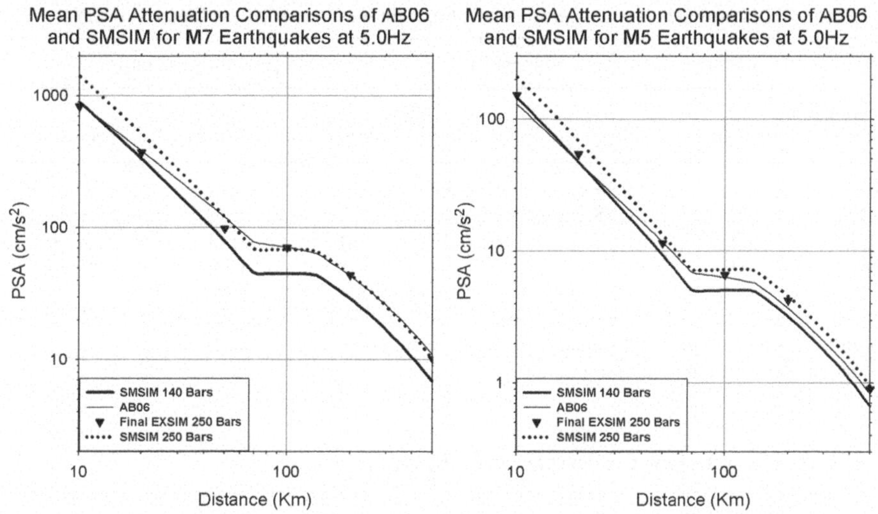

Fig. 7.2 Comparison of 5 Hz spectral acceleration estimates in ENA showing the impact of using stress drops of 140 and 250 bars in SMSIM [6]. All other seismological parameters are from Atkinson and Boore [8]

Atkinson et al. [6] studied the problem and concluded that a stress drop of 200 bars was required in SMSIM to match ground motion estimates from EXSIM using a stress drop of 140 bars for M 5 and 7 events at a distance of 200 km. A lower stress drop was required at shorter distances. Boore [15] concluded that the need for different stress drops was largely due to differences in scaling relations assumed in the two programs and that once these differences were corrected that the two programs predict similar response spectra for small magnitudes and large distances. However, he found that there were still differences in predicted response spectra at large magnitudes and large distances due to the different way that source duration is calculated. Both of these latter two studies found that a stress drop of around 250 bars was needed in SMSIM to approximately match the predictions of Atkinson and Boore [8], confirming my earlier results that it was necessary to approximately double stress drop in order to match the predictions from the GMPE (Fig. 7.2).

Boore et al. [18] used the seismological model of Atkinson and Boore [8] with SMSIM to determine the stress drops needed to fit the high-frequency response spectra of eight well-recorded earthquakes in ENA. These are the same earthquakes that Atkinson and Boore [8] used to derive a stress drop of 140 bars with EXSIM. Four different assumptions regarding spreading are used (Table 7.2). The calculated stress drops using the same spreading as Atkinson and Boore [8] are 235 and 172 bars with and without the Saguenay earthquake, respectively. Other attenuation models with less near-source spreading lead to significantly smaller stress drops, indicating that stress drop is inexorably tied to assumptions regarding spreading.

Table 7.2 Estimates of stress drop in bars for eight well-recorded earthquakes in ENA

Attenuation Model[a]	Stress Drop (without Saguenay) Mean	Std. Dev.[b]	Range	Stress Drop (with Saguenay) Mean	Std. Dev.[b]	Range
A04	172	0.19	107–336	235	0.41	107–2,161
AB95	42	0.18	23–78	57	0.39	23–467
BA92	44	0.18	23–81	60	0.41	23–603
1/R	61	0.16	38–108	82	0.38	38–650

[a]A04, Atkinson [2]; AB95, Atkinson and Boore [7]; BA92, Boore and Atkinson [16]; 1/R, hypothesized r^{-1} geometrical spreading of Boore et al. [18].
[b]Standard deviation in natural log units.

7.3.3 Near-Source Geometrical Spreading

Campbell [21, 22] used $r^{-1.0}$ near-source geometrical spreading in both WNA and ENA determined from regional seismological studies to develop regional adjustment factors. As a result, the near-source ground motions predicted by this HE-GMPE are consistently larger than those predicted by the WNA E-GMPEs. Atkinson [2], using an expanded weak motion database, revised the ENA near-source geometrical spreading to $r^{-1.3}$. Campbell [23] found that this higher rate of attenuation leads to the prediction of similar or lower ground motions in ENA than in WNA at near-source distances using the seismological parameters listed in Table 7.1 with a stress drop of 140 bars. Of course, as noted previously, the higher stress drop needed in SMSIM to match predictions from EXSIM removes this inconsistency, but the difference in near-source spreading is still noticeable.

Atkinson and Wald [12] show that MMI estimates from small-to-moderate earthquakes in the central and eastern US (CEUS) are one unit higher at distances less than 30 km and 1.5 to 2 units higher at distances greater than 100–200 km compared to earthquakes of similar magnitude in California. Interestingly, their overall regression results do not appear to support differences in near-source attenuation between the two regions. These investigators conclude that the higher overall intensities in CEUS are consistent with a higher stress drop and that any regional differences in near-source attenuation are likely "obscured by other shape parameters in the prediction equations".

As noted previously, it is the *relative* difference and not the *absolute* values of ground motion and, therefore, spreading in ENA and WNA that is important in applying the HEM, a point overlooked by Tavakoli and Pezeshk [46, 47]. Therefore, the issue is whether the regional difference in geometrical spreading in Table 7.1, which is based on seismological studies in southern California [40] and southeastern Canada and the northeastern US [2], is both scientifically justified and transferable to other regions of WNA and ENA. In order to address this issue, Campbell [23] reviewed the scientific literature to see if he could find other seismological studies that could either support or refute a difference in spreading between or within these two regions. He concluded from the diverse range of observed and theoretical geometrical spreading functions available in the literature (primarily published and unpublished studies by R. Herrmann and his colleagues, see [23]) that there are large regional differences in spreading that are not easily quantified in terms of simple tectonic environments, such as ENA and WNA. These studies show that near-source spreading is a complicated function of focal mechanism and crustal structure as well as other factors. The importance of crustal structure has also been noted by Douglas et al. [30].

Since Campbell [23] performed his review, several additional seismological studies have become available that further show how spreading can vary significantly within a given tectonic region. Zandieh and Pezeshk [48] found $r^{-1.0}$ near-source geometrical spreading in the New Madrid seismic zone from a study of small-to-moderate earthquakes, confirming a similar result found by Samiezade-Yazd et al. [41]. This result is significant considering that the New Madrid seismic zone is an important source of hazard in the CEUS and that this result might imply a similar rate of attenuation in other regions of the CEUS. Fatehi and Herrmann [32] found near-source geometrical spreading that ranged from $r^{-0.7}$ to $r^{-1.5}$ for small-to-moderate earthquakes in central and northern California and the Pacific Northwest. Malagnini et al. [35], Atkinson and Morrison [10], and Chiou et al. [28] found noticeable differences in spreading and stress drop between small-to-moderate magnitude earthquakes in southern and northern California. Boore et al. [18] demonstrate the relatively large impact that assumptions regarding spreading can have on estimates of near-source ground motion and stress drop in ENA (Fig. 7.3).

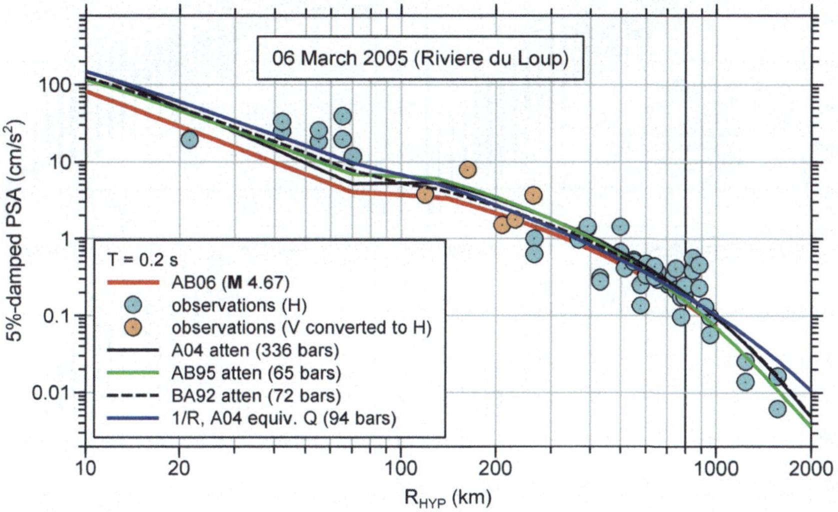

Fig. 7.3 Differences in near-source attenuation for the 2005 Riviere du Loup earthquake inferred from the geometrical spreading terms listed in Table 7.1. All other seismological parameters are held constant except for stress drop, which is fit to match the response spectra data. Also shown are the predictions from Atkinson and Boore [8]. Modified from Boore et al. [18]

7.3.4 Small-Magnitude Scaling

Atkinson [4] and Atkinson and Morrison [10] compiled a set of small-magnitude intensity (MMI) and ground motion data collected from the US. Geological Survey "Did You Feel It?" and ShakeMap databases and compared them to the empirical ground motion predictions of the NGA E-GMPEs [39]. MMI was converted to ground motion using the relationships between MMI and PGA, PGV, and PSA of Atkinson and Kaka [9]. They found that there is a general attenuation discrepancy between the MMI-based and the ShakeMap ground motion predictions and the NGA E-GMPE ground motion predictions at small magnitudes ($M < 5.5$), with the latter having less attenuation. Similar discrepancies at small magnitudes have been observed in the NGA models by Chiou et al. [28] and in several of the European E-GMPEs by Cotton et al. [29]. These latter investigators give a theoretical basis for greater distance attenuation at small magnitudes. Atkinson and Morrison [10] suggest that this attenuation discrepancy is likely a generic issue with GMPEs derived from strong motion data.

I confirmed this discrepancy for the Campbell and Bozorgnia [27] NGA E-GMPE used in the example application presented in this study, after comparing it with the data compiled by Atkinson [4] and Atkinson and Morrison [10] (Fig. 7.4).

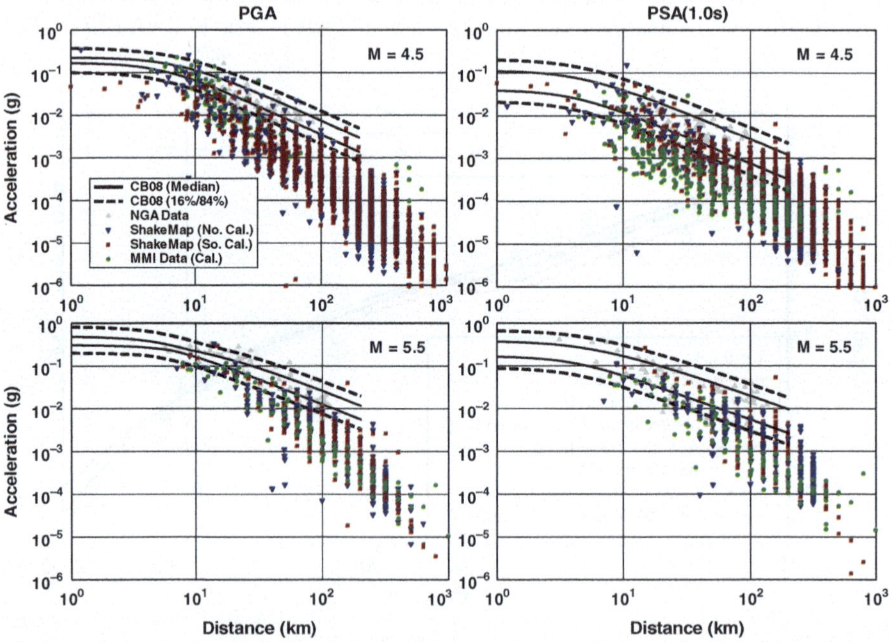

Fig. 7.4 Comparison of the Campbell and Bozorgnia [27] E-GMPE with small-to-moderate magnitude NGA and ShakeMap ground motion recordings and MMI data that have been converted to ground motion parameters. Data courtesy of G. Atkinson (personal communication, [4])

The attenuation discrepancy suggests that the WNA empirical model predicts a near-source attenuation rate that is too low compared to the small-magnitude MMI and ShakeMap data. Atkinson [4] and Chiou et al. [28] suggest that an adjustment factor could be used to correct this discrepancy. Unless such a correction is made, the HE-GMPE will overestimate ground motions at moderate-to-large distances. Fortunately, all of these studies find that the attenuation discrepancy disappears at the larger magnitudes of greatest engineering significance. This suggests that the problem might be a bias in the way that strong motion recordings are selected for processing, with small-to-moderate magnitude events with large ground motions preferentially being selected because of their greater significance.

7.3.5 Large-Magnitude Scaling

One major assumption in the example application of the HEM that has not been thoroughly validated is whether the magnitude-saturation characteristics predicted by a WNA E-GMPE are directly transferable to ENA. The key issue is whether the observed and modeled near-source saturation of ground motion and the possible physical mechanisms that are used to explain this saturation can be expected to occur in ENA as well as in WNA. This issue is closely related to the stress-drop issue noted previously. The ENA finite-source stochastic simulations of Atkinson and Boore [8] predict that near-source saturation occurs at a larger magnitude than the WNA empirical model of Campbell and Bozorgnia [27] if a stress drop of 140 bars is used in ENA. For a stress drop of 280 bars, the Atkinson-Boore and HE-GMPE estimates derived in the example application in this paper are generally similar at large magnitudes for distances greater than 10 km (Fig. 7.1).

Campbell [23] reviewed this issue and noted that there is a lack of understanding on what physical mechanisms might lead to the ground motion saturation observed for large earthquakes in active tectonic regimes and, therefore, whether these mechanisms are also valid in stable tectonic regimes. He concluded from these observations that, although the assumption that the magnitude saturation predicted by WNA E-GMPEs is transferable to ENA is a viable hypothesis, it is a significant source of epistemic uncertainty and needs further study. Atkinson [5] came to a similar conclusion, noting that "it may be that saturation effects not modeled in the [finite-source] stochastic predictions, but inferred from observations in other regions, cause overestimation of near-source amplitudes from large events in Atkinson and Boore [8]".

7.4 Conclusions

The widespread application of the HEM has made it a viable approach for developing GMPEs in regions where there are few strong motion recordings but ample weak motion data from small-magnitude earthquakes. The method has been successfully applied in ENA, the US Pacific Northwest, central Europe, southern Spain, and southern Norway. In this paper, I demonstrate the use of the HEM with an example application that was initially intended to revise the HE-GMPE developed by Campbell [21, 22] using an updated E-GMPE for WNA [27] and an updated seismological model for ENA [2, 8]. This tentative revision identified several issues that need to be addressed before a HE-GMPE can be reliably developed for ENA. These issues are (1) whether a Brune omega-square SCPS source spectrum is appropriate for estimating regional adjustment factors for large-magnitude earthquakes, (2) what value of stress drop should be used for ENA earthquakes and whether this stress drop is model dependent, (3) what rate of near-source spreading should be used for ENA and WNA stochastic ground motion simulations, (4) whether WNA E-GMPE ground motion predictions are valid at small magnitudes, and (5) whether the magnitude-saturation characteristics of ground motion at large magnitudes predicted by WNA E-GMPEs is transferable to ENA.

The issues identified in this paper are generic to the application of the HEM in any region and should be adequately addressed through careful study to ensure that reliable estimates of ground motion are obtained. If these issues cannot be adequately resolved, their epistemic uncertainty should be incorporated in the development of the HE-GMPE. As I have shown in this paper, there has been some progress made in addressing these issues, but much more work needs to be done. Fortunately, these and other issues concerning the estimation of ground motion in ENA are being addressed in a comprehensive PEER NGA-East project, which is expected to be concluded in 2012. Until these issues are adequately resolved, they will continue to be a significant source of epistemic uncertainty.

Acknowledgments Research supported by the U.S. Geological Survey (USGS), Department of the Interior, under award number 05HQGR0032. The views and conclusions contained in this document are those of the author and should not be interpreted as necessarily representing the official polices, either expressed or implied, of the US Government. I thank Dave Boore and Gail Atkinson for their insightful discussions and their interest and assistance in attempting to understand and resolve the issues discussed in this paper. John Douglas provided helpful comments.

References

1. Assatourians K (2008) Stress parameter distribution on an earthquake fault based on a stochastic modeling approach. Ph.D. thesis, Carlton University, Ottawa, ON, Canada
2. Atkinson GM (2004) Empirical attenuation of ground motion spectral amplitudes in southeastern Canada and the northeastern United States. Bull Seismol Soc Am 94: 1079–1095
3. Atkinson GM (2005) Ground motions for earthquakes in southwestern British Columbia and northwestern Washington: crustal, in-slab, and offshore events. Bull Seismol Soc Am 95: 1027–1044
4. Atkinson GM (2007) Analysis of "Did you feel it?" intensity data to determine ground motion characteristics for the central/eastern United States. National Earthquake Hazard Reduction Program, U.S. Geological Survey Award 07HQGR0071, final report
5. Atkinson GM (2008) Ground motion prediction equations for eastern North America from a referenced empirical approach: implications for epistemic uncertainty. Bull Seismol Soc Am 98:1304–1318

6. Atkinson GM, Assatourians K, Boore DM, Campbell K, Motazedian D (2009) A guide to differences between stochastic point-source and stochastic finite-fault simulation. Bull Seismol Soc Am 99:3192–3201
7. Atkinson GM, Boore DM (1995) New ground motion relations for eastern North America. Bull Seismol Soc Am 85:17–30
8. Atkinson GM, Boore DM (2006) Earthquake ground-motion prediction equations for eastern North America. Bull Seismol Soc Am 96:2181–2205
9. Atkinson G, Kaka S (2007) Relationships between felt intensity and instrumental ground motions for earthquakes in the central United States and California. Bull Seismol Soc Am 97:497–510
10. Atkinson GM, Morrison M (2009) Observations on regional variability in ground-motion amplitudes for small-to-moderate earthquakes in North America. Bull Seismol Soc Am 99:2393–2409
11. Atkinson GM, Silva W (2000) Stochastic modeling of California ground motions. Bull Seismol Soc Am 90:255–274
12. Atkinson GM, Wald DJ (2007) "Did you feel it?" intensity data: a surprisingly good measure of earthquake ground motion. Seismol Res Lett 78:362–368
13. Boore DM (2003) Prediction of ground motion using the stochastic method. Pure Appl Geophys 160:635–676
14. Boore DM (2005) SMSIM – Fortran programs for simulating ground motions from earthquakes: version 2.3 – a revision of OFR 96-80-A. U.S. Geological Survey Open-File Report 00–509
15. Boore DM (2009) Comparing stochastic point-source and finite-source ground-motion simulations: SMSIM and EXSIM. Bull Seismol Soc Am 99:3202–3216
16. Boore DM, Atkinson GM (1992) Source spectra for the 1988 Saguenay, Quebec, earthquakes. Bull Seismol Soc Am 82:683–719
17. Boore DM, Atkinson GM (2008) Ground-motion prediction equations for the average horizontal component of PGA, PGV and 5%-damped PSA at spectral periods between 0.01 s and 10.0 s. Earthquake Spectra 24:99–138
18. Boore DM, Campbell KW, Atkinson GM (2010) Determination of stress parameters for eight well-recorded earthquakes in eastern North America. Bull Seismol Soc Am 100: 1632–1645
19. Boore DM, Joyner WB (1997) Site amplification for generic rock sites. Bull Seismol Soc Am 87:327–341
20. Campbell KW (2001) Development of semi-empirical attenuation relationships for the CEUS. National Earthquake Hazard Reduction Program, U.S. Geological Survey Award 01HQGR0011, final report
21. Campbell KW (2003) Prediction of strong ground motion using the hybrid empirical method and its use in the development of ground-motion (attenuation) relations in eastern North America. Bull Seismol Soc Am 93:1012–1033
22. Campbell KW (2004) Erratum: prediction of strong ground motion using the hybrid empirical method and its use in the development of ground-motion (attenuation) relations in eastern North America. Bull Seismol Soc Am 94:2418
23. Campbell KW (2007) Validation and update of hybrid empirical ground motion (attenuation) relations for the CEUS. National Earthquake Hazard Reduction Program, U.S. Geological Survey Award 05HQGR0032, final report
24. Campbell KW (2008a) Hybrid empirical ground motion model for PGA and 5% damped linear elastic response spectra from shallow crustal earthquakes in stable continental regions: example for eastern North America. In: Proc 14th World Conf Earthq Eng, paper S03-001, Beijing
25. Campbell KW (2008b) Comment on "Empirical-stochastic ground-motion prediction for eastern North America" by Behrooz Tavakoli and Shahram Pezeshk. Bull Seismol Soc Am 98:2094–2097

26. Campbell KW, Bozorgnia Y (2006) Next generation attenuation (NGA) empirical ground motion models: can they be used in Europe? In: Proc 1st Eur Conf Earthq Eng Seismol, paper 458, Geneva
27. Campbell KW, Bozorgnia Y (2008) NGA ground motion model for the geometric mean horizontal component of PGA, PGV, PGD and 5% damped linear elastic response spectra for periods ranging from 0.01 to 10 s. Earthquake Spectra 24:139–171
28. Chiou B, Youngs R, Abrahamson N, Addo K (2010) Ground-motion attenuation model for small-to-moderate shallow crustal earthquakes in California and its implications on regionalization of ground-motion prediction models. Earthquake Spectra 26:907–926
29. Cotton F, Pousse G, Bonilla F, Scherbaum F (2008) On the discrepancy of recent European strong-motion observations and predictions from empirical models: analysis of KiK-net accelerometric data point-sources stochastic simulations. Bull Seismol Soc Am 98:2244–2261
30. Douglas J, Aochi H, Suhadolc P, Costa G (2007) The importance of crustal structure in explaining the observed uncertainties in ground motion estimation. Bull Earthq Eng 5:17–26
31. Douglas J, Bungum H, Scherbaum F (2006) Ground-motion prediction equations for southern Spain and southern Norway obtained using the composite model perspective. J Earthquake Eng 10:33–72
32. Fatehi A, Herrmann RB (2008) High-frequency ground-motion scaling in the Pacific Northwest and in northern and central California. Bull Seismol Soc Am 98:709–721
33. Frankel A, Mueller C, Barnhard T, Perkins D, Leyendecker E, Dickman N, Hanson S, Hopper M (1996) National seismic hazard maps: documentation June 1996. U.S. Geological Survey Open-File Report 96–532
34. Lin PS (2007) A comparison study of earthquake strong-ground motions in California and Taiwan. Report PEER 2006/12, Pacific Earthquake Engineering Research Center, University of California, Berkeley
35. Malagnini L, Mayeda K, Uhrhammer R, Akinci A, Herrmann RB (2007) A regional ground-motion excitation/attenuation model for the San Francisco region. Bull Seismol Soc Am 97:843–862
36. Motazedian D, Atkinson G (2005) Stochastic finite-fault modeling based on a dynamic corner frequency. Bull Seismol Soc Am 95:995–1010
37. Peruš I, Fajfar P (2009) Ground-motion prediction by a non-parametric approach. Earthq Eng Struct Dyn 39:1395–1416
38. Petersen M, Frankel A, Harmsen S, Mueller C, Haller K, Wheeler R, Wesson R, Zeng Y, Boyd O, Perkins D, Luco N, Field E, Wills C, Rukstales K (2008) Documentation for the 2008 update of the United States national seismic hazard maps. U.S. Geological Survey Open-File Report 2008-1128
39. Power M, Chiou B, Abrahamson N, Bozorgnia Y, Shantz T, Roblee C (2008) An overview of the PGA project. Earthquake Spectra 24:3–21
40. Raoof M, Herrmann R, Malagnini L (1999) Attenuation and excitation of three-component ground motion in southern California. Bull Seismol Soc Am 89:888–902
41. Samiezade-Yazd M, Herrmann RB, Malagnini L, Liu W (1997) A regional comparison of vertical ground motion in North America., Report 1434-94-G-2403, Saint Louis University, Saint Louis, Missouri, www.eas.slu.edu/People/RBHerrmann/GroundMotion.
42. Scasserra G, Stewart JP, Bazzurro P, Lanzo G, Mollaioli F (2009) A comparison of NGA ground motion prediction equations to Italian data. Bull Seismol Soc Am 99:2961–2978
43. Scherbaum F, Bommer JJ, Bungum H, Cotton F, Abrahamson NA (2005) Composite ground-motion models and logic trees: methodology, sensitivities, and uncertainties. Bull Seismol Soc Am 95:1575–1593
44. Shoja-Taheri J, Naserieh S, Ghofrani H (2010) A test of the applicability of NGA models to the strong ground motion data in the Iranian plateau. J Earthquake Eng 14:278–292
45. Stafford PJ, Strasser FO, Bommer JJ (2008) An evaluation of the applicability of the NGA models to ground-motion prediction in the Euro-Mediterranean region. Bull Earthq Eng 6: 149–177

46. Tavakoli B, Pezeshk S (2005) Empirical-stochastic ground-motion prediction for eastern North America. Bull Seismol Soc Am 95:2283–2296
47. Tavakoli B, Pezeshk S (2008) Reply to "Comment on 'Empirical-stochastic ground-motion prediction for Eastern North America' by Behrooz Tavakoli and Shahram Pezeshk" by Kenneth W. Campbell. Bull Seismol Soc Am 98:2098–2100
48. Zandieh A, Pezeshk S (2010) Investigation of geometrical spreading and quality factor functions in the New Madrid seismic zone. Bull Seismol Soc Am 100:2185–2195

Part II
Accelerometric Data Repositories

Chapter 8
Record Processing in ITACA, the New Italian Strong-Motion Database

R. Paolucci, F. Pacor, R. Puglia, G. Ameri, C. Cauzzi, and M. Massa

Abstract The development of the new Italian strong-motion database ITACA (ITalian AC-celerometric Archive, http://itaca.mi.ingv.it) has been performed under the funding of the National Department of Civil Protection (DPC) within Project S4, in the framework of DPC-INGV 2007–2009 research agreement. This work started from the alpha version of ITACA [11], where 2,182 3-component records from 1,004 earthquakes, mainly recorded by the National Accelerometric Network, RAN, operated by DPC, were processed and included in the database. Earthquake metadata, recording station information and reports on the available geological-geophysical information of 452 recording sites, corresponding to about 70% of the total, were also included. Subsequently, ITACA has been updated and has reached its final stage by the end of Project S4, around mid-2010, with additional features, improved information about recording stations, and updated records, including the M_W6.3 L'Aquila earthquake. All records were re-processed with respect to the alpha version [12], with a special care to preserve information about late-triggered events and to ensure compatibility of corrected records, i.e., that velocity and displacement traces obtained by the first and second integral of the corrected acceleration should not be affected by unrealistic trends. After a short introduction of ITACA and its most relevant features and statistics, this paper mainly deals with the newly adopted processing scheme, with reference to the problems encountered and the solutions that have been devised.

8.1 Introduction

The development of the new Italian strong-motion database ITACA (ITalian ACcelerometric Archive, http://itaca.mi.ingv.it) has been performed under the funding of the Italian Department of Civil Protection (DPC) within Project S4, in the framework of DPC-INGV 2007–2009 research agreement. This Project has

---

R. Paolucci (✉)
Department of Structural Engineering, Politecnico di Milano, Milano, Italy
e-mail: paolucci@stru.polimi.it

S. Akkar et al. (eds.), *Earthquake Data in Engineering Seismology*, Geotechnical, Geological, and Earthquake Engineering 14, DOI 10.1007/978-94-007-0152-6_8,

continued the activity originally developed by Project S6, within the previous 2004–2006 DPC-INGV agreement, in which the alpha version of ITACA was originally developed [11].

The main goal of the S6 and S4 Projects has been to organize into a comprehensive, informative and reliable database (and related webtools) the wealth of strong-motion records, obtained in Italy during the seismic events occurred starting from the Ancona earthquake sequence in 1972, up to the L'Aquila 2009 sequence.

The published version 1.0 of ITACA includes several improvements and additional features, namely: – strong motion records from other local and/or temporary networks, and from recent seismic events, in primis the April 6 2009 M_W6.3 L'Aquila earthquake and its main aftershocks; – updated reports, with an improved format, on the available geological/geophysical information of recording stations, including average HVSR from microtremors and earthquakes where available; – identification of stations and records showing distinctive features, either due to geological/topographic irregularities or due to seismic source effects; – online tools for selection of spectrum-compatible records.

To date, ITACA contains 3,955 three-component waveforms: 3,562 of them were recorded during 1,802 earthquakes starting from the M_W4.7 Ancona earthquake, up to the M_W5.4 2008 Parma (Northern Italy) and the M_W6.3 2009 L'Aquila (Central Italy) earthquakes and related $M_W > 4$ aftershocks. The introduction of records from minor earthquakes from 2008 to present has not been completed yet.

The recordings mainly come from the National Accelerometric Network (RAN, Rete Accelerometrica Nazionale), now operated by DPC. RAN presently consists of 334 free-field digital stations and 84 analogue stations, the replacement of which with digital instruments is currently in progress. The goal is to achieve a final configuration of more than 500 digital stations installed throughout the Italian territory, with an average inter-station spatial distance of about 20–30 km in the most seismically active regions of Italy (Gorini et al., 2010). Further records are provided by the Strong Motion Network of Northern Italy (*Rete Accelerometrica dell'Italia Settentrionale*, RAIS http://rais.mi.ingv.it, [4]), consisting of digital instruments, installed around the Garda lake area, and by sparse stations (analogue and digital) operated by ENEA (Ente per le Nuove tecnologie, l'Energia e l'Ambiente (*Italian energy and environment organization*)), over the time span 1972–2004. In addition to these, waveforms recorded during the L'Aquila seismic sequence by the accelerometer installed on the very broad band AQU station (http://mednet.rm.ingv.it) are also present.

All ITACA records were re-processed with respect to the alpha version [11], with a special care to preserve information about late-triggered events and to ensure compatibility of corrected records, i.e., velocity and displacement traces obtained by the first and second integral of the corrected acceleration should not be affected by unrealistic trends.

This paper mainly deals with the newly adopted processing scheme, with reference to the problems encountered and the solutions that have been devised.

8.2 Characteristics of the ITACA Dataset

Figures 8.1 and 8.2 summarize the main characteristics of the ITACA dataset in terms of focal parameters and distance ranges. As shown in Fig. 8.1, magnitude (either M_w or M_L) ranges from 2 to 6.9 with the best sampled distance interval from 5 to 100 km. The epicentral distance (R_{epi}), for M < 5.5 events, and the Joyner-Boore distance (R_{jb}) for stronger earthquakes are considered, based on the fault geometry data available in the DISS database [5, 8]. Nine 3-component records with epicentral distance $R_{epi} \leq 10$ km are available in the range $5.9 \leq M_w \leq 6.3$ (5 from the L'Aquila earthquake, 3 from the Friuli aftershocks of September 1976, and 1 from the Umbria-Marche September 1997 mainshock). The strongest events in ITACA, i.e., the M_w 6.4 May 6 1976 Friuli and the M_w 6.9 November 23 1980 Irpinia earthquakes, were recorded at $R_{epi} > 10$ km.

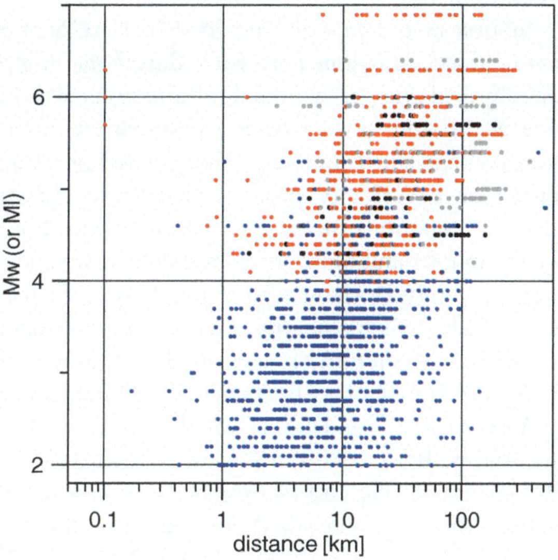

Fig. 8.1 Magnitude vs. distance (either Joyner-Boore for M > 5.5 or epicentral distance otherwise) distributions for the ITACA dataset. The records are grouped by focal mechanism (*red*: normal; *gray*: reverse, *black*: *strike*, *blue*: unknown)

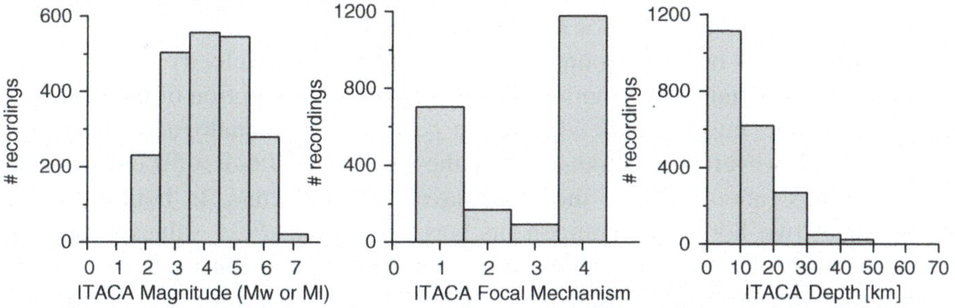

Fig. 8.2 Distribution of ITACA records plotted as a function (*left*) Magnitude; (*middle*) focal mechanism (1: normal; 2: reverse, 3: *strike*, 4: unknown); (*right*) focal depth

Distributions of records as a function of magnitude, focal depth and focal mechanism are plotted in Fig. 8.2. Most events with magnitude less than 4 have unknown focal mechanisms. For the strongest earthquakes, the focal mechanisms were assigned following the classification of Zoback [14], as described in Luzi et al. [11]. Among the strongest earthquakes, most of them were caused by normal faults in Central and Southern Apennines (namely, the Irpinia, Umbria-Marche and L'Aquila earthquakes), with focal depths less than 10 km. Earthquakes in NE Italy, including the Friuli seismic sequence, and in the Northern Apennines, are deeper and mainly characterized by a compressional tectonic regime. Finally, strike-slip events mainly occurred in Southern Italy, including the M_w 5.7 October 31 2002 Molise earthquake, at focal depths generally between 20 and 30 km.

As a whole, waveforms collected in ITACA were recorded by 665 strong-motion stations. Among these stations, 287 are presently not in operation, since they were either part of temporary networks or equipped with old analogue instruments, which were removed.

Station metadata were included in ITACA after collection of pre-existing data and field investigations performed during the S6 project and the ongoing S4 project. Geophysical and geotechnical information at the ITACA recording stations is available at different levels: from the simple geological description up to a complete geotechnical site characterization, including stratigraphic logs, V_S (S-wave) and V_P (P-wave) velocity profiles, dispersion curves, fundamental frequencies, site response functions, noise measurements etc. For most sites, based on strong and weak motions and noise measurements, it was possible to apply spectral ratio techniques, mainly Horizontal to Vertical (HVSR) and, in few cases, Standard Spectral Ratio (SSR), when a nearby reference station was available.

All ITACA stations are classified according to the EC8 [7] site classes, i.e., class A: $\mathbf{V}_{S30}$ 800 m/s, class B: $\mathbf{V}_{S30}$=360–800 m/s, class C: $\mathbf{V}_{S30}$ = 180–360 m/s, class D: $\mathbf{V}_{S30}$ <180 m/s and class E: 5–20 m of C – or D-type alluvium underlain by stiffer material with $\mathbf{V}_S > 800$ m/s. However, since $\mathbf{V}_{S30}$ will be available only for about 100 stations at the end of Project S4, it was decided to denote by a star (∗) those stations that were classified only based on the geological/geophysical information available (S4 project – http://esse4.mi.ingv.it – Deliverable D4, 2009), but not on a direct measurement of $\mathbf{V}_{S30}$. Among stations with $\mathbf{V}_{S30}$ available at present, 8% were classified as A, 42% B, 27% C, 2% D and 21% E.

The distributions of peak ground acceleration (*PGA*) and velocity (*PGV*) values reflect the event-distance distribution (Fig. 8.3). With the exception of the L'Aquila seismic sequence, most records with largest peaks are from analogue instruments. A total of 360 waveforms (about 20% of the total) have $PGA > 50$ cm/s^2 while 160 recordings (about 10% of the total) have $PGV > 5$ cm/s. In both cases the maximum of two horizontal components was considered. *PGA* values exceeding 400 cm/s^2 were recorded at stations in the epicentral area, during the L'Aquila, Umbria Marche and Friuli earthquakes. The 1980 Irpinia earthquake generated the largest *PGV* (70 cm/s) at Sturno station (STR).

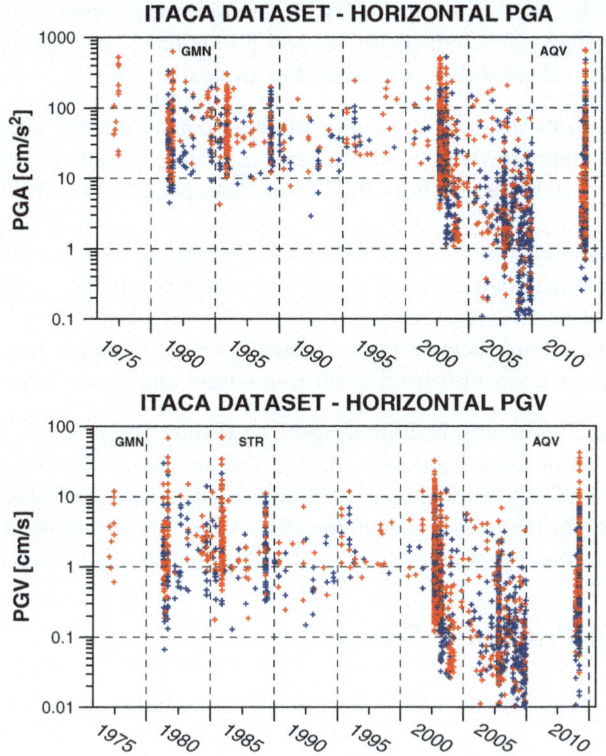

Fig. 8.3 Distribution with time of maximum horizontal *PGA* (*top*) and *PGV* (*bottom*) of ITACA records. The *blue* symbols represent values recorded at rock site (class A); the *red* ones at all the other site classes

8.3 Record Processing

The problem of defining a procedure to process acceleration time series recorded by analogue and digital instruments has been tackled since the first appearance of ITACA database. The proposed correction scheme involves the processing of analogue and digital records in different ways, with particular attention to the treatment of analogue data, as most of the strongest Italian events were recorded by analogue instruments.

The main steps of the processing procedure are described in Massa et al. [12] and involve: mean removal, baseline correction, instrument correction (for analogue data), band-pass filtering (with acausal filters) and integration of the processed acceleration in order to obtain velocity and displacement waveforms. This scheme was applied to each individual record, with the aim of preserving the low frequency content of the signals. Although the ITACA waveforms were treated by following the worldwide accepted techniques that aim to remove low and high frequency noise, the compatibility among acceleration, velocity and displacement was not guaranteed in the alpha version of ITACA. Within the revision activities to publish the beta version of the database, several points have been addressed, dealing with the quality and reliability of corrected records, namely:

- to check the accuracy and reliability of the frequency range of the corrected records and compare them with the corresponding records available in other international databases, such as PEER and European Strong Motion Database (ESMDB);
- to ensure the compatibility of corrected accelerograms, so that no further correction is required to obtain the velocity and displacement traces by single and double integration, respectively;
- to identify the late-triggered records, typically on the S-phase, that form a large portion of analogue records from small-to-medium magnitude earthquakes.

Based on the above discussions a novel procedure for processing the ITACA strong-motion records has been devised, with the objectives of providing a rational solution to the previous problems and of being robust as well as reliable enough to be effectively used for reprocessing of all the ITACA records, including the most recent ones from the Parma (December 2008) and L'Aquila (April 2009) earthquakes.

8.3.1 ITACA Processing Scheme

The diagram block of the new procedure is illustrated in Fig. 8.4. Its basic steps are the followings:

- baseline correction (constant de-trending);
- application of a cosine taper, based on the visual inspection of the record (typically between 2% and 5% of the total record length); records identified as late-triggered are not tapered;
- visual inspection of the Fourier spectrum to select the band-pass frequency range; whenever feasible, the same range is selected for the 3-components;
- application of a 2nd order two-passes time-domain acausal Butterworth filter to the acceleration time-series;
- double-integration to obtain displacement time series;
- linear de-trending of displacement;
- double-differentiation to get the corrected acceleration.

Note that zero-pads are added at the beginning and end of the signal before the acausal filter is applied [6]. However, this may pose several problems when using the corrected accelerograms, especially for engineering applications. As a matter of fact, very long initial zero-pads would most likely be removed by those end-users who are interested in using the waveforms for time-consuming non-linear time history analyses of dynamic response of soils and structures. As a consequence, the numerical simulations may start from non-zero initial conditions and present spurious trends in terms of input velocity and displacement, with the risk to compromise the reliability of results. To overcome this problem, it was decided to re-establish after filtering the original initial time-scale, whenever feasible. This is done by

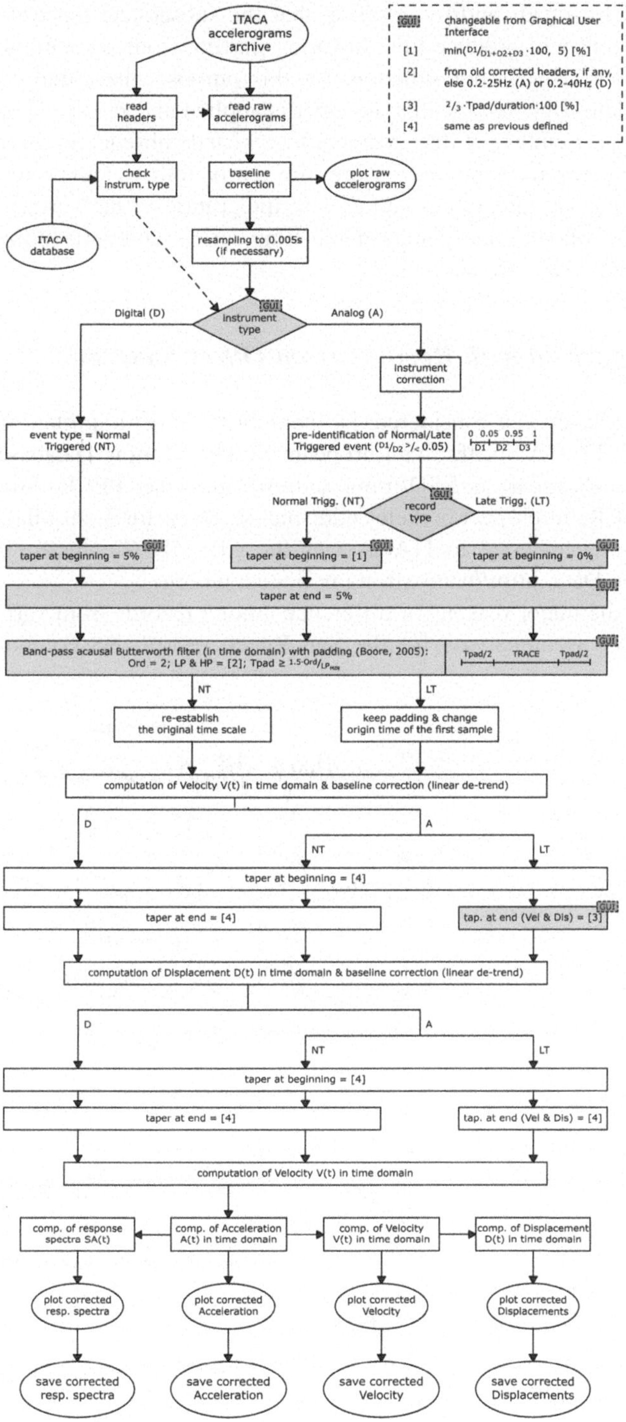

Fig. 8.4 ITACA data processing scheme

removing the zero-pads and by ensuring that the subsequent tapering of velocity and displacement will produce time histories starting from zero initial conditions. Otherwise, if tapering is not sufficient for this purpose, the initial zero-pads are retained. For late-triggered records, no taper is applied and zero-pads are kept.

The linear de-trending of displacement traces, and subsequent differentiation to obtain the corrected accelerations, ensures the compatibility of all corrected records, in the sense that the integration and double integration of the corrected accelerograms produce velocity and displacement time series with zero initial conditions and without unrealistic trends.

8.3.2 Comparison with Records from Other Sources

Three sources have been considered that contain the most important records from Italy, namely ITACA itself, the European Strong Motion Database (ESMDB, http://www.isesd.cv.ic.ac.uk/ESD/frameset.htm) and the PEER Strong motion database (PEER, http://peer.berkeley.edu/smcat). Only for L'Aquila 2009 earthquake the source external to ITACA was the CESMD (Center for Engineering Strong Motion Data, http://www.strongmotioncenter.org).

To clarify the major reasons of difference among records from various sources, Fig. 8.5 shows a comparison for the San Rocco record, NS component, of the

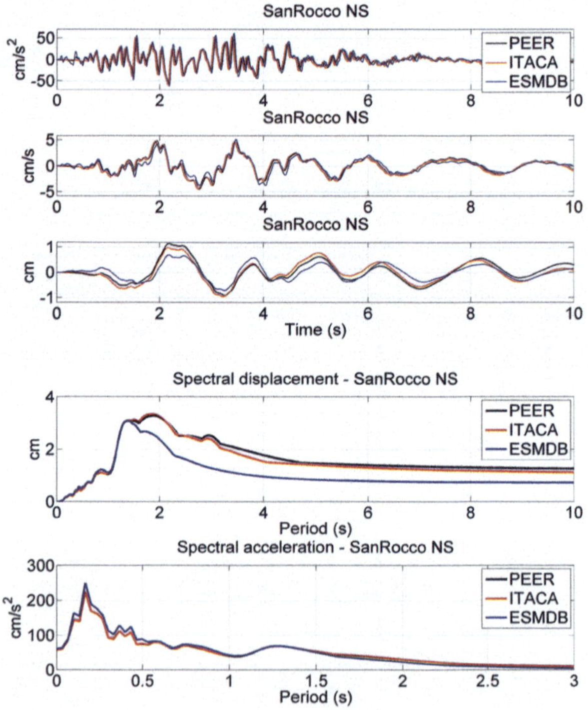

Fig. 8.5 Comparison of San Rocco corrected record, NS component, from the M_W6.1 Friuli aftershock of Sep 15 1976, 03:15 GMT, as available from ITACA, ESMDB and PEER databases. From *top* to *bottom*: corrected acceleration, velocity, displacement, spectral displacement and spectral acceleration

M_W6.1 Friuli aftershock of September 15 1976 (03:15 GMT). In this case, PEER and ITACA records are similar, with similar high-pass (HP) filter corners (0.1 and 0.15 Hz, respectively). None of these records have zero-pads at the beginning, but the tapering allows one to obtain compatible velocity and displacement time series.

On the other hand, the ESMDB record is not tapered, it is HP filtered at 0.45 Hz and keeps zero-pads at the beginning (not shown in the plot). If zero-pads were removed to re-establish the original time scale, the displacement would be affected by a trend.

As a second example, Fig. 8.6 illustrates the comparison of the corrected Bagnoli NS record of the M_W6.9 Irpinia earthquake in 1980. In this case, the HP corner

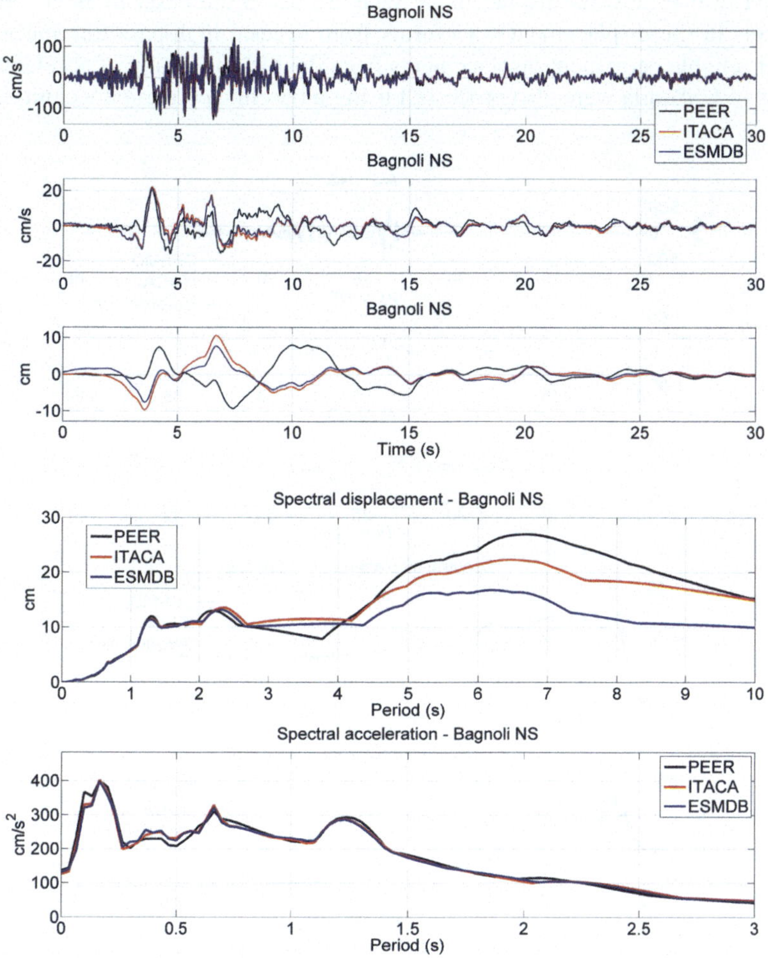

Fig. 8.6 As Fig. 8.5, for the NS component of Bagnoli corrected record, from the M_W6.9 Irpinia earthquake, 1980

frequency of corrected records are similar (0.1 Hz for both ITACA and PEER and 0.15 Hz for the ESMDB), but the PEER velocity and displacement traces are different from the other two.

Such a difference could be due to causal filtering of the record, affecting the phase of the signal. ITACA and ESMDB time series, both processed by acausal filter, are quite similar in this case, although the ESMDB record has zero pads at beginning that are not shown in the plot.

As a further example, Fig. 8.7 illustrates a case of corrected ground motion from digital records. Reference is made to the NS component of the AQV record of the M_w6.3 L'Aquila earthquake and the alternative source is the CESMD. In this case the HP frequency is 0.1 Hz for ITACA and 0.05 Hz for CESMD. The difference in the HP frequency is the reason of the clearer evidence of the acausal filter transient in the CESMD displacement trace. To avoid the onset of such spurious transients in the displacement waveforms from acausal high-pass filtering and to recover reliable permanent displacements from double integration of accelerations, records of L'Aquila were also processed using a baseline correction technique that

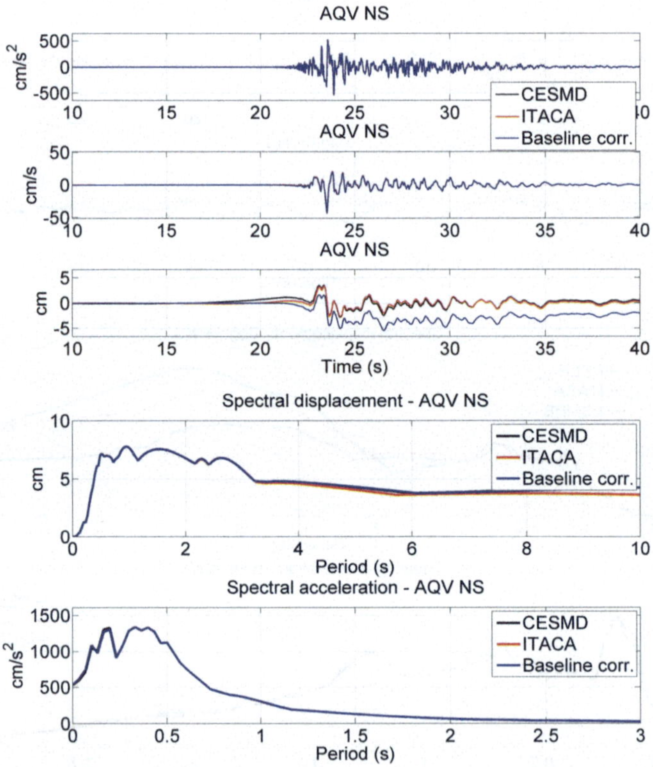

Fig. 8.7 As Fig. 8.5, for the AQV corrected record, NS component, from the M_w6.3 L'Aquila earthquake, 2009. Superimposed is the record corrected with a piecewise baseline on velocity to retrieve permanent displacements

consists of least-squares fitting the velocity time histories by three consecutive line segments, and subsequently removing these trends from the velocity time histories [1]. The resulting permanent displacements were found to be consistent with the GPS and INSAR findings (Anzidei et al., [2]; Atzori et al., [3]). Note that long period response spectral ordinates are practically unchanged using the three different processing techniques, confirming the findings by Paolucci et al. [13] regarding the reliability of long period response spectral ordinates from digital accelerograms.

Due to the space limitations of the paper, instead of documenting similar comparisons on a much larger set of records, we summarize here the most significant outcomes of such comparisons:

- for digital records, results of ITACA, PEER, ESMDB and CESMD processing are similar;
- for analogue records, ITACA and ESMDB provide similar results except for (i) a more conservative selection of the ESMDB band-pass frequency range in several cases, (ii) tapering on a longer portion of records in ITACA and (iii) the retention of zero-pads in the ESMDB records;
- ITACA and PEER analogue records practically coincide whenever the PEER records are processed by acausal filters.

8.3.3 Processing of Late-Triggered Records

A significant portion of analogue strong-motion records of ITACA consists of accelerograms triggered by the S-phase arrival (*late-triggered records*). Processing such records faces several major difficulties, especially because tapering of the initial part of the signal would inevitably cancel out some important portions of the signal itself. In the new version of ITACA, late-triggered records are identified by a specific field, so that the end-user may decide to query the database without considering such records.

To support the identification of late-triggered (LT) records in the processing stage, a criterion was introduced based on the cumulated Arias intensity function, $I(t)$. For this purpose, each record is subdivided into three portions, as shown in Fig. 8.8, where D_1 is the time between the starting of the record and the time t_{05} for which $I(t_{05})=0.05$, and $D_2 = t_{95}-t_{05}$, where $I(t_{95})=0.95$. It was found that most of the LT records in ITACA could be identified by the condition $D_1/D_2 < 0.05$, although visual inspection of the records is always required.

Once the LT record has been identified, the procedure for correction is similar to the one for NT records, except for the following:

- the initial part of the record is not tapered;
- the zero-pads are always retained.

We can gain an interesting insight about the quality of LT records, by considering two co-located stations in Nocera Umbra, an analogue one (denoted by NCR in ITACA) and a digital one (denoted by NCR2). Table 8.1 lists the events for which

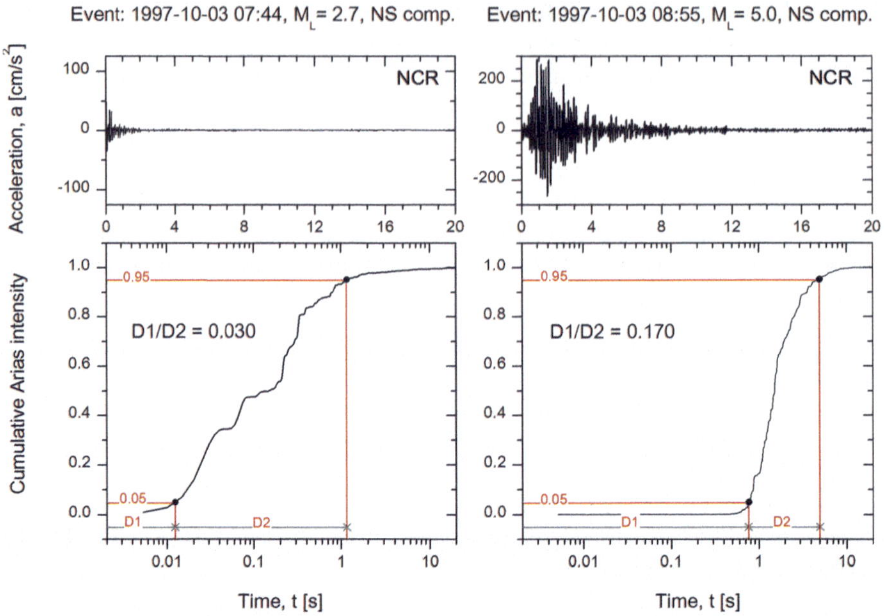

Fig. 8.8 Two analogue records from the same station NCR, identified as *late-triggered* (LT, *left*) and *normally-triggered* (NT, *right*)

both digital and analogue records are available, as well as the corresponding D1/D2 ratios and the N_d(0–0.5 s) parameter between the NCR and NCR2 response spectra normalized by NCR2. The latter parameter (N_d) measures the average difference of the response spectral ordinates in the 0–0.5 s period range. Therefore, N_d=0 means

Table 8.1 List of events and parameters associated to analogue records at NCR station

			D_1/D_2		N_d (0–0.5 s)		
ID	M_w	M_L	NS	EW	NS	EW	Class. rec.
19971003_074404		2.7	0.030	0.019	0.55	0.25	LT
19971003_121624		2.9	0.003	0.006	0.81	0.42	LT
19971003_124844		3.1	0.011	0.017	0.56	0.39	LT
19971007_012434	4.2	4.1	0.005	0.108	0.02	0.17	LT
19971007_050956	4.5	4.3	0.036	0.028	0.68	0.49	LT
19971012_110836	5.2	5.1	0.024	0.031	0.16	0.10	LT
19971014_075405		3.3	0.008	0.011	0.42	0.01	LT
19971014_152309	5.6	5.5	0.059	0.099	0.10	0.06	LT
19971108_153153		4.1	0.014	0.034	0.85	1.32	LT
19980405_155221	4.8	4.5	0.059	0.050	0.05	0.07	LT
19971002_105956	4.7	4.1	0.073	0.082	0.12	0.07	NT
19971003_085522	5.2	5.0	0.170	0.164	0.09	0.10	NT
19971006_232453	5.4	5.4	0.346	0.398	0.07	0.09	NT
19971011_032057		3.7	1.163	0.526	0.12	0.04	NT

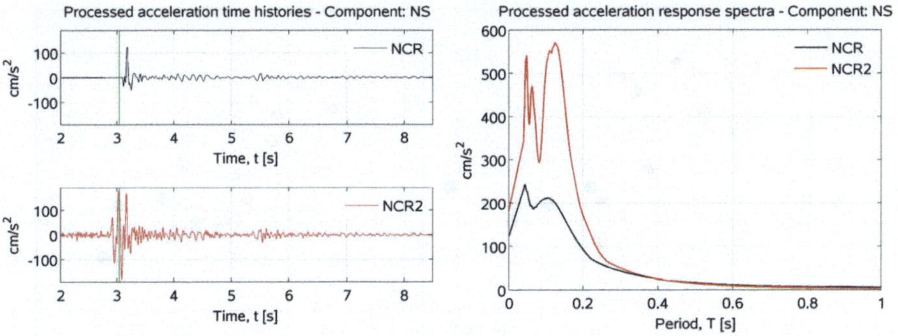

Fig. 8.9 Analogue (NCR) and digital (NCR2) corrected accelerograms of event 19971014_075405 (NS component in Table 8.1)

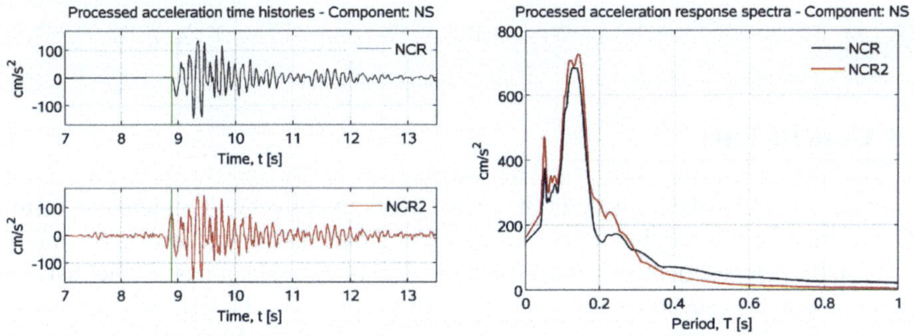

Fig. 8.10 Same as Fig. 8.9 for the NS component of event 19980405_155221

that the analogue and digital spectra coincide, while $N_d=1$ means that the average difference is 100%.

Examples of the corrected LT records at NCR, with the corresponding digital co-located records of NCR2 and the corresponding 5% damped response spectra of acceleration are shown in Fig. 8.9 and Fig. 8.10. It is clear that the case plotted in Fig. 8.9 illustrates a very poor quality record ($N_d = 0.47$, according to Table 8.1), while the corrected analogue accelerogram in Fig. 8.10 ($N_d = 0.05$) approaches the spectral ordinates of the digital record and can be considered usable for engineering applications.

Another interesting illustration about the quality of the LT records and their relationship with the proposed parameter D_1/D_2 is shown in Fig. 8.11 that shows the plot of N_d vs. D_1/D_2. This plot suggests that the proposed rule-of-thumb $D_1/D_2 < 0.05$ to identify LT records is rather satisfactory, but it is difficult to use the same parameter D_1/D_2 to discriminate between "good" and "poor" quality LT records. A similar conclusion was drawn by Douglas [9], when considering a similar criterion to check the quality of LT records, based on the bracketed duration for acceleration values larger than 0.005 g.

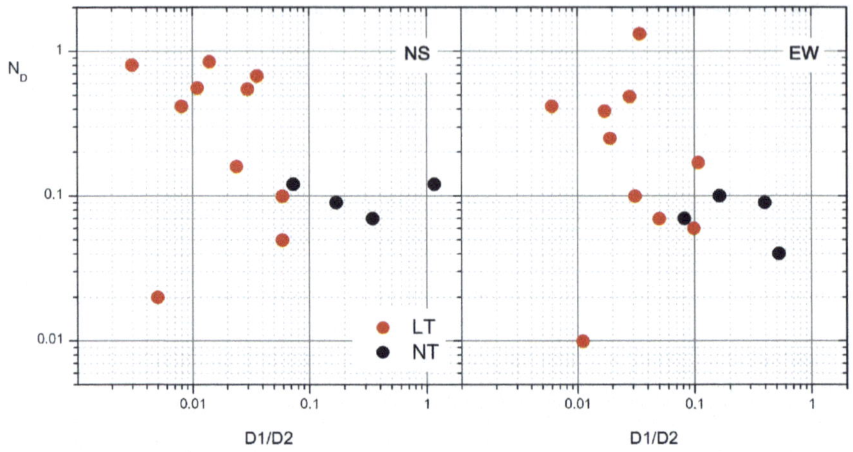

Fig. 8.11 Variation of the index N_d(0–0.5 s) as a function of the ratio D_1/D_2 for the records of NCR station

8.4 Conclusions

A notable effort has been made in the recent years to collect and organize in a single, informative and reliable Italian strong-motion database by the joint cooperation of the Italian Department of Civil Protection, the Istituto Nazionale di Geofisica e Vulcanologia, and several University research groups. The ITalian ACelerometric Archive ITACA contains most of the strong motion accelerograms recorded in Italy since 1972. The published version 1.0 of ITACA has been released in June 2010 at the end of Project S4.

The quality and the level of station and event metadata were appreciated by many researchers and professionals who accessed ITACA after the L'Aquila earthquake, as well as the rapid response of ITACA for collecting, processing and disseminating the data of this earthquake from Italian networks.

Among different topics addressed in Project S4 to improve ITACA, this paper illustrated the main issues that were faced to provide reliable corrected accelerograms from a large set of records with a wide variation in quality and amplitude that are usable both for the engineering and research communities.

Acknowledgments The authors are indebted to the DPC referees, Antonella Gorini and Adriano De Sortis, and to all partners of Project S4 within the DPC-INGV Project S4, 2007–2009. The invitation by Sinan Akkar to attend and contribute to the 2nd Euro-Mediterranean meeting on Accelerometric Data Exchange and Archiving, Ankara, 10–12 Nov 2009, is gratefully acknowledged, as well as his very careful revision of the manuscript.

References

1. Ameri G, Massa M, Bindi D, D'Alema E, Gorini A, Luzi L, Marzorati S, Pacor F, Paolucci R, Puglia R, Smerzini C (2009) The 6 April 2009, Mw 6.3, L'Aquila (Central Italy) earthquake: strong-motion observations. Seismol Res Lett 80(6):951–966

2. Anzidei M, Boschi E, Cannelli V, Devoti R, Esposito A, Galvani A, Melini D, Pietrantonio G, Riguzzi F, Sepe V, Serpelloni E (2009), Coseismic deformation of the destructive April 6, 2009 L'Aquila earthquake (central Italy) from GPS data, Geophys. Res. Lett., 36, L17307, doi:10.1029/2009GL039145
3. Atzori S, Hunstad I, Chini M, Salvi S, Tolomei C, Bignami C, Stramondo S, Trasatti E, Antonioli A, Boschi E (2009), Finite fault inversion of DInSAR coseismic displacement of the 2009 L'Aquila earthquake (central Italy), Geophys. Res. Lett., 36, L15305, doi:10.1029/2009GL039293
4. Augliera P, D'Alema E, Marzorati S, Massa M (2009) A strong motion network in northern Italy: detection capabilities and first analysis. Bull Earthq Eng doi: 10.1007/s10518-009-9165-y, on-line
5. Basili R, Valensise G, Vannoli P, Burrato P, Fracassi U, Mariano S, Tiberti MM, Boschi E (2008) The Database of individual seismogenic sources (DISS), version 3: summarizing 20 years of research on Italy's earthquake geology. Tectonophysics 453:20–43
6. Boore DM, Bommer J (2005) Processing of strong-motion accelerograms: needs, options and consequences. Soil Dyn Earthq Eng 25:93–115
7. CEN (2004) Eurocode 8: design of structures for earthquake resistance – Part 1: general rules, seismic actions and rules for buildings. Bruxelles
8. DISS Working Group (2009) Database of individual seismogenic sources (DISS), Version 3.1.0: a compilation of potential sources for earthquakes larger than M 5.5 in Italy and surrounding areas. http://diss.rm.ingv.it/diss, © INGV 2009 – Istituto Nazionale di Geofisica e Vulcanologia – All rights reserved.
9. Douglas J (2003) What is a poor quality strong-motion record? Bull Earthq Eng 1:141–156
10. Gorini A, Nicoletti M, Marsan P, Bianconi R, De Nardis R, Filippi L, Marcucci S, Palma F, Zambonelli E (2010) The Italian strong motion network. Bull Earthquake Eng (2010) 8: 1075–1090.
11. Luzi L, Hailemikael S, Bindi D, Pacor F, Mele F (2008) ITACA (Italian accelerometric archive): a web portal for the dissemination of Italian strong motion data. Seismol Res Lett doi:10.1785/gssrl.79.5
12. Massa M, Pacor F, Luzi L, Bindi D, Milana G, Sabetta F, Gorini A, Marcocci S (2009) The Italian accelerometric archive (ITACA): processing of strong motion data. Bull Earthq Eng DOI 10.1007/s10518-009-9152-3, on-line
13. Paolucci R, Rovelli A, Faccioli E, Cauzzi C, Finazzi D, Vanini M, Di Alessandro C, Calderoni G (2008) On the reliability of long-period response spectral ordinates from digital accelerograms. Earthq Eng Struct Dyn 37:697–710
14. Zoback ML (1992) First and second-order patterns of stress in the lithosphere: the world stress map project. J Geophys Res 97(B8):11703–11728

Chapter 9
The European-Mediterranean Distributed Accelerometric Data-Base

A. Roca, P. Guéguen, S. Godey, X. Goula, T. Susagna, C. Péquegnat, C.S. Oliveira, J. Clinton, C. Pappaioanou, and C. Zülfikar

Abstract We created an archive for European acceleration data, based on distributed database of accelerogram waveforms, accessed through the new European Earthquake Data Portal (www.seismicportal.eu). Data are open to the scientific and engineering community. Currently the 6 core partners contribute data from 1,379 earthquakes with magnitudes ranging from M1.0 to M7.4. Strong Motion Data are available with epicentral distances up to 1,000 km. Additionally, agencies are encouraged to contribute data. Waveforms included in the database are uniformly processed to create a set of engineering parameters that are used to search the database. In addition to the database, we compiled a survey of the existing accelerometric stations in the Euro-Mediterranean region. We expect this platform to be the basis for growing sharing of European Strong Motion Data in an open environment, in as near to real-time as is possible from network operators.

9.1 Introduction

In Europe, strong motion earthquake waveforms are made available by numerous individual European networks [6, 12, 16, 17, 21, 24, 25] and from independently created international compilations [1, 2, 8]. In the absence of a general strong motion archive for all European strong motion data, the NERIES project [19] included the initiative to create a distributed database for accelerometric waveforms in order to provide waveforms to the scientific and engineering communities, similar to what is available from the existing major strong motion distributors (e.g. COSMOS [3], K-Net [13] and CWB of Taiwan [23]). The infrastructure described in this paper is the result of this initiative.

The Strong Motion Networks who constitute the core implementation group are: Institut Geològic de Catalunya, IGC; Instituto Superior Técnico, IST; Laboratoire de Geóphysique Interne et Tectonophysique, LGIT; Kandilli Observatory and

A. Roca (✉)
Institut Geològic de Catalunya, Barcelona, Spain
e-mail: aroca@igc.cat

S. Akkar et al. (eds.), *Earthquake Data in Engineering Seismology*, Geotechnical, Geological, and Earthquake Engineering 14, DOI 10.1007/978-94-007-0152-6_9,

Earthquake Research Institute, KOERI; Institute of Engineering Seismology and Earthquake Engineering, ITSAK; and the Swiss Federal Institute of Technology, ETHZ.

The group first compiled a survey of the existing accelerometric stations in the Euro-Mediterranean region, which included a detailed description of the recording sensor and datalogger and any existing site response information (geotechnical and site type e.g. free-field or building installation).

A strong motion processing software was developed such that key strong motion parameters can be determined uniformly for all waveforms. The core group runs a wide variety of networks operating in very different tectonic regimes so waveform quality and event sizes are very different. This means our software tool needs to be flexible enough to process both weak motions as well as the strong ground motions. We compute well known engineering parameters including peak amplitudes, spectral values and duration and ground motion intensity estimates.

These computed parameters are linked to the event information (including epicentral distance of the recording), and associated with the global EMSC identifier. All this information is searchable via the European Earthquake Data Portal (www.seismicportal.eu), where selected data can be accessed in a variety of standard seismological and engineering formats (ASCII, SEED and SAC).

There are over 25,000 recordings from events ranging from M1.0 to M7.4 recorded at epicentral distances varying less than 1 km and up to 1,000 km. All data are recorded by digital instrumentation and they span a period range between 1995 and 2009. They are made available on the aforementioned web site starting from December 2009.

9.2 State of Strong Motion Instrumentation in European-Mediterranean Region

In the period 2006–2008 we conducted an exhaustive survey of existing accelerometric infrastructure in the European Mediterranean region in order to obtain an overview of accelerometer sensor density and quality across the region. In addition to producing a comprehensive inventory of strong motion network architecture we also compiled information on operational procedures. Individual station metadata includes geographic location, location type (e.g. free-field, inside buildings, on dams), soil conditions (preferably EuroCode classification or $\mathbf{V}_{S30}$ – the average shear wave velocity over the top 30 m) and instrument technical characteristics.

This survey identified a total of 51 networks that operate strong motion stations in the Euro-Mediterranean region. Those networks are distributed across 39 countries spanning from Yemen to Norway. By April 2008 we determined a total of 3,695 stations under operation in the Euro-Mediterranean region, of which 78% were reported to be digital instruments. The 6 agencies that comprise of the core development group compiling this European database operate 498 stations. From the total number of operational stations, 52% are reported to be installed in buildings and 38% are deployed in the free-field; 86% of the instruments that are installed

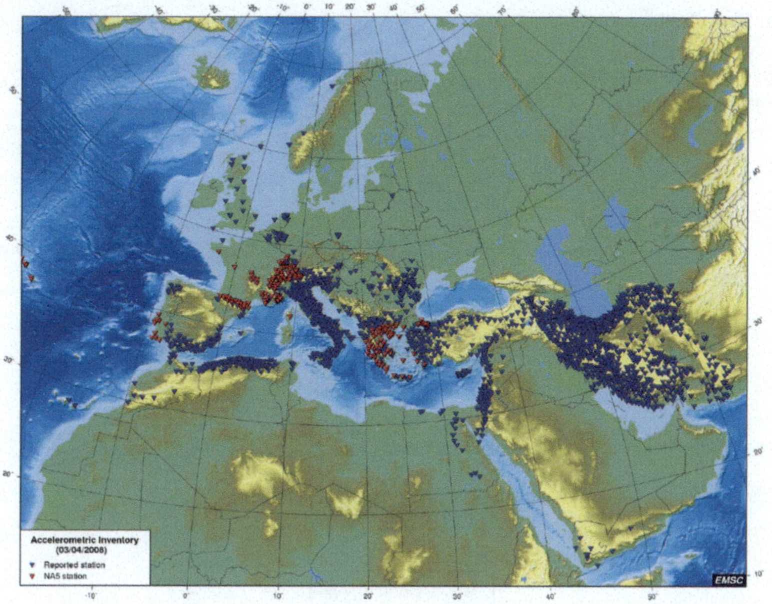

Fig. 9.1 Distribution of accelerometric stations in Euro-mediterranean region (April 2008). Stations in *red* contribute to this database

in buildings are located at the basement level. The distribution of accelerometric stations in the European-Mediterranean region is presented in Fig. 9.1.

9.3 Standard Processing Procedures: Parameter Definition and Software Description

In addition to the acceleration time series, earthquake engineers typically use various simplified waveform parameters that allow the characterisation of ground motions for use in the analysis of the expected structural response against earthquakes. Therefore, our strong motion database not only provides the raw accelerograms, but also typical engineering parameters, such as the response spectrum. In our website, queries on these parameters are used to find appropriate earthquake events and records in the dataset. To ensure consistency across the various data sets it is important to process all records in a homogeneous way. Therefore, a standard computation procedure was developed [18] and used by all the contributing accelerometer networks. A simplified flowchart of this process is shown in Fig. 9.2.

Each of the computed engineering parameters [7, 15], and some key steps in the computation process are summarised below:

- Raw acceleration, *a(t)*: acceleration time-history, in cm/s^2, base-line corrected by approximation straight line fitted by the least squares (i.e. the linear trend is removed).

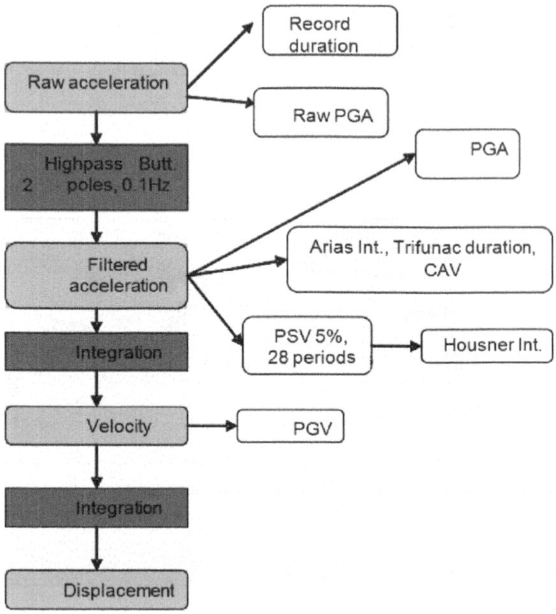

Fig. 9.2 Flowchart for the data processing to obtain accelerogram parameters for the European distributed database

- Raw PGA: Peak Ground Acceleration, in cm/s^2, from raw acceleration record *a(t)*.
- High-pass filter (acausal): a 2 pole, 2 pass Butterworth IIR high-pass filter, with constant cut-off frequency of 0.1 Hz. This value was selected as a compromise reflecting the wide variety of instrument types (in terms of frequency response and dynamic resolution) operated by the networks. Acausal filtering was selected in order to avoid phase distortion. Zero padding is introduced to avoid low frequency distortion, with zero length equivalent to 5% of the duration applied at both the beginning and at the end of signal [5]. No tapering is applied.
- Filtered acceleration: acceleration time-history obtained after the application of high pass filter.
- PGA: Peak Ground Acceleration, in cm/s^2, from filtered record. It is directly obtained from the maximum absolute value of the filtered acceleration time-history.
- AI: Arias Intensity [4], in cm/s. It is a specific function related to the energy content, calculated as:

$$AI = \frac{\pi}{2 \cdot g} \int_0^\infty a^2(t)dt$$

- TD: Trifunac duration is the time interval, in *seconds*, between the 5 and 95% of the Husid function [11]:

$$Husid(t) = \frac{\int_0^t a^2(t)dt}{\int_0^\infty a^2(t)dt}$$

- CAV: Cumulative Absolute Velocity, in cm/s. It is the area under the absolute accelerograms and has been found to correlate well with structural damage potential [14], computed according to the expression:

$$CAV = \int_{0}^{\infty} |a(t)|\, dt$$

- PSV (5%): Pseudo-Velocity Response Spectrum, in cm/s, computed for 28 logarithmically spaced frequencies from 0.15 to 39 Hz.
- HI: Housner intensity [10] or response spectrum intensity, in cm. This is the time integral of a 5% damped PSV calculated between 0.1 and 2.5 s, and gives a measure of the damage potential of the accelerogram for typical engineered structures.

$$I_{Housner}(\xi) = \int_{0,1}^{2,5} PSV(\xi, T)dT, \text{with } \xi = 5\%$$

- Integration: the trapezoidal method (time domain) was used to obtain velocity and displacement time-histories.
- Velocity time history (cm/s): integration of the filtered acceleration time-history.
- PGV: Peak Ground Velocity, in cm/s. It is directly obtained from the maximum absolute value of the calculated velocity time-history, *v(t)*.
- Displacement time history (cm): integrated velocity time history.

Note that the cut-off filtering frequency is constant for all the records, irrespective of amplitude, duration or frequency content, in order to ensure homogeneous and transparent data processing. As earthquake records from small earthquakes often do not contain signal above the microseismic peaks or the long period site/sensor noise floor, it was decided that the peak ground displacement (PGD) should not be included as a parameter. Records from larger earthquakes, which do contain seismic signals at long periods, would require a variable filter, possibly beyond 10 s, to produce correct displacement estimates. So, it is beyond the scope of this project to compute automatically PGD.

The processing software program was written in Matlab® and can be download from the European Earthquake Data Portal (www.seismicportal.eu). A snapshot of a processed record, using the visualization module of the software is shown in Fig. 9.3 to process large datasets the program can be run in batch mode.

Prior to running the software on the records, the contributor is required to set up a summary table documenting key parameters of each of the records to be processed, including information that associate the record to a universal earthquake ID. In practice, we use the UNID EMSC identifier [9], and simple tools are available to link approximate event origin times and locations with the relevant UNID.

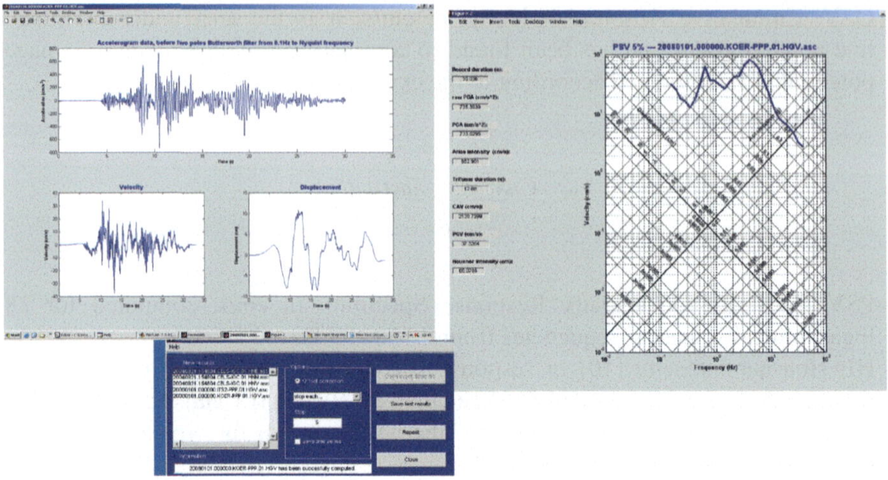

Fig. 9.3 The software visualization tool for examining processed records

Table 9.1 Number of accelerometric records (as of December 2009) by agency, dates, magnitude range and epicentral distance

Agency	Dates	# of events	Mag.	Epicentral distance (km)	# of records
IST	1996–2006	238	2.1–5.9	1–490	1,158
IGC	1996–2008	71	1.0–5.2	3–240	345
LGIT	1995–2007	378	3.0–6.8	1–863	5,253
KOERI	1999 Izmit sequence	7	5.2–7.4	13–273	369
ETHZ	2003–2009	286	2.5–5.5	0–495	15,536
ITSAK	2003–2008	399	2.8–6.9	2–697	2,379
Total	1995–2009	1,379	1.0–7.4	0–863	25,040

9.4 Assembled Core Contributor Dataset

The datasets contributing to the distributed European database by the core implementation group are summarised in Table 9.1. By the end of 2009, a total of 25,040 single component records from 1,374 different events were processed.

The dataset includes records spanning a wide range of event sizes, recorded essentially at epicentral distances between 0 and 200 km and from several different tectonic settings. Consequently they can be considered as a representative sample of the Euro-Mediterranean Region. The type of the dataset is illustrated in Fig. 9.4, where the peak ground acceleration from only RAP (LGIT) and ITSAK accelerograms recording events with magnitudes between M4-5 are plotted versus epicentral distance [20]. One can note the consistency in terms of PGA between the two agencies that provided data to this database. Another observation is that the recording threshold of instruments from ITSAK is much higher than those from RAP.

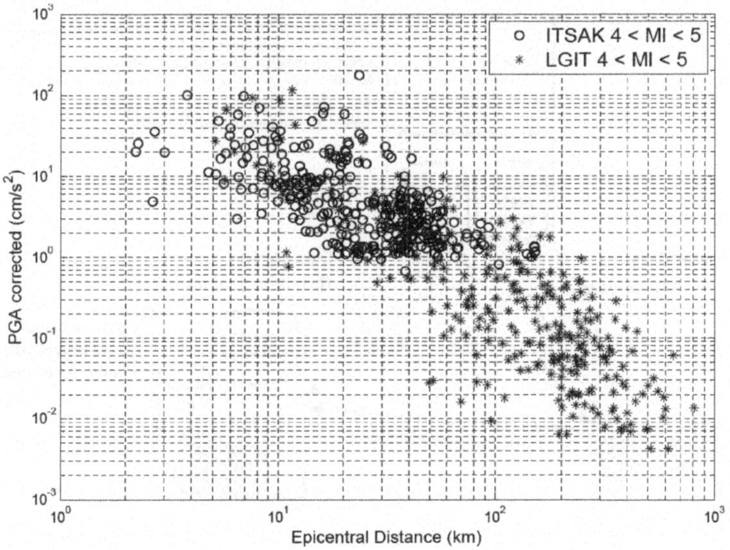

Fig. 9.4 PGA values corresponding to a range of magnitudes between 4 and 5, recorded by LGIT and ITSAK plotted versus epicentral distance

Better understanding of the characteristics of the datasets may be obtained when we consider the data that are based on regions with similar tectonic sources. Oliveira et al. [20] perform such an investigation on our dataset and include uncertainty data analysis, influence of magnitude and distance in spectral shapes, and the correlations between the calculated Engineering parameters. As this database is extended to include datasets from other regional agencies the additional information will aid future analysis of European-Mediterranean strong motion data.

9.5 Technical Description of the Database

The European-Mediterranean database structure and data server architecture are initially based on the experience of the French Accelerometric database [21]. Structure and formats of the tables are fully described in the companion article in this publication [22].

Specifications of protocols and formats for waveforms access were agreed and defined by the core implementation group. Though the data server is not visible to the data user, it is essential for collecting, processing and storing the data and information that allows data dissemination through the Earthquake Portal.

In order to encourage the dissemination of the data to a larger community, it was decided that the strong motion datasets available on the Earthquake Portal would be available in a variety of standard formats: engineering seismology standard formats

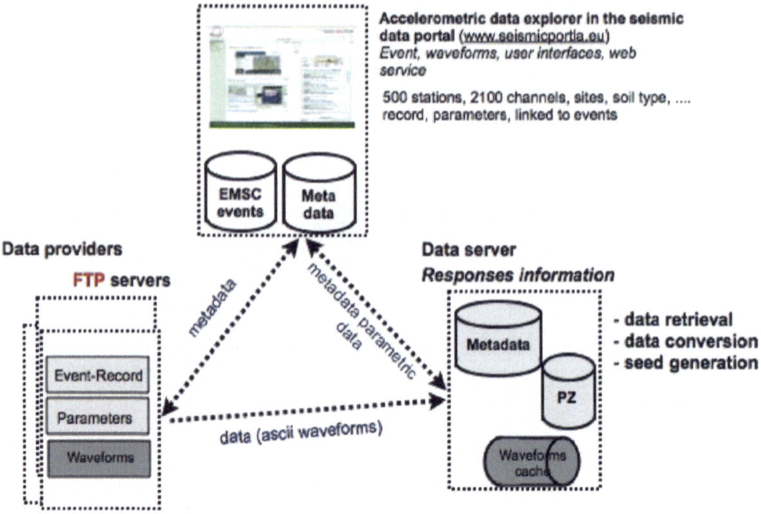

Fig. 9.5 Summary of distributed database architecture. Note relationship between the three nodes of the system – the accelerometric data explorer, the data providers and the data server [22]

(ASCII) and seismology standard formats (SEED, SAC). To provide this functionality and also to ensure that data are primarily stored on the local server of each agency, the data server architecture (Fig. 9.5) was defined to support and connect 3 separate systems (i) the data provider (i.e. the individual accelerometric networks) who remain in control of their data and metadata by maintaining and archiving the basic waveform data; (ii) portal access to the data, for the management of the end-user's requests, and (iii) the data server, which focuses on the processing of requests from the accelerometric portal and converting the raw data to the required download format.

Of particular importance was the development of a metadata stream protocol between the seismic Data Portal and data server. A data request mechanism allows end-users to collect data from different databases maintained by the core participants, thus creating our "distributed archive" (Fig. 9.5).

9.5.1 Waveform Data

The original waveform data reside on with each contributing agency for an efficient operability and a better updating procedure. No central archive was planned. The data contain both, waveforms (accelerograms) and engineering parameter data, computed using the standard software. Moreover, the data are "event based", that means that they are explicitly linked to the events. The basic station descriptions and characteristics, developed during the Survey described in Section 9.2 are stored at the Accelerometric Portal. Data provider nodes are typically *ftp servers* offering protected access to primary waveforms and partial metadata in two tables which describe in turn the events that are linked to waveforms, and the parameters calculated for each waveform.

9.5.2 Data Access

Waveform data can be browsed and selected for downloading from the Earthquake Data Portal using two kinds of criteria: event/station selectors (e.g. event magnitude, recording epicentral distance, station local soil condition) and engineering parameters (e.g. PGA, PSV). As referred above, data formats available for download, are ASCII, SEED and SAC. Data in ASCII format are split into two volumes: the first volume contains the time series accelerograms and the second contains the computed parameters. Each file has a common header. Figure 9.6 shows a screenshot of the Acceleration Data Explorer in the Earthquake Data Portal with the Euro-Mediterranean map zoomed in the Pyrenees region, with the epicenters and stations resulting from a request, the list of records available and the status of the download process, in three different windows.

9.5.3 Data Server

The inventory of acceleration infrastructure provided by each agency contains sufficient information on the sensors and site to allow complete description of the station metadata. For the instrument, specific properties such as type of instrument, orientation, sampling, trigger procedure and local data access are included. For site information, the basic geotechnical properties and the type of installation facility are given. Standard nominal poles and zeros for each instrument type have been assembled to allow a generic transfer function to be assembled (in dataless SEED format) for each of the sensors providing waveforms to the database.

9.6 Long Term Vision

The number and quality of strong motion accelerometer stations in Europe is increasing. A significant number of accelerometers are now being placed in buildings and other engineered structures such as dams, industrial facilities, bridges and lifelines, in addition to the free-field sites. Moreover, new instruments capabilities are becoming commonplace, including sensor systems with very high dynamic range and frequency resolution capable of recording weak motion in an extended frequency band, and new acquisition technologies including real-time communications and high volume data storage. Many networks, including those in

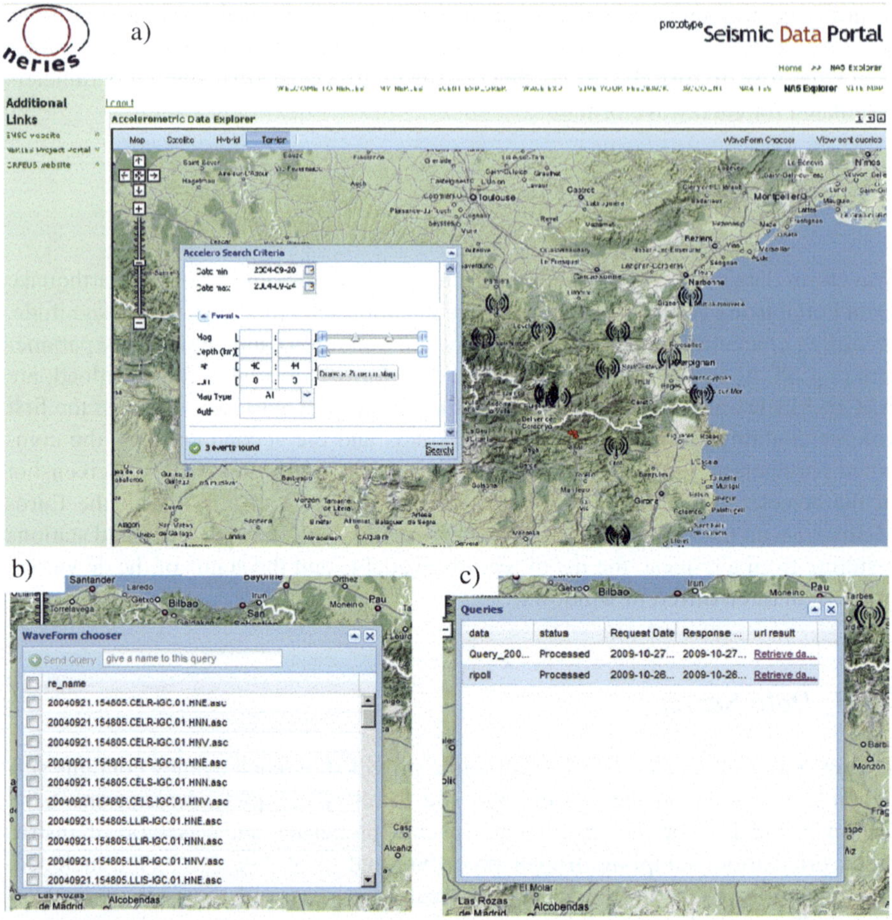

Fig. 9.6 Example of a request to the acceleration data explorer in the seismic data portal with: **a** the Euro-mediterranean map zoomed in the Pyrenees region, with the epicenters and stations resulting from a query "Accelero Search Criteria"; **b** the list of records available ("Waveform Chooser"); and **c** status of the download process ("Queries")

the core group, are now recording continuously strong motion data at high sample rates.

As the architecture of strong motion networks changes, it is clear that accelerometric data are now capable of being used for many non-traditional applications. We try to summarise and describe the uses of strong motion data by making the following separations for strong motion data: (i) different ground motion parametric values; (ii) different recording site types; and (iii) different recording access, storage systems and formats. In Table 9.2 we summarise the type of data required by different end-users according to the type of applications.

Table 9.2 Summary of type of data typically required by different end-users

Application	PGA thresholds	Recording site			Access to Data		Storage		Data Format	
		Free Field	Down-Hole	Bldg. or structure	Offline	Real time	Event recording + parameters	Continuous	ASCII	SEED
Engineering Practice	>0.05 g	X	X	X	X		X		X	
Engineering Research	>0.001 g	X	X	X	X		X		X	X
Seismology: Network Event Detection; Research	$>10^{-6}$g (or) continuous	X	X		X	X	X	X		X
Rapid alert, Rapid shake maps, Rapid damage scenarios	>0.01 g	X	X	X		X	X			

It is important to take into consideration the different user requirements shown in Table 9.2 in order to introduce future developments in the Accelerometric data exchange systems toward providing enhanced user oriented service.

To continue the advancements made in generating the database and developing the infrastructures, and to encourage the growth of the Accelerometric Portal in the Earthquake Data Portal, we propose the following measures:

- Gathering user suggestions from different communities to improve and optimize portal functionalities.
- Continuing our efforts to enlarge the current contributing community by engaging networks who are in a position to contribute (e.g. INGV/DPC in Italy are currently converting their archives to make them available over the Earthquake Portal).
- Improving station metadata information with site conditions and additional information for accelerometers on structures.
- Ensure European – Mediterranean accelerometric networks are involved in an International Community to organize periodic meetings among partners, support data interchange protocols, define international rules for network, station codes and formats consensus.

9.7 Conclusions

We have created a distributed database of European-Mediterranean accelerograms available to the community on the Earthquake Data Portal at www.seismicportal.eu. Using a standard software developed by the core group, accelerogram waveform parameters from our dataset are computed for their use as searchable parameters to find the requested data on the portal. The software and a description of the procedures required to incorporate strong motion data are freely available for candidate agencies to join this distributed data system. It is expected that this database will grow with additional agencies contributing data to the system in the coming years.

Acknowledgments This paper has been partially supported by EC Project NERIES, Sixth Framework Programme, Contract number: RII3-CT-2006-026130. Special thanks are due to all author's Institutions for financial contributions and to Torild Van Eck for reviewing the manuscript.

References

1. Ambraseys NN, Smit P, Berardi R, Rinaldis D, Cotton F, Berge-Thierry C (2000) European Strong-Motion Database. CD-Rom. Collection European Council, Environment and Climate Research Programme, ENV4-CT97-0397. European Database.
2. Ambraseys NN, Smit P, Douglas J, Margaris B, Sigbjörnsson R, Ólafson S, Suhadolc P, Costa G (2004) Internet site for European strong motion data. Bolletino di Geofisica Teorica ed Applicata 45:113–129
3. Archuleta RJ, Steidl J, Squibb M (2006) The COSMOS virtual data center: a web portal for strong motion data dissemination. Seismol Res Lett 77(6):651–658

4. Arias A (1970) A measure of earthquake intensity. In: RJ Hansen (ed) Seismic design for nuclear power plants. MIT Press, Cambridge, MA, pp 438–483
5. Boore D, Akkar S (2003) Effect of causal and acausal filters on elastic and inelastic response spectra. Earthq Eng Struct Dyn 33:1729–1748
6. Çagnan Z, Akkar S, Gülkan P (2009) A predictive ground-motion model for Turkey and its comparisons with recent local and global GMPEś. In: S Akkar, P Gülkan, T Van Eck (eds) Earthquake data in engineering seismology: networks data management and predictive models, Chapter 4. Springer, pp 39–53
7. Chen WF, Scawthorn C (2003) Earthquake Engineering Handbook (New Directions in Civil Engineering). 1512.
8. Douglas J, Guéguen P, Chaljub E, Cotton F, Suhadolc P, Costa G, Faeh D, Spühler E, Gosar A, Priolo E, Barnaba C, Paolucci R, Cauzzi C, Eva C (2006) Alpine Accelerometric Data-Base. CDRom.
9. Gilles S, Godey S, Merrer S (2009) Unique Identifier. Deliverable Dx1 of the Networking Activity 5, NERIES project (EC project number 026130), Sixth Framework Proogramme EC, 37 pages.
10. Housner GW (1952) Spectrum intensities of strong-motion earthquakes. In Proceedings of Symposium on Earthquake and Blast Effects on Structures. Earthquake Engineering Research Institute, Berkeley, CA
11. Husid R (1973) Terremotos: análisis spectral y características de acelerogramas como base de diseño sísmico. Andres Bello, 447.
12. IGN (1999) Base de datos de aceleración 1984–1997. CR-Rom.
13. Kinoshita S (2003) Kyoshin Net (K-Net), Japan. In: Lee W, Kanamori H, Jennings P, Kisslinger C (eds) International handbook of earthquake and engineering seismology. Academic Press, pp 1049–1056
14. Kramer SL (1996) Geotechnical earthquake engineering. Prentice Hall Inc, Upper Saddle River, NJ, 651 pp
15. Lee W, Kanamori H, Jennings P, Kisslinger C (2002) International handbook of earthquake and engineering seismology. Academic Press, International Association of Seismology and Physics of the Earth's Interior, 1942 pp
16. Luzi L, Hailemikael S, Bindi D, Pacor F, Mele F, Sabeta F (2008) ITACA (Italian Accelerometric Archive): a web portal for the Italian strong-motion data. Seismol Res Lett 79:5
17. Margaris B, Skarlatoudis A, Savvaidis A, Theodulidis N, Kalogeras I, Koutrakis S (2010) Strong-motion networks in Greece and their efficient use in the derivation of regional ground motion prediction models. In: S Akkar, P Gülkan, T Van Eck (eds) Earthquake data in engineering seismology: networks data management and predictive models, Chapter 6. Springer, pp 73–83
18. Marsal A, Susagna T, Goula X, Oliveira CS (2008) Implementation of Accelerometric Parameters. Computation and Exchange. NA5-D4 Report, NERIES.
19. NERIES Project (2006) Network of Research Infrastructures for European Seismology (http://www.neries-eu.org/). Accessed 27 Jan 2010.
20. Oliveira CS, Gassol G, Susagna T, Goula X, Papaioannou C, Guéguen P, Clinton J, Zulfikar C, Godey S (2010) Statistical analysis of European accelerograms parameters homogeneously computed from NERIES Database (submitted to the 14ECEE).
21. Péquegnat C, Guéguen P, Hatzfeld D, Langlais M (2008) The French accelerometric network (RAP) and national data center (RAP-NDC) (http://www.rap.obs.ujf-grenoble.fr/). Accessed 27 Jan 2010.
22. Pequegnat C, Jacquot R, Gueguen P, Godey S, Frobert L (2010) Distributed archive and single access system for accelerometric event data: a NERIES initiative. In: S Akkar, P Gülkan and T Van Eck (eds) Earthquake data in engineering seismology: networks data management and predictive models, Chapter 10. Springer, pp 131–146

23. Shing TC, Tsai YB, Yeh YT, Lin CC, Wu YM (2003) Strong motion instrumentation programs in Taiwan. In: Lee W, Kanamori H, Jennings P, Kisslinger C (eds) International handbook of earthquake and engineering seismology, Academic Press, pp 1057–1062
24. Vilanova SP, Ferreira MA, Oliveira CS (2009) PAD-1.0 Portuguese accelerometer database, CD-Rom. Seismol Res Lett 80(5):839–844
25. Wyss A (2004) Swiss national strong motion network. Strong motion bulletin January 2004 – December 2004. Available at http://seispc2.ethz.ch/strong_motion/download/Bulletin04.pdf (last Accessed 15 Mar 2010).

Chapter 10
Distributed Archive and Single Access System for Accelerometric Event Data: A NERIES Initiative

C. Péquegnat, R. Jacquot, P. Guéguen, S. Godey, and L. Frobert

Abstract We developed a common access facility to homogeneously formatted accelerometric event data and to the corresponding sheet of ground motion parameters. This paper is focused on the description of the technical development of the accelerometric data server and the link with the accelerometric data explorer. The server is the third node of the 3-tier architecture of the distributed archive system for accelerometric data. The server is the link between the data users and the accelerometric data portal. The server follows three main steps: (1) Reading and analysis of the end-user request; (2) Processing and converting data; and (3) Archiving and updating the accelerometric data explorer. This paper presents the description of the data server and the data explorer for accessing data.

10.1 Introduction

One activity of the NERIES project is to improve the access to the European accelerometric data. The necessity is the development of common access to homogeneously formatted accelerometric data and to the corresponding sheet of ground motion parameters. First, the input and output formats, the protocols for exchanging data, the structures of the database and the description of acquisition channels were discussed and agreed on with the other partners. A preliminary version of the accelerometric database structure was defined based on the experience of the French Accelerometric Network database [5]. Structures and formats of the database tables were fixed in relation to the description of soil conditions and EMSC inventory of the accelerometric stations in Europe [2]. Based on the inventory, the description of the accelerometric channels found in Europe were implemented in conformity with the tool used for building SEED volume.

The distributed system for accelerometric event data was described in 2007 in Gueguen et al. [3], the prototype data server implemented in 2009 and the

C. Péquegnat (✉)
LGIT/CNRS/LCPC – Joseph Fourier University, Grenoble, France
e-mail: pequegna@obs.ujf-grenoble.fr

S. Akkar et al. (eds.), *Earthquake Data in Engineering Seismology*, Geotechnical, Geological, and Earthquake Engineering 14, DOI 10.1007/978-94-007-0152-6_10,

accelerometric data explorer as a portal in the seismic data portal in 2009. The data server (invisible for end users as well as for data providers) is an important piece of the system: in charge of collecting, processing and archiving the data, it will allow their dissemination through a unique and standard system. Unlike the broad-band community which has defined standards and protocols for exchanging data a long time ago, accelerometric networks are less structured and the data server was a good opportunity to define and fix some specifications.

Based on the French Accelerometric Network experiences, conversion tools from earthquake engineering standardized format (ASCII) to seismology standardized format (SAC, miniseed) was chosen in order to increase the dissemination of these data to a large community. For this reason, the accelerometric system has been defined following a 3-tiers architecture: data providers (accelerometric networks) in charge of giving access to their data and metadata; data explorer to the data for the management of the end-users requests; and the data server only focused on the processing, the conversion and the download of the data, following the request file transmitted by the accelerometric data explorer.

The accelerometric data explorer and the data server are described in this document. The main objectives of the data server are first described, as well as the structure chosen for its development. During the entire period of the project, the main concern was to select and choose solutions as close as those existing for the other system devoted to the data access of the seismological data. The metadata stream between accelerometric data explorer and data server is then described in the second part of this document. Finally, the specific tools developed are listed and described, from the XML files reader to the data converter.

10.2 Distributed System

As described in Guéguen et al. [3], the specifications of the system for accessing the data have been defined in three parts:

* Data are waveforms (accelerograms) and engineering parametric data. Moreover, the data are "event based", that means that they are explicitly linked to events by data providers.
* Data selectors and data extraction tools will allow the expression of two kinds of criteria for retrieval: seismological criteria (event, magnitude, station to event distance) and engineering criteria (PGA, PGV etc).
* Accelerometric data are delivered by tools in 2 formats:
  - in a specific ASCII format for the engineering community,
  - in SEED format in order to share and disseminate it widely.

10.2.1 Specific ASCII Format for Accelerometric Data

Different formats have been examined from Strong Motion (SM) databases: COSMOS (The Consortium of Organizations for Strong-Motion Observation

Systems, www.cosmos-eq.org) in California, KNET in Japan [4], and the RAP in France [5]. All of them are composed by (i) headers containing information related to event, station and record and (ii) acceleration time histories.

Some basic ideas have guided our choices:

* a simple header with the most important information concerning station, event and record. A similar format to the one used by RAP seems to be the most adequate, between the very simple content of KNET and the comprehensive content of COSMOS (result of a long history of US formats).
* waveforms will be present by one sample per line with a first column for time and a second for acceleration.
* SM parameters and PSV response spectra (5% damping) at 28 frequencies computed and stocked by each participant in a homogeneous way [7] and available in the data explorer are also provided.

In conclusion, data in ASCII format will be split into 2 volumes: the first one for accelerations and the second one for parameters (SM parameters and PSV (5%) response spectra at 28 frequencies) with the same header. A common header for metadata will be added to each volume.

10.2.2 Main Nodes

The distributed system is based on a 3-tier architecture, the nodes of which are (1) data providers' nodes, (2) data explorer and metadata node and (3) data server node (Fig. 10.1).

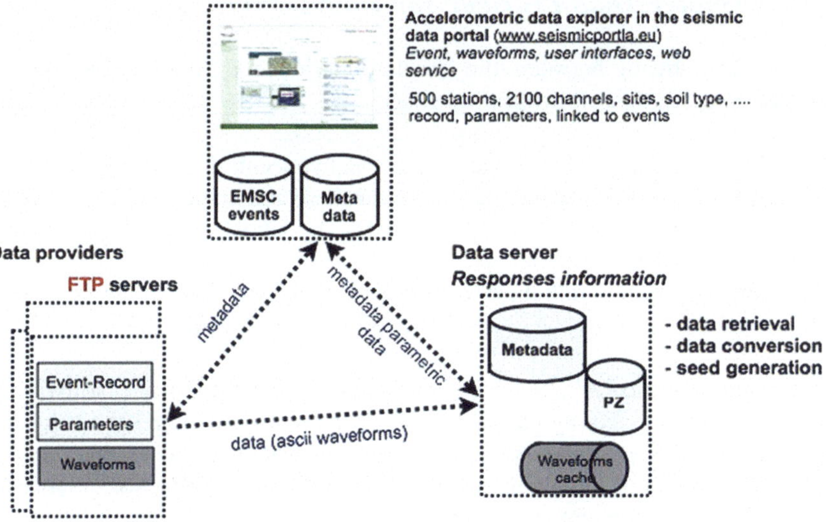

Fig. 10.1 Description of the three nodes of the system, between the accelerometric data explorer, the data providers and the data server

* Data providers' nodes do set up ftp server offering protected access to primary waveforms (without any header) and partial metadata in two ASCII tables. The first table is for event-record association, the second is for record parameters. The syntax and the semantic of these two ASCII tables: *event-record.txt* and *parameters.txt,* as well and the directory structure for the initial raw data files have been defined in Guéguen et al. [3].
* EMSC node provides User Interfaces for data queries expression. EMSC operates two databases: the EMSC event database and the station metadata database, which contains all the metadata about networks, i.e. site conditions, stations, channels, records and parameters, except the specific information about instrumental responses. The structure of the station metadata database (which is derived from the RAP database) has been described in Guéguen et al. [3]. The content of this database is based on the EMSC report [1].
* The last node is the data server, which processes queries and data. The main resources of this node are:
* A local metadata database, the structure of which is very close to the metadata database. This local database will be progressively filled up with and during requests processing.
* The PZ database (poles and zeroes database) which contain the generic instrumental responses of the NA5 instruments (8 types of digitizer configurations, 7 types of generic sensor). The structure of the PZ database, its content as well as the tools used to manage it have been described in details in Guéguen et al. [3].

10.2.3 Metadata Stream Between the Accelerometric Data Explorer, Data Provider and Data Server

The metadata stream between the accelerometric data explorer and the data providers' nodes is the following (Fig. 10.2):

- Providers should use a Webservice to retrieve hypocentral localisations (in XML format) from the EMSC database. In this database and in the retrieved XML file, events are identified by a key called UNID. The UNID implementation by the EMSC is described in [2] and [1].
- Providers have then to link their waveforms to this UNID using their own association methods and store the result in the formatted ASCII file: *event-record.txt.*

The *event-record.txt* ASCII table is the input of the PARAMACC software [7], which computes engineering parameters for each record, and stores the values in a second ASCII table: *parameters.txt.* Those two tables must be accessible on the ftp server of the providers, known by the data explorer. It downloads ASCII tables and integrates them into the metadata database.

The metadata stream between data explorer and data server is the following (Fig. 10.3):

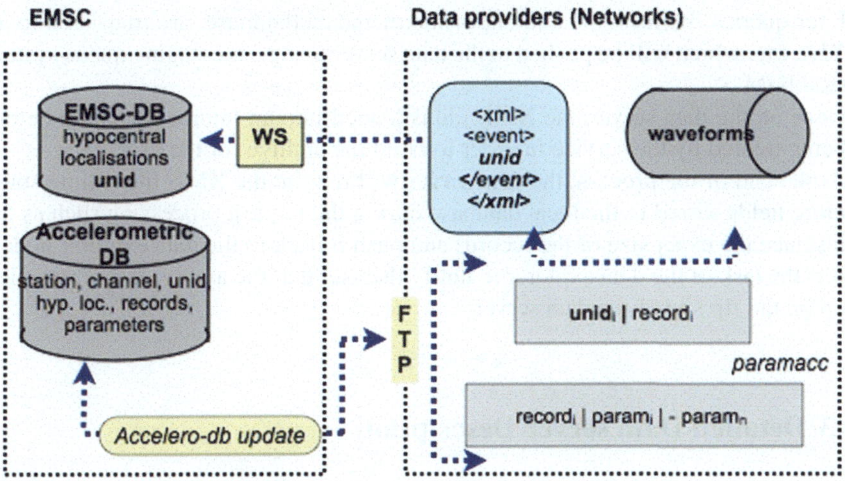

Fig. 10.2 Description of the stream between the accelerometric data explorer in the seismic data portal (www.seismicportal.eu)

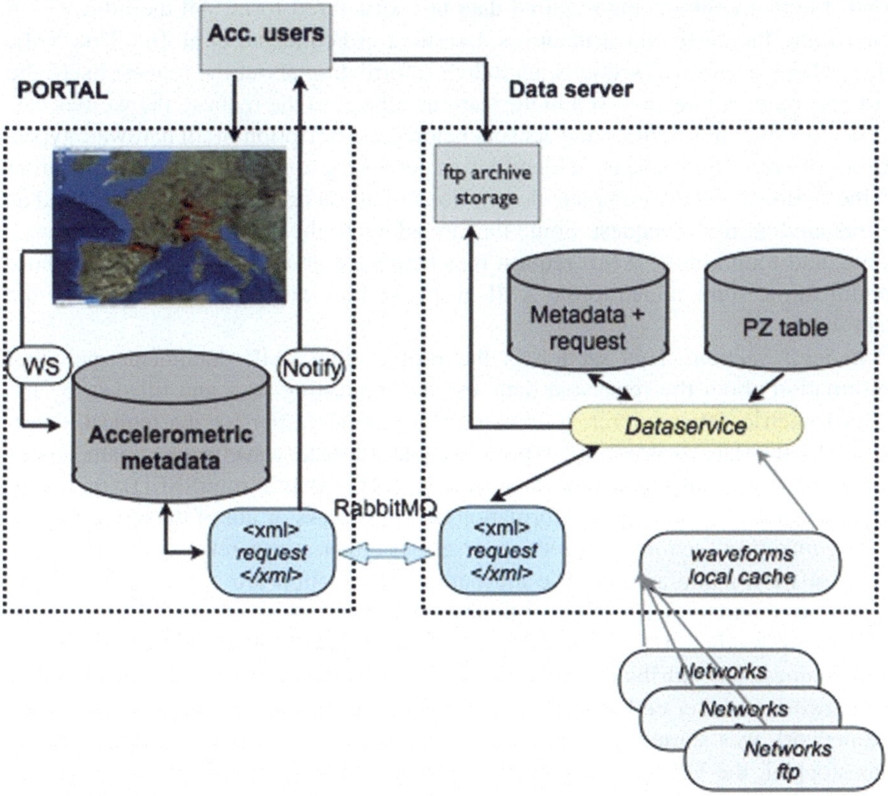

Fig. 10.3 Description of the stream between the data explorer and the accelerometric data server

- User queries, built on the content of the metadata database, are translated to an *XML* file, which will be pushed to the data server using a messaging queue system (RabbitMQ).
- Once on the data server, the XML file is loaded into an internal database before being treated by the service in order to build the archive for the user.
- At the end of the process, the data server will rewrite the XML file (filling some more fields stored in the local database during the request processing such as for instance the exact size of the record) and push it back to the data explorer node.
- It is the task of the data explorer to notify the user that the archive is ready to pick up on the ftp site of the data server.

10.3 Detailed Data server Description

10.3.1 Structure of the Request

The accelerometric data explorer sends requests to the data server in the form of an XML file that contains the required data in a structured form. All the information concerning the XML file structure is described in Péquegnat et al. [6]. This XML file contains a series of sections, including information about the request itself, the various organizations involved in the stations related to the request, the various networks to which the stations and sensors belong, a description of all hardware types, the list of events that have been identified as pertaining to the request, the description of the stations, and the complete descriptions of all records that have been found as corresponding to the request. Some flag are added to the XML file of the response, some field found in the XML request files being preserved in the response without modification, some added to the XML response files calculated or filled in by the data server.

Request section: This section is the root of the XML document, including information about the requested data and the requesting user and filled in by the accelerometric data explorer: it contains a unique identifier for the request, determined by the data explorer, the types of requested data (waveforms, parameters or both), the format of the output data (ASCII, SAC, SEED, miniSEED), the email address of the user and some information on the success or not of the request.

Organization section: This section describes either a research facility, or a hardware/software manufacturer. The attribute of the section are a unique ID for the organization and the URL of the organization website.

Network section: This section describes the networks, owner of the data, found in the request. The attribute is a unique ID for the network, the FDSN unique ID for the network (2-letter codes used in SEED data), a time stamp that describes when the network was started, an optional time stamp that describes when the network was stopped, the URL to the network's website (which should contain a description of the network), the organization ID pointing to the organization that manages the network. The FDSN code attribute is optional, because all the accelerometric networks do not have such a code. In that case, a default code will be added by the data server tool in order to deliver seed volume anyway.

Equipment section: This section has no attributes and contains one or more "sensor", "preamp" and "adc" elements. The "sensor" element describes each type of sensor found in the request. It contains a unique ID for the sensor type, given by the data explorer, the ID of the manufacturer that refers to the ID of an "organization" element, the URL to the file with the PZ description and values for the sensor (pz_url), the corner frequency of the sensor, the seed code for the sensor band and the sensor sensitivity. In the incoming XML file, the pz_url attribute is empty. It will be filled up by the data server tools. The rule to derive the pz_url base name file is to concatenate the manufacturer and the sensor ID. If the pz_url element remains empty in an outgoing XML file, it signals an inconsistency in the PZ database. The "preamp" element describes each type of preamplifier found in the request, in relation with the sensor. It contains a unique ID for the preamplifier type, generated by the data explorer application, the URL to the file with the PZ values for the preamplifier, the value of the analogue gain for the preamplifier. The "adc" element describes each type of digitizer found in the request. It is formed by an unique ID for the pair type of digitizer code – sampling frequency, generated by the data explorer application, the ID of the manufacturer (refers to the ID of an "organization" element), the digitizer type code, the URL to the file with the PZ values for the digitizer, the sampling frequency, in Hz, and the digital gain as set on this digitizer type. In the station metadata database, *adc id* are integer (numeric keys) and a foreign key give access to the data type code via the dastype table. ADC entries have 3 dependent fields: the analogue device converter code (ADC), the sampling frequency and the numerical gain. In case of an unknown ADC, the "XXXX" word will be used.

Event section: This section describes the events that are requested. It contains the UNID of the event provided by EMSC/CSEM, the date of event, the longitude, latitude and depth of the epicenter and the standardized name of the region where the event occurred.

Complementary to the event section, the *magnitude section* is defined for each event element. It contains the type of the magnitude value and the value of the magnitude of the event, each event may have several magnitude values (M_L, M_w, mb etc).

Station section: This section describes the stations found in the request. It contains the station ID used in seed data, the URL at which the station is described, the station's owner ID (referring to one of the "organization" elements above), the station's manager ID (referring to one of the "organization" elements above), the soil type where the station is located ("S" or "R"), the Eurocode definition for the soil where the station is located, the latitude and longitude of the station.

Channel section: This section describes all the information linked with the data. For each channel, there's a "site" description, an "equipment" description and a list of "record". The attributes are the channel code used in SEED data (HNN for example), the channel's location code (01 for example), the channel's azimuth, a time stamp to identify the day when the channel was set up, a time stamp to indicate the day when the channel was stopped, a distance in meters where to the north the channel sensor is located with respect to the station, a distance in meters where to the east the channel sensor is located with respect to the station, a distance in meters of how much higher the channel sensor is located with respect to the station (ground level), the depth at which the sensor is located with respect to the above mentioned ground level, the URL of the web page where the channel is described, an optional URL to the channel's response file, an optional URL to the channel's dataless file (containing the metadata), the channel's manager (refers to an organization ID) and the channel's SEED network ID.

Site section: This section describes the building where the channel is located. The attributes are the type of the site, the total number of floors in the building, the floor on which the sensor is located and the depth at which the sensor is located with respect to the floor. The site types are:

- R: rock based on surface geological observations
- A: rock or stiff geological formation ($\mathbf{V}_{S30} > 800$ m/s)
- B: stiff deposits of sand, gravel or over consolidated clays ($\mathbf{V}_{S30} > 360$ m/s and $\mathbf{V}_{S30} < 800$ m/s)
- C: deep deposit of medium dense sand, gravel or medium stiff clays ($\mathbf{V}_{S30} > 180$ m/s and $\mathbf{V}_{S30} < 360$ m/s)
- D: loose cohesionless soil deposits ($\mathbf{V}_{S30} < 180$ m/s)
- E: soil made up of superficial alluvial layer, with a thickness ranging from 5 to 20 m with $\mathbf{V}_{S30}$ value in class C and D ranges covering stiffer deposits (class A)
- BHd: Borehole sensors. d = depth in meters
- Bm.n: Sensors in building (m = total number of stories of the building including ground floor, n=floor where the instrument is installed)
- O: other

Record section: This section contains the information on the record in relation with the channel and the event. The attributes are the event ID (refers to the UNID of one of the events that is requested), the URL of the source data for this record, the URL for the corresponding miniSEED file, the size of the data in bytes, the time stamp of the record, the time stamp of the first sample in the record that corresponds to the event, the time stamp of the last sample in the record, the time stamp of when the record was created, the PGA_uncorrected parameter, the PGA_corrected parameter, the Arias intensity, the Trifunac parameter, the CAV parameter, the PGV parameter, the 28 values defining the PSV parameter and the Housner parameter.

10.3.2 Local Database

A database is used locally to store the information from the requests and responses. This database contains all the fields defined in the XML files described above. When a processing task has finished, it updates the database to reflect the status of the data.

At the end of the request, the last process extracts the information from the database to create the response XML file that is sent back to the data explorer.

10.3.3 Application Architecture

The application running on the data server, that is responsible for generating the request response files containing the requested data, is designed as a pair of processes that run in parallel as system daemons (Fig. 10.4). A request-reader task watches over a directory, waiting for request description XML files to come in, then reads it into the database, and signals via station metadata the second module, the data-engine, to do the actual work. Each module is a thread (that is, it runs in parallel to other modules if the data it requires is ready at the time). The data-engine starts the various modules, which start doing their work and signal each other to

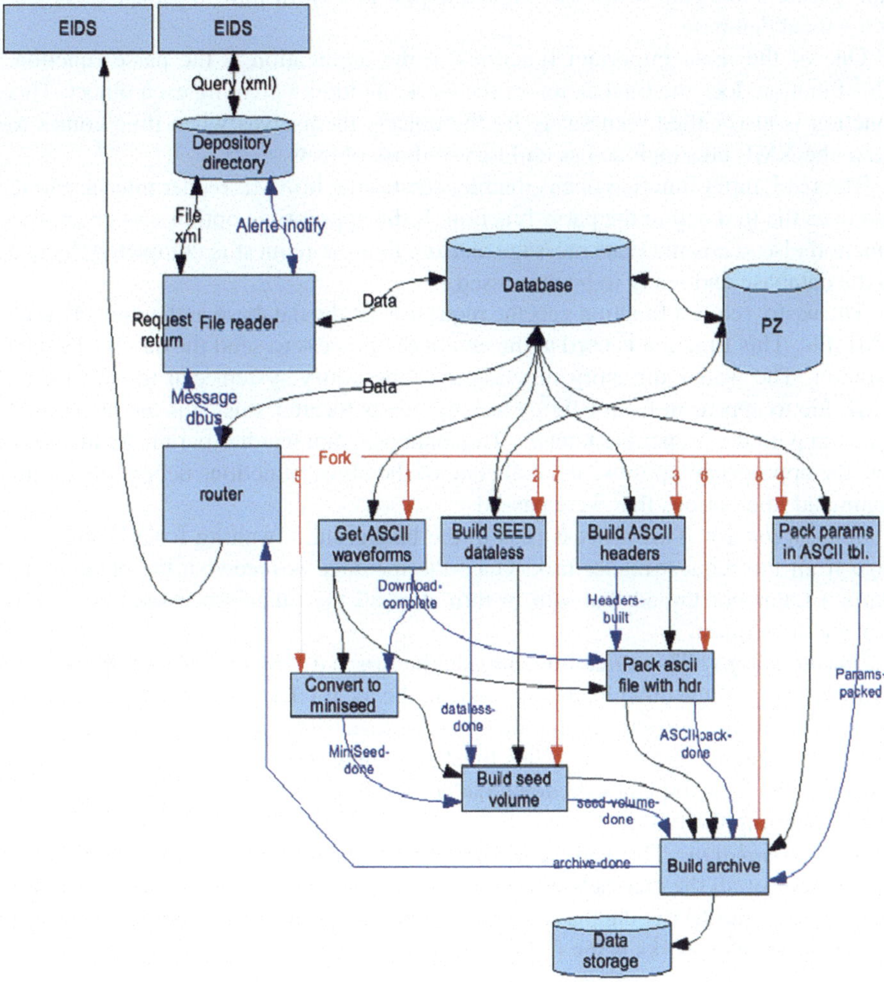

Fig. 10.4 Description of the accelerometric data server designed for processing the request coming from the accelerometric data explorer

synchronize the work to be done. All the information concerning the application architecture is described in Péquegnat et al. [6].

Request-reader.py: The request-reader.py application is used at two locations in the system. It is used as the loader application. When started with the proper arguments, this application uses the notify system call to watch a directory in which RabbitMQ is configured to drop the XML request files when they arrive. The application then opens the XML file, and uses the libXML reader interface to parse the XML file into its components, and create specific python objects for each XML tag. The parsing follows instructions contained in a python dictionary data structure (NA5REQUEST). Those instructions include information about the fact that the parameter is required, the parameter type and list of valid values. It also contains pointers to a pair of functions (get and put) that know how to persist the object in the local database.

One of the most important functions in the application is the parse function. This function does the final action of the work, including creating each object. This function is also called recursively by the objects themselves when time comes to parse the XML tags included as children of those objects.

The read_input function opens the file, creates the libXML reader interface to it, and does the first call to the parse function. If the appropriate option was given, this function also sends the dbus message alerting that the request is completely loaded in the database and ready to be processed.

The write_request function gets the request from the database and outputs it as an XML file. This function is used at the end of the process to send the data to the data explorer. The watch_directory function uses the notify system call to wait for an XML file to appear in from EIDS. Once the file is located, this calls the read_input function with the proper parameters. The main program handles parameter analysis, sets the appropriate options, and calls one of the above functions depending on the command line options that were passed.

data-engine.py: This process runs in the background, waiting for a DBus message from the request-reader task. Once the message is received, the data-engine starts a Launcher thread that will in turn start all the other processes required to execute the request.

downloader.py: This thread downloads the original ASCII files that are not yet cached locally. If the download is successful, the local URL is stored in the database (which signals the presence of the file for subsequent runs). When all files are downloaded, the thread signals that it has finished, which in turn allows for subsequent tasks to start. If a file can't be downloaded, then the URL field in the database is empty, warning the failure.

datalessmaker.py: This task generates a SEED dataless file from the data sent in the request for all the channels of each station it contains. It widely uses some tools coming from the BDsis database system set up for the RAP network [5]. The way the datalessmaker works is the following:

1. An (ASCII) intermediary file is built for each station, using some pieces of the specific channels information stored in the internal database (and particularly but not only, the pz_url attributes of the *sensor*, *preamp* and *adc* elements). Such a file (called "dbird", for "database instrumental responses description")

establishes all the links between the different channels elements for a station, and the corresponding (relative) PZ files of the PZ database. An example is given further.
2. The SEED writer itself reads a dbird file, loads the entire PZ database, checks that the PZ files names which are referenced in the dbird file are available, performs some controls in those PZ files, and build the dataless volume. The seed writer first builds Stations header logical records, then builds Abbreviations header logical records, and finally builds Time header logical records.

sacmaker.py: Once the ASCII data is downloaded, this thread will transform each file into a file in the SAC format. It is run twice, the first time to generate ASCII format SAC files, the second time, to generate Binary format SAC files. This thread uses a pure python implementation of SAC writing that can be found in pysac.py.

mseedmaker.py: This task converts the binary SAC files built above into miniSEED files using the sac2mseed program. The main part is identifying which SAC files are there that haven't been yet transformed into mini-seed, then calling sac2mseed for each of them. Once this is done, the task alerts whichever other task that is interested that it is finished and that they can start their own work.

seedmaker.py: This task build a large SEED file for each station with all the miniSEED info previously set up. The miniSEED files are "cat-ed", and the seed volume for each station is built using the RDSEED program.

This task generates the appropriate ASCII data files according to the 2 volumes format as shown in Fig. 10.5.

volume 1

```
---- EVENT
unid : <unid>
source : EMSC/CSEM
localisation : NW SPAIN
1996-02-18 01:45:45.5 2.54 42.79 8 5.2 <type_mag1> 6.3 <type_mag2> <mag3> <type_mag3>
---- STATION
code : <stationcode> network : IGC
latitude : 42.730 longitude : 7.320 elevation : 250.0
depth : 0.000 site : <R/S/A/B/C/D> sensor : EST
building_code : XX
---- EVENT-STATION
distance_to_event : 68.0
event_to_station_azimuth : 206.16
station_to_event_azimuth : 25.87
---- RECORD
julian_day : 327/2006 time : 06:58:01
sample_rate : 0.00800 nb_points : 5632
component : ENE unit : cm/s2
azimuth : 90.0 incident : 90.0
------------------------------------------------------------
VERSION : 2008-07-24T12:35:21Z – Neries project
------------------------------------------------------------
0.0000 1.527225e-02
0.0080 6.927599e-03
0.0160 -6.662256e-03
```

volume 2

```
---- EVENT
unid : <unid>
source : EMSC/CSEM
localisation : NW SPAIN
1996-02-18 01:45:45.5 2.54 42.79 8 5.2 <type_mag1> 6.3
---- STATION
code : <stationcode> network : IGC
latitude : 42.730 longitude : 7.320 elevation : 250.0
depth : 0.000 site : <R/S/A/B/C/D> sensor : EST
building_code : XX
---- EVENT-STATION
distance_to_event : 68.0
event_to_station_azimuth : 206.16
station_to_event_azimuth : 25.87
---- RECORD
julian_day : 327/2006
component : ENE
azimuth : 90.0 incident : 90.0
---- RECORD-PARAMETERS
pgau : 4.573729 cm/s2
pga : 4.597282 cm/s2
al : 0.011488 cm/s
td : 8.310000 s
cav : 5.622970 cm/s
pgv : 0.198258 cm/s
hl : xxxxx cm
f(hz) psv(cm/s)
0.15 0.594359
0.19 0.297106
0.23 0.219786
0.28 0.185797
0.34 0.220771
0.42 0.298527
0.52 0.109993
[... 28 values total]
------------------------------------------------------------
VERSION : 2008-07-24T12:35:21Z – Neries project
------------------------------------------------------------
```

Fig. 10.5 Description of the volume1 and volume 2 of the ASCII format

10.4 The Accelerometric Data Explorer

The accelerometric data explorer is part of the seismic data explorer developed within the NERIES project and is accessible at www.seismicportal.eu. This portal is an aggregation of different portal (thought as different applications) running on the server side from different Web servers but visually on the same Web Portal, accessible at an unique web address (Fig. 10.6).

Fig. 10.6 The NERIES web portal access (www.seismicportal.eu). Access to the accelerometric portal is available for registered users

This Portal currently includes:

- The Event Explorer to search for event information;
- The Wave Form Explorer for retrieving Broad band waveform data;

– Access to historical events' data;
– The accelerometric data explorer as presented here.

This last portal is a visual front end and easy tool to search for accelerometric data using specific criteria (upon specific seismic events or accelerometric parameters). This portal helps the end-user to have access to accelerometric data using a single access point giving access to accelerometric data providers (currently six). The back end of the portal consists of a database of the station metadata (network, hardware, sensors installation…), a database of event records with accelerometric parameters and the events catalog from the data explorer.

The portal is used to query these databases to found existing record event and retrieving accelerometric data in an easy user interface. An example of the user interface is given on Fig. 10.7.

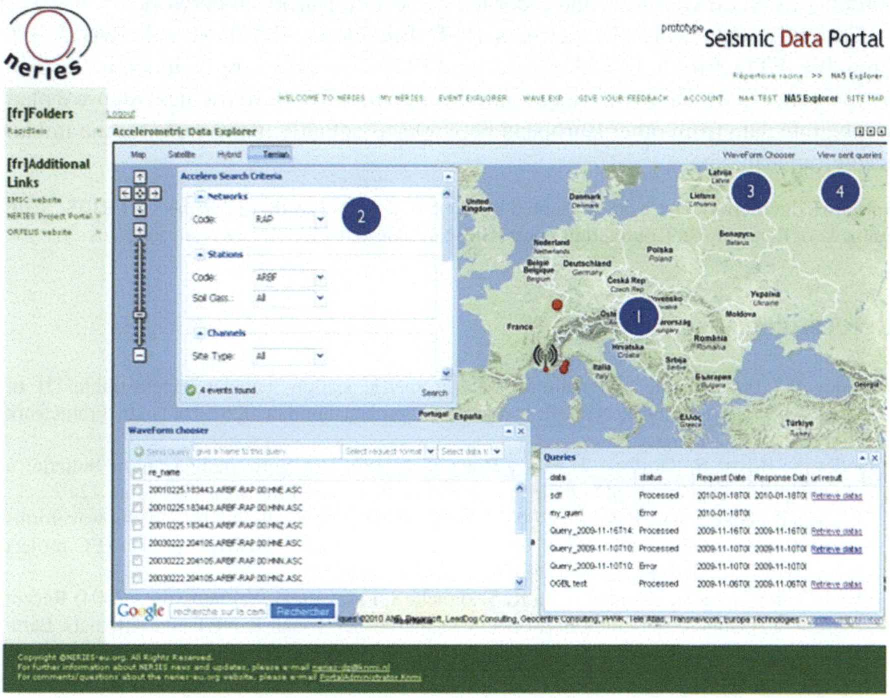

Fig. 10.7 The accelerometric data explorer portal shows a map (1) displaying the result of the user query. A user query is made using the Accelero Search Criteria (2) The Waveform chooser (3) displays all the records available for the selected events, the user chooses the desired records, give a name to the query, selects the result format and sends the query. In the Queries panel (4) the user can see the different queries made and the current status of each query. When a query has been processed a link to the result is provided and the user downloads the result

10.5 Conclusion

A first and preliminary version of the database used for the accelerometric data server and the accelerometric data explorer has been defined, as a portal in the NERIES seismic data portal. The data server was designed for uploading data (waveforms and ground motion parameters) in several format (SAC, ASCII and MiniSEED). The tier architecture of the system designed for accelerometric data was defined for seperating the data providers, the data server and data explorer system. The stream between the data explorer and the data server is based on an XML formatted file containing all the informations.

The inventory and the description of the accelerometric stations in the Euro-Med region is used by the data explorer for helping the end-users of the accelerometric data to perform request. The data server and the data explorer collect, archive and provide data from several European accelerometric networks (data providers) in the homogeneous format and ground motion parameters. By this way, it is now possible to download accelerometric data corresponding to seismic events localized at the borders of several countries and recorded by several European networks.

Currently, data from the partners (IGC Barcelona, IST Portugal, RAP-LGIT Grenoble, ETH Zurich, ITSAK Greece and KOERI Turkey) are included in the data server, concerning data with magnitude over 2 since 2000. In the next step we plan to integrate data from other European networks as actually in progress for the Italian accelerometric data.

Acknowledgements The work presented in this article is part of the EC Project NERIES, Sixth Framework Programme, Contract number: RII3-CT-2006-026130.

References

1. Godey S (2007) Define and maintain accelerometric station metadata, Deliverable D1 of the Networking Activity 5, NERIES project (EC project number 026130), Sixth Framework Programme EC, 36 pages
2. Godey S, Bossu R, Guilbert J, Mazet-Roux G (2006) The Euro-mediterranean bulletin: a comprehensive seismological bulletin at regional scale. Seismol Res Lett 77(4):460–474
3. Guéguen P, Péquegnat C, Revilla J (2007) Specifications of protocols and formats for waveforms access. Task C – Deliverable D3 of the Networking Activity 5, NERIES project (EC project number 026130), Sixth Framework Programme EC, 38 pages
4. Okada Y, Kasahara K, Hori S, Obara K, Sekiguchi S, Fujiwara H, Yamamoto A (2004) Recent progress of seismic observation networks in Japan (Hi-net, F-net, K-NET and KiK-net). Earth Planets Space 56:xv–xxviii
5. Péquegnat C, Guéguen P, Hatzfeld D, Langlais M (2008) The French accelerometric network (RAP) and national data centre (RAP-NDC). Seismol Res Lett 79(1):79–89
6. Péquegnat C, Jacquot R, Guéguen P (2009) Development and implementation of unified waveform request. Task C – Deliverable D5 of the Networking Activity 5, NERIES project (EC project number 026130), Sixth Framework Programme EC, 91 pages
7. Tapia M, Susagna T, Goula X (2007) Task B – Specifications for PSA and PSV – Definition and computation of parametric data. Deliverable D2 of the Networking Activity 5, NERIES project (EC project number 026130), Sixth Framework Programme EC, 140 pages

Part III
Arrays and Networks

Chapter 11
Euroseistest 3D Array for the Study of Complex Site Effects

K. Pitilakis, D. Raptakis, K. Makra, M. Manakou, and F.J. Chávez-García

Abstract Euroseistest is currently the longest running instrumented test site in the world. It was originally defined as the 2D (N–S) cross section of the Mygdonian basin, N-E from Thessaloniki Greece, epicenter area of the M6.4 1978 earthquake. In this paper, we present the effort to extend the test site to a larger portion of the whole sedimentary structure, i.e., from 2D to a 3D structure. To this end we have compiled available geological and geotechnical information. We have analyzed microtremor and earthquake data. We present the results of the analysis of all available information and data. The synthesis of all data allowed us to propose reliable image of the geometry and the properties of the basin. We have also obtained a reliable estimate of the site response throughout the basin and we have discussed several aspects of site effects in complex geologic structures, including the increase of spectral amplification compared to 1D site amplification.

11.1 Introduction

Site conditions affect strong ground motion during destructive earthquakes. They have been shown to affect the intensity of ground motion by a value between one and two units (e.g. [4]). The importance of site effects led to the creation of test sites; sites where detailed observations on the variations of site response could be related to the subsoil structure. The first test sites were those of Turkey Flat USA [29] and Ashigara valley Japan [9]. In Europe, Euroseistest was established as a test site for seismology and earthquake engineering studies in 1993. During its operation, a large number of universities and institutes throughout the world have been involved in both field measurements and research studies.

EUROSEISTEST project is actually a very powerful multidisciplinary European experimental site to perform integrated studies in earthquake engineering, engineering seismology, seismology, and soil dynamics. It generates a set of high quality

K. Pitilakis (✉)
Department of Civil Engineering, Aristotle University of Thessaloniki,
GR-54124 Thessaloniki, Greece
e-mail: kpitilak@civil.auth.gr

S. Akkar et al. (eds.), *Earthquake Data in Engineering Seismology*, Geotechnical, Geological, and Earthquake Engineering 14, DOI 10.1007/978-94-007-0152-6_11,

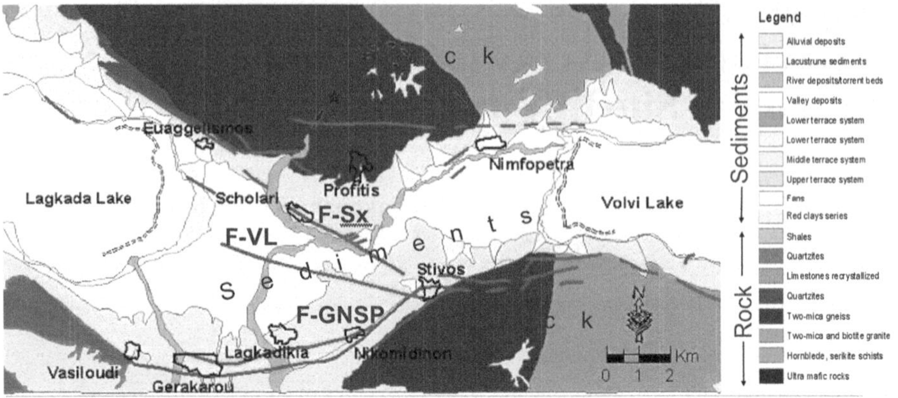

Fig. 11.1 Geological map and main tectonic features (seismic fault F-GNSP: 12 km length, E–W direction, 70–80° deep and 2 branches: F-Sx: 8 km length and F-VL: 5 km length) of the central part of the Mygdonian basin. The *star* depicts the epicenter of the 1978 Ms=6.4 Volvi earthquake

data relevant to the above scientific topics. Scientists, researchers, engineers working in any of the above fields have access to valuable data to check their models and methodologies, to validate and to improve models and processes, and to develop new ones. The core of the research activities performed during the last 15 years is threefold: (a) to construct a reliable model of the geological structure of the Mygdonian basin (Fig. 11.1), adequate as input model in site response studies, (b) to conduct experimental and theoretical research for understanding the physics of ground motion variations due to 1D/2D/3D site effects for engineering applications, (c) to contribute to the ongoing elaboration of the new generation of Eurocode 8 in terms of site specific response spectra, soil amplification, etc.

At present, from different test sites whose data were opened to the scientific community a decade ago, only Euroseistest continues to be operational. This makes it one of the longest running test sites in the world. During these years, much insight has been gained from the detailed study of a 2D cross section of a 3D sedimentary structure. The detailed structure of this cross section is validated [6, 24, 25]. Moreover, some papers have already been published [3, 18, 15] that explore the possible use of the lessons learned in Euroseistest to the more general framework required by building codes. For all these reasons, it has been decided to extend Euroseistest from 2D to a 3D test site. So much has been gained from the 2D Euroseistest, that it is considered worthwhile to pursue further the idea of this test site, into the current problems posed by 3D geometries.

For these reasons, we report the 3D model of the valley and an evaluation of its seismic response. In order to constrain the 3D structure, a huge amount of geophysical and geotechnical data was synthesized. Concerning site response, we present an evaluation of many datasets distributed throughout the basin, from earthquake and noise records at temporary weak motion networks and the 3D permanent accelerograph network. All these results are correlated with the detailed information available at the 2D cross section and the 3D soil model as well providing preliminary constraints to the modeling of the 3D seismic response at this site.

11.2 Definition of a Complex Geologic Structure

11.2.1 Background, Aims and Workplan

The importance of site effects led to the establishment of test sites where detailed observations of the variations of site response could be related to the subsoil structure determined from a thorough geophysical and geotechnical exploration program. This actually implies the knowledge of the geometry of the main geologic formations in the studied area as well as their main dynamic properties from free surface down to the sediments bedrock interface and their spatial distribution in the whole basin [11, 14, 30], Shin Aoi et al. 2002, [7], Guoqing [13].

During the operation time of Euroseistest, a large number of field measurements and research studies were performed. At the beginning, a 2D cross section of the Mygdonian basin was studied (Raptakis 1995 [10, 28, 25]). At that time, the significance of lateral variations in the subsoil structure relative to the large amplification caused by impedance contrasts in the vertical direction was still being debated. During the last years, our interest was extended from the 2D cross-section to the whole basin in three dimensions.

At the beginning of Euroseistest, an extensive seismic campaign with conventional prospecting and a detailed geotechnical in-situ and laboratory survey were conducted, oriented to the determination of the main geologic formations in Mygdonian basin. The results of all these measurements were correlated and compared with each other. Shear wave velocity, Vs, values played the most important role for the configuration of 1D and 2D profiles for seismic ground response analysis. Surface Wave Inversion (SWI) method provided the majority of the Vs values. The resulting 2D soil model is presented in Fig. 11.2 and consists of 7 soil layers that can be divided in two units according to their geological age. Surface layers A, B, C and D belong to Mygdonian System and have an average Vs of 330 m/s, while layers E, F and G* belong to the Premygdonian one with average Vs of 750 m/s [15, 25]. This model was used as a reference point to extend our knowledge from a 2D cross-section to a 3D structure.

We present herein the 3D model of the valley and an evaluation of its seismic response. We first define the region of the basin that can be realistically constrained. The selected portion is not a closed basin, but it is the general case in sedimentary basins. In order to construct the 3D structure, we have reappraised the available geological data. Based on that, we have evaluated the 3D geometry of the Mygdonian basin proposed in an old study by BRGM (1971) concentrated on the determination of the interface between sediments and bedrock, using P-wave refraction tests, electrical soundings, and data from few hydrogeological boreholes. By that time, no data concerning Vs and lateral variations were available. For this reason, a large testing program was designed and performed including, mainly, seismic prospecting

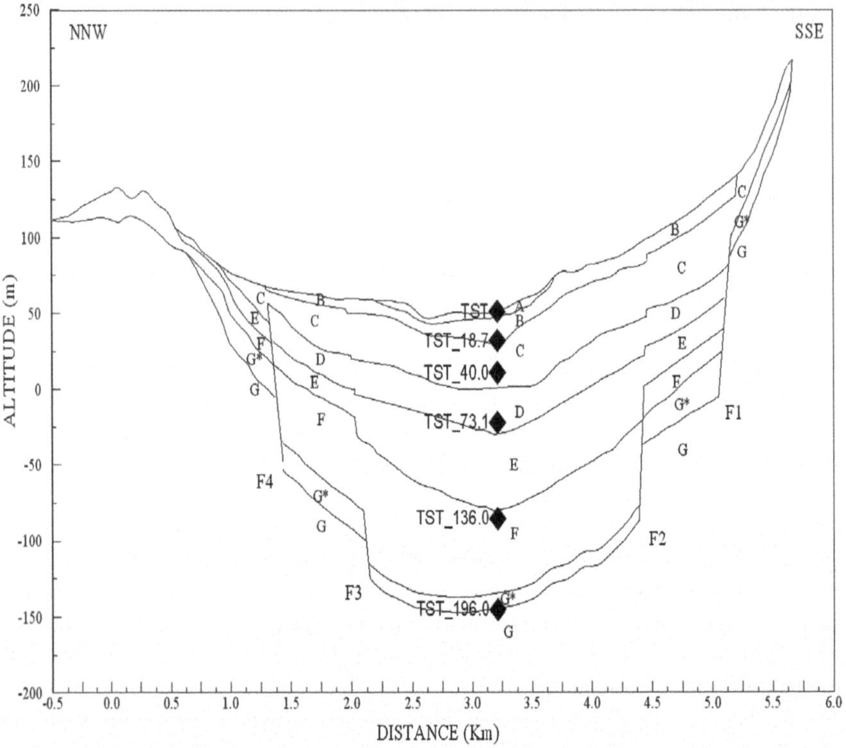

Fig. 11.2 2D NS (Profitis – Stivos) soil model and the TST downhole array [25]

lines in the EW direction and numerous microtremor array measurements well distributed in the whole basin (Fig. 11.3). Additional information was obtained from the recently performed array microtremor measurements and seismic prospecting together with electrical tomographies at the edges of the basin, and from near surface in situ geotechnical tests (Standard Penetration Test, SPT, and Cone Penetration Test, CPT). Large effort has been carried out during the validation step to bridge Vs profiles with borehole and electrical data, in situ geotechnical tests, and earthquake recordings, as well.

11.2.2 3D Model Configuration

To investigate the Vs at sites within the basin that was poorly known or completely unknown especially at larger depths, a huge program of microtremor array measurements was performed. In a first stage, few ambient noise array recordings, conducted along the NS cross-section, gave comparable results [1, 12] with seismic prospecting results [26, 28, 25, 10]. Results from large-scale refraction experiments (using big explosions and of total length around 8 km) in the middle of the basin along the EW

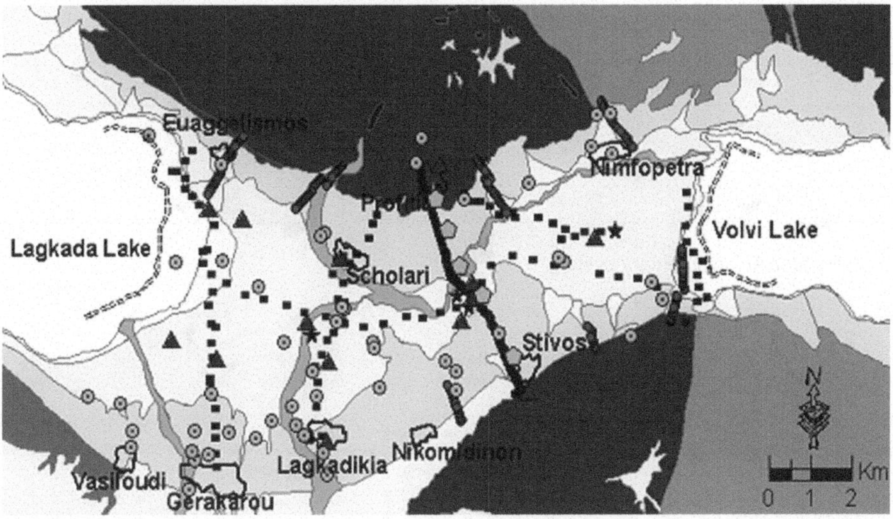

Fig. 11.3 Geological map of the central part of the Mygdonian basin; with (*circles*) hydrological and geotechnical boreholes (*squares*) single station microtremor measurements (*hexagons*) array microtremor measurements (*small circles*) electrical tomographies at the boundaries of the basin (*triangles*) temporary accelerograph network operating during 1997 and (*black stars*) linear refraction experiments

direction [21, 27], indicated that a constant average value for the Vs does not apply at the Mygdonian basin. For this reason, in a second stage, Spatial Autocorrelation Coefficient (SPAC) method was applied in more than 35 sites [19, 21].

In general, the Vs from surface to bedrock have values between 90 and 800 m/s (Fig. 11.4). The interface between the two main sedimentary systems, Premygdonian – Mygdonian has been determined at all sites of the noise measurements. The depth of this interface at the western part of the basin is of about 120 m, while at the central and eastern part it is about 80 m with an average velocity of the Mygdonian system between 290 and 330 m/s. In some sites, the penetration depth did not reach the bedrock, but Vs at large depth reached values greater than 600 m/s indicative of the stiff formations of the Premygdonian system. Thus, it seems that the depth variations of sediments-bedrock interface are due to thickness variations of the Premygdonian system that is found to present Vs between 500 and 800 m/s.

Especially, at the TST site where a downhole array with 6 3D accelerometers installed at different depths (0, 18, 40, 73, 136, and 196 m), the Vs vertical profile was determined using earthquake recordings. The average Vs profile between the downhole stations was computed from the cross correlation of a set of earthquake recordings and found to be in agreement with estimates from SWI and CH (Fig. 11.5 [5, 17]).

An important issue to construct the 3D soil model was the definition of the interface between sediments and bedrock [27], which plays a significant role in site response studies since its shape governs the wave field propagation. In the middle of

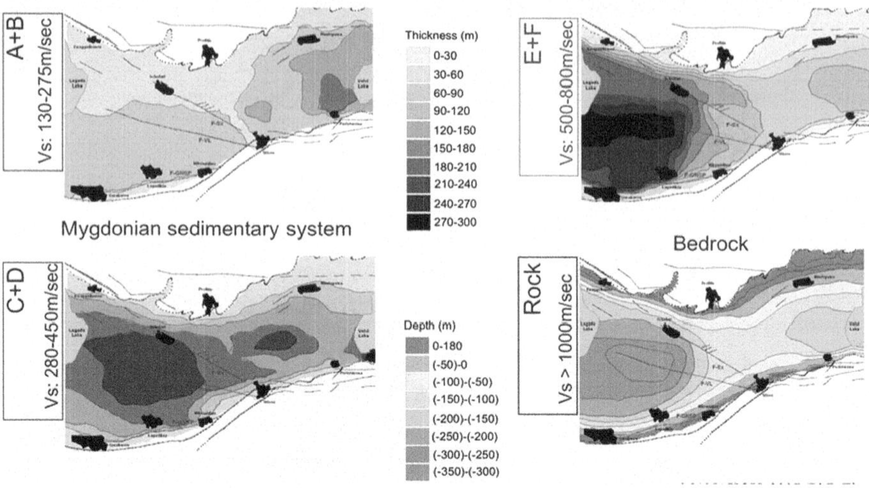

Fig. 11.4 Contour-maps showing the geometry of the Mygdonian (A + B, C + D) and Promygdonian (E + F) sedimentary formations and bedrock interfaces, with their thickness (in m) and depth (in m referred to sea level), together with their range of Vs velocity. *Solid thick lines* indicate the main faults in the region. The sediments bedrock boundary is shown with *solid circles* [21]

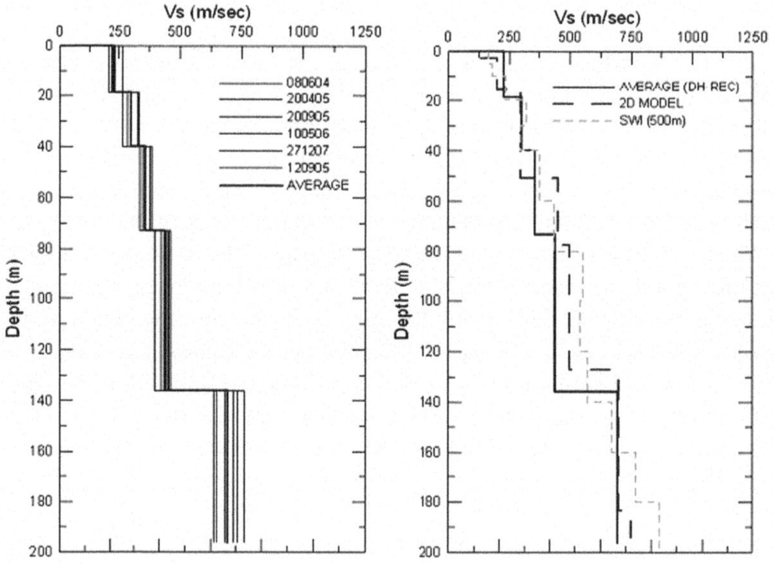

Fig. 11.5 Vs profiles determined from cross-correlation functions computed from the recordings of 6 events at TST downhole stations and their average value (*left*). Comparison between the aforementioned average Vs profile at TST with the Vs profile at TST obtained from Surface Wave Inversion and the revised version of the Vs distribution with depth for TST site [17]

the basin, the bedrock depth and Vs were determined mainly on the basis of SPAC results and in cases where this was not feasible (smaller penetration depths) through the correlation between empirical and theoretical transfer functions (for details see next paragraphs). At the western part of the basin, the depth of the bedrock with Vs larger than 1,000 m/s, is about 400 m while at the central and eastern part it is about 200 m (Fig. 11.4). The basement successively deepens, more than 200 m, close to the offshore line of Volvi Lake. At the northern and southern edges of the basin, the sediments-bedrock interface was determined based on results from array microtremor recordings and P-wave refraction tests [20], together with many electrical tomographies and soundings [32]. These measurements were performed perpendicular to the contact of bedrock formations with sediment deposits and constrained the geometry of the basin at the edges. Almost all of them showed smooth sediments – bedrock interface, with a gentle dip towards the centre of the basin. Some bedrock discontinuities were detected, probably because of the activity of several branches of the faulting system of the area. All the aforementioned results were confirmed by a series of hydrological and geotechnical boreholes and as well as electrical soundings [8].

In addition to the above, a series of CPT (> 30 along EW and > 10 along NS direction) and SPT tests were performed to obtain the geotechnical classification of the near surface soil materials. For the interpretation of CPT tests and soil classification, a thorough examination of geotechnical conditions is undertaken based on a considerable number of geotechnical boreholes and adjacent Vs profiles. The central part of the basin is essentially flat and consists mainly of loose to medium silty-clayey soil, at the Western part, and loose to medium silty-sands with organics at the Eastern part. The correlation between tip cone resistance (q_c) and shear modulus at very low shear stain levels (G_{max}) is rather difficult due to the inherent differences of the two methods (as cone resistance is a punctual measurement while Vs (or G_{max}) represents larger volume of soil). Nevertheless, in general those values seem to follow the relationship proposed by Mayne and Rix [22] as a first approximation.

Finally, we put into the procedure of integrating the 3D soil model other indirect information such as fundamental frequencies from empirical (ambient noise measurements) and theoretical transfer functions (based on the determined Vs profiles) [21]. The fundamental resonant frequencies in combination with amplification factors, for the whole sedimentary basin from all data sets (i.e. single station and array microtremor measurements at more than 300 sites, and a large number of earthquake recordings at more than 30 stations, during the operation of temporary and permanent networks), and their correlation with other direct (measured) parameters of soil profiles, lead to useful conclusions for the 3D geometry of the basin [21, 27]. For example, at most sites close to the centre of the basin, we observed the maximum amplification ratios in a wide resonant frequency band in both horizontal components. This implies that, for example at the centre of the basin, lateral constrains are significant or that different layers contribute to ground motion at different frequencies. In any case the characteristics of site response in frequency domain allow us to propose a map of dominant frequency throughout Mygdonian basin (Fig. 11.6).

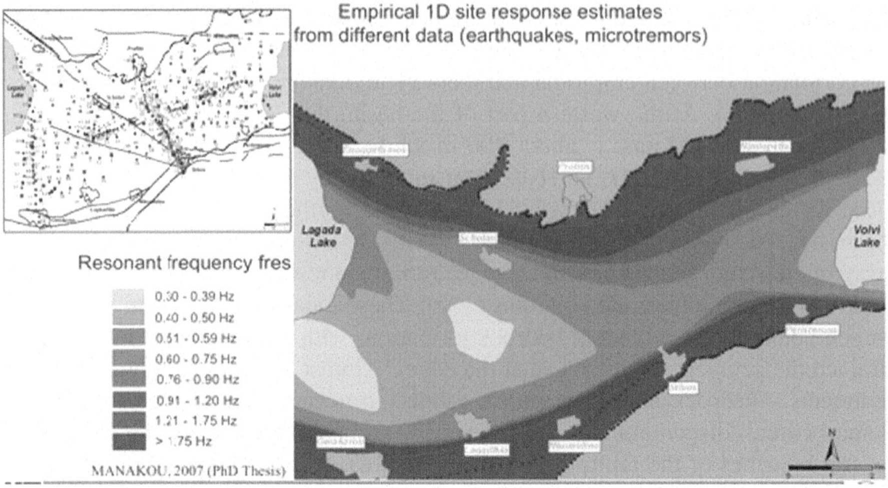

Fig. 11.6 Map obtained through the interpolation and smoothing of all available estimates of dominant frequency (together with information provided by the electrical measurements and borehole data). The dominant frequencies in this map correspond to the amplification of the whole sediment volume, Mygdonian plus Premygdonian systems [19, 21]

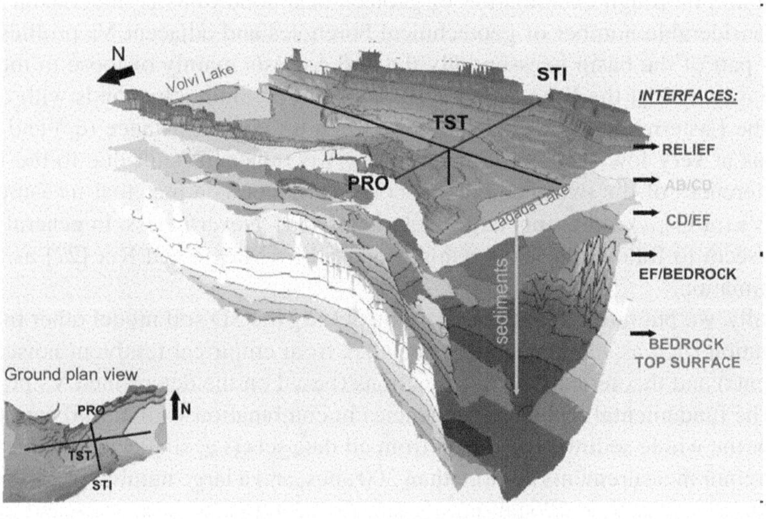

Fig. 11.7 3D model of the Mygdonian basin geological structure

All these results, including amplification factor, are well correlated with Vs profiles, proving the sedimentary volume of the basin (Fig. 11.7).

Concluding, field investigations and their integration lead to a rather complex geological model (Fig. 11.7) useful for empirical and theoretical studies of site response. It is a rather shallow basin according to its shape ratio. The 3D soil model of the Mygdonian sedimentary basin has a rather asymmetrical semi-cylindrical shape, which deepens at the open edges. This benchmark model is composed of two main sedimentary geological systems (Mygdonian – Premygdonian) according to their geological age in consistency with the initial 2D NS cross section [15, 25]. The characteristic Vs values of main systems and the bedrock basement are also given as average velocities of the sublayers from top to the bedrock. The bedrock is deeper at the western part of the basin than at the eastern part. At the north and south borders of the basin the bedrock appears at the surface. At the southwestern part of the basin the extension of the Premygdonian sediments is greater than in southeastern part and for this reason the sediments – bedrock surface contact extends more towards the south.

11.3 Strong Motion Array and Site Effect Studies in Complex Media

Taking advantage of the knowledge of the subsoil conditions in the Mygdonian basin and the existence of earthquake recordings at a 3D strong motion array, many site response studies, using conventional and advanced tools, revealed the effects of the local complex geology in ground motion characteristics. In this section we present briefly the main scientific conclusions from these studies combining both instrumental (based on the total length of recordings or their time windows using Standard Spectral Ratio, SSR, Horizontal to Vertical Spectral Ratio, HVSR, and Generalized Inversion Scheme, GIS) and theoretical approaches (1D and 2D modeling), both in frequency and time domain, in order to show the specific characteristics of site effects, regarding the dominant frequency, the amplification level, and the shaking duration and their dependencies from the geologic complex conditions.

11.3.1 Strong Motion Array

At the central part of the Mygdonian basin, a well-designed 3D permanent strong motion array was deemed of major importance to record local and regional seismicity in the area, in order to study the effect of local complex geology in ground motion characteristics. The effort to establish such a network started in early 1993 under the financial support of a European project EUROSEISTEST. After the continuous support of European funds (EUROSEISMOD 1995 and EUROSEISRISK 2002), we were able to expand and upgrade the strong motion array. Today, Euroseistest is the longest-running strong motion arrays in Europe. It covers an area of about 6 km $\times$ 8 km and consists of 21 surface and downhole 3D digital accelerographs, installed in a cross shaped scheme (Fig. 11.8), covering different geological conditions and depths down to the bedrock. It is supported with GPS units and permanent remote monitoring useful to transfer and archiving of earthquake recordings. In the middle of the basin, at TST site, a dense downhole array of accelerometers has been installed at depths of 18, 40, 73, 136, 196 m-bedrock, connected to a

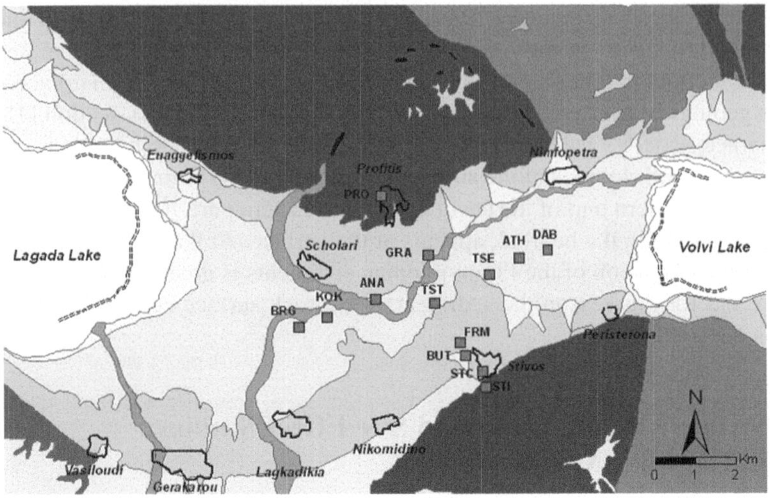

Fig. 11.8 Plan view of the free field instruments (*squares*) of the Euroseistest strong motion array

18-channel recording system, as well as a pore-pressure probe installed at 17 m depth (Fig. 11.2). A smaller downhole array of 2 sensors (surface and 33 m depths) is installed in outcrop rock conditions at the northern part of the basin (PRO site). More than 300 local and regional earthquakes have been recorded in the array, most of them with peak ground accelerations less than 130 gals.

11.3.2 Empirical and Theoretical Site Response Along 2D Cross-Sections

A comprehensive study of site effects was based on a well-known 2D soil model (Fig. 11.2) using observations together with 1D and 2D analyses [6, 25]. These studies gave the basic and at the same time crucial achievements for the *identity* of the details of site response in this test site. Many other posterior studies confirmed its specific achievements using different approaches and data sets. The main conclusions of these studies are summarized in the following paragraphs.

The study of site response, in frequency domain, of the entire seismograms of two events, one recorded at the permanent accelerograph network and the other at a temporary seismograph array along the 2D NS cross-section showed that the lateral discontinuities play an important role in the amplification pattern both in their vicinity and at the central part of the valley. The examination of three different time-windows (P, S, and surface wave, SW) of the accelerograms showed that the SW time-window contributes significantly to the peak resonance amplitude (Fig. 11.9). Love and Rayleigh waves observed in all components in the centre of the valley are locally generated at the lateral variations. The existence of Rayleigh waves in the

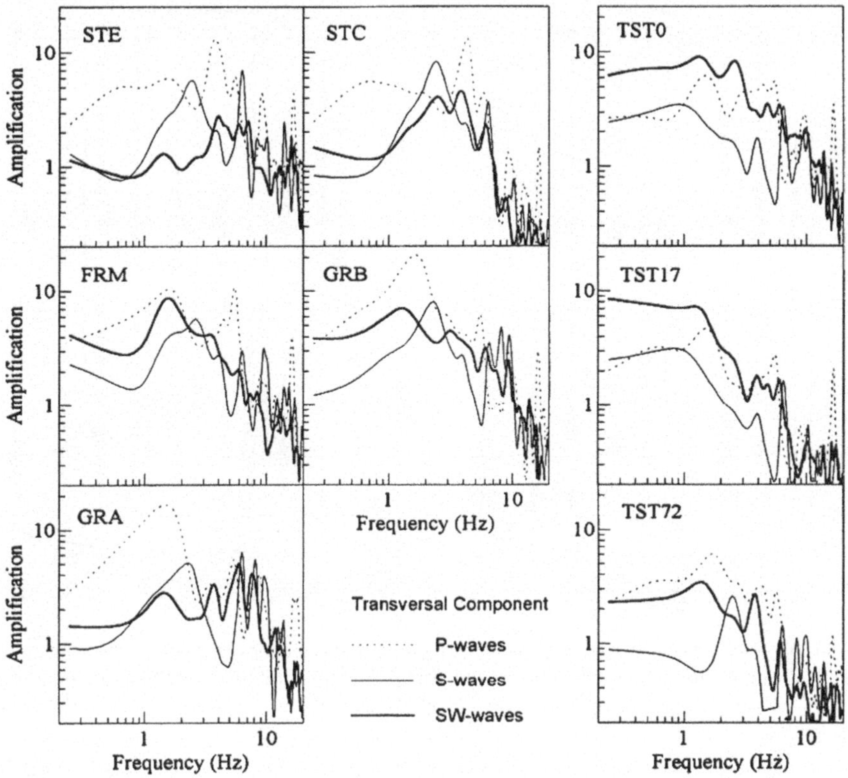

Fig. 11.9 Transfer functions obtained for P, S and SW time-windows of the transversal accelerograms for GRA, GRB, FRM, STC, STE, TST0, TST17 and TST72 accelerographs. Reference records obtained at PRO. The transfer functions from SW time-window (*thick line*) show amplification larger than those of S time-window (*thin solid line*) at the stations within the central faults (F2 and F3) for frequencies up to 2 Hz [25]

vertical component may also justify the disparity between the amplitudes of HVSR and SSR [28]. Additionally, it has been shown that 1D theoretical transfer functions do not reproduce the complexity of the empirical ones. Seismograms and accelerograms from the first S wave arrival to the end showed that wave trains with frequency content up to 3.5 Hz are distributed either in the S-wave or SW time-window. Thus, it is difficult to distinguish the contribution of different type of waves in resonance. This underscores the fact that the study of site effects in frequency domain gives only a partial image. Synthetic time histories obtained by the convolution of 1D transfer function (TF) with records at the reference site, show large disparities, with respect to the recorded traces concerning the duration of shaking within the valley (Fig. 11.10). It is clear that the use of the best 1D TF to model ground motion in time domain under complex soil conditions is not very efficient since the underestimated duration of shaking may lead to erroneous predictions of ground motion.

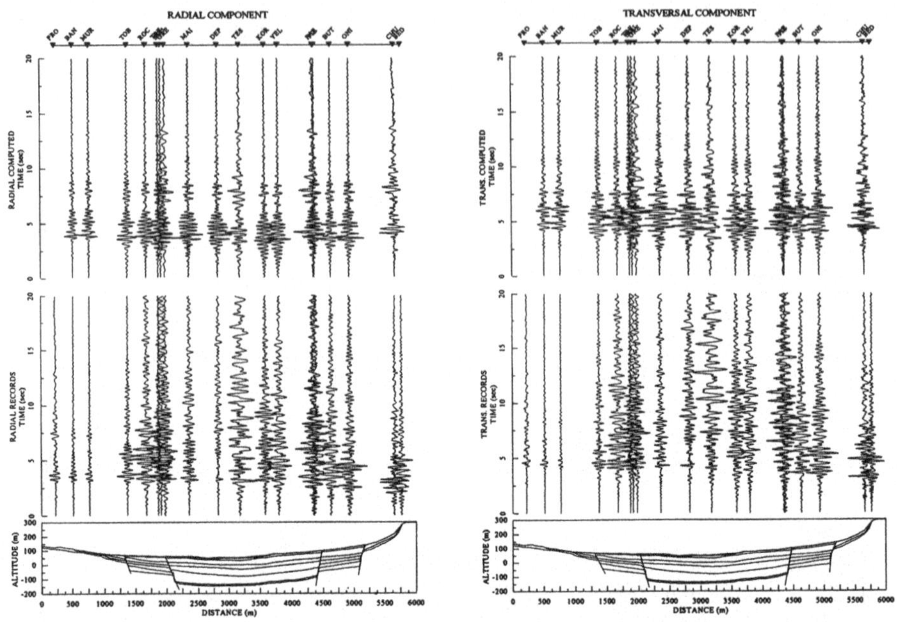

Fig. 11.10 Comparisons between filtered seismograms and synthetics obtained by the convolution of 1D TF with the records at the reference site. The radial (*left*) and transversal (*right*) component of both recordings (*bottom*) and synthetics (*top*) are shown together with respect to the cross-section [25]

In addition to 1D computations and their comparison with the observations, we have computed the response of the 2D structure to vertically incident SH waves. The results confirm that the dominant feature of site response is the locally generated surface waves (Fig. 11.11). We have investigated the evidence of these surface waves in earthquake data recorded at TST site including both seismograph and accelerograph data. We observed that locally generated surface waves contribute significantly to ground motion, and that they appear at the same frequencies as the resonance frequency of vertically propagating shear waves (Fig. 11.11). Similar conclusion on the predominant phenomena of the lateral propagation have been conducted with a different approach, which combines the analysis of the classical spectral ratio with the sonogram approach applied on simultaneous recordings at the site of interest and a reference site as well [23].

In a later step, a parametric study of the seismic response was carried out [15] to give a practical answer from engineering point of view to the necessary degree of detailing of subsoil structuring of a site, to provide a reasonable evaluation of its site response. Thus, we have computed the seismic response on a series of 2D models representing the existing 2D one with different degrees of detail. We computed the response at the surface of 3 2D models with the same geometry but different velocity

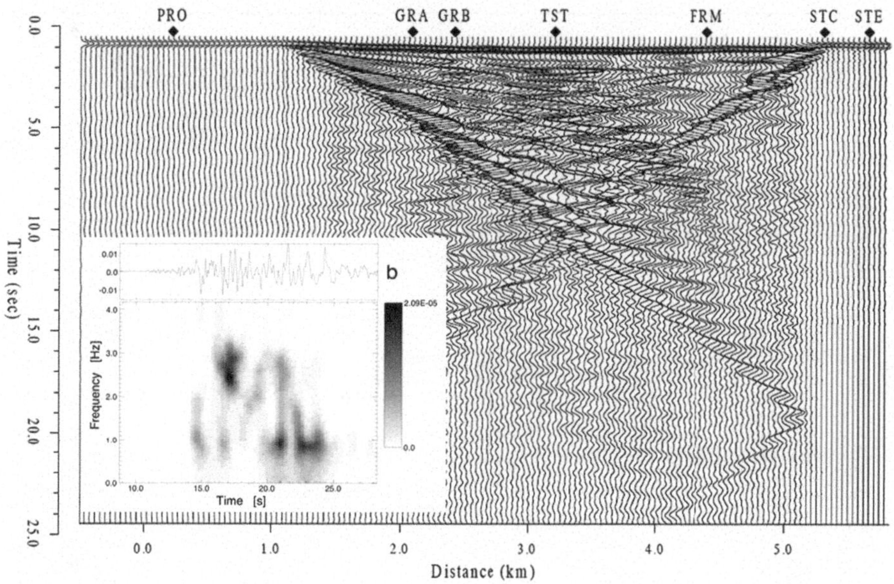

Fig. 11.11 Seismic section computed at the surface of the 2D model shown in Fig. 11.2 for vertical incidence of SH waves. The positions of the surface accelerographs of the permanent network have been indicated for reference. Traces have been low-pass filtered with a 3.5 Hz frequency cutoff [6]. *Inset* figure shows the evolutionary spectrum of the accelerogram recorded at station TST during the event of 5.03.95, transverse component (SH motion). Above the spectrum is shown the corresponding observed accelerogram. Before computing the spectra, traces were band-pass filtered between 0.1 and 5 Hz [6]

structure in the sediments. Finally, two additional models were considered representing different possible geometries of the sediment/bedrock interface (Fig. 11.12). We then calculated the differences among these models in terms of frequency transfer functions, and time domain synthetics. Our results show clearly that, in the case of Euroseistest, site response depends fundamentally on its closed basin shape because it is largely controlled by locally generated surface waves (Fig. 11.13). Thus, in terms of predicting site response, a rough idea of its shape ratio and of the average mechanical properties of the sediments, are sufficient for an accurate estimation of the response of 2D valley cross section, even with a poor knowledge of the 1D profile at the centre of the valley. Although the details of ground motion may vary significantly between the models, the relative amount of surface waves generated in the 2D models seems to be relatively constant. Similar conclusions were also reached from another parametric study on the same 2D model, however, using boundary element method for SH and SV wave fields [31].

In a strict sense, the aforementioned conclusions are valid only for Euroseistest. However, in most sedimentary valleys site effects are due to the combination of

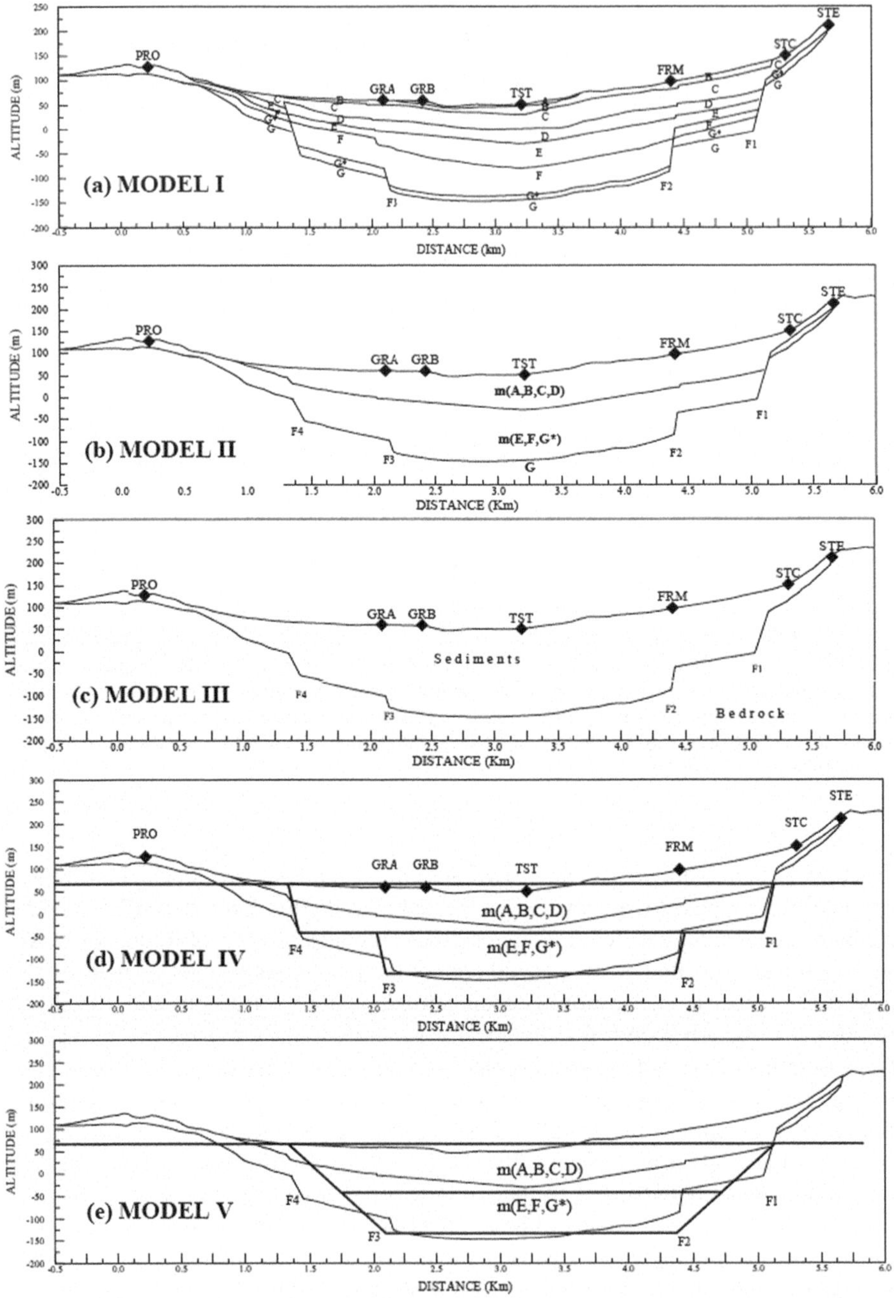

Fig. 11.12 (Contd.)

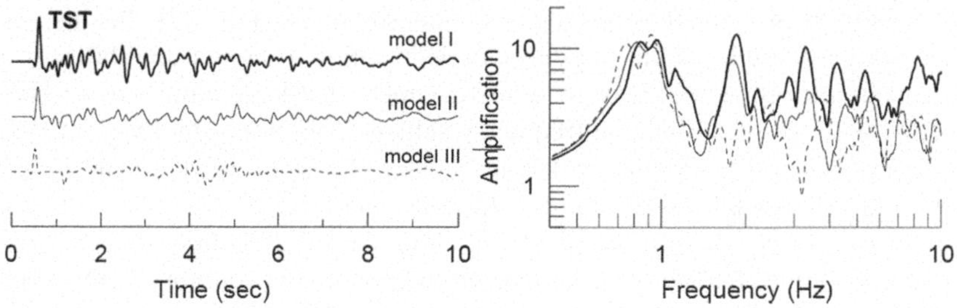

Fig. 11.13 *Left panel*: synthetic seismograms computed using a finite differences code at TST site for the three different 2D models (I, II, and III) shown in Fig. 11.12. *Right panel*: corresponding transfer functions for each synthetic seismogram. Different *line* styles identify the transfer functions with the corresponding synthetic seismogram [15]

1D resonance and the lateral propagation of locally generated surface waves [2]. Thus, if the seismic response of Euroseistest is representative of these conditions for shallow alluvial valleys, then our results may apply to a large number of sites.

11.3.3 Empirical and Theoretical Site Response in the Whole Basin

Concerning the experimental site response studies in the Mygdonian basin, we present an evaluation of many datasets: single station noise measurements at almost 300 points (distributed throughout the basin), earthquake records at a temporary seismic network (operated in 1997), recordings at the permanent accelerograph array as well as microtremor array recordings at 30 additional sites at the basin. All results are correlated with the detailed information available for the 2D cross section, providing preliminary constraints to the modeling of the 3D seismic response at this site. The analysis of available earthquake and microtremor recordings throughout the basin provides a good general idea of the spatial variation of the site response characteristics. The main conclusions of these studies are briefly summarized in the following paragraphs.

Fig. 11.12 Five possible 2D models describing Euroseistest valley: **a** 2D cross section [25]. F1, F2, F3, and F4 indicate the four normal faults. **b** Simplification of the model shown in (**a**). The *shape* of the interfaces is kept, but each of four topmost (A, B, C, and D) and the three lowermost formation (E, F, and G*) have been replaced by a single *homogeneous layer*. The properties of the replacing layer are the average (weighted by thickness) of the properties of the replaced layers at the *centre* of the valley. **c** Further simplification of the sediment velocity structure. The interface sediment/bedrock is the same as in (**a**) but the sediments are considered homogeneous. The soil properties are the average values of the properties at the centre of the valley. **d** and **e** simplifications of the geometry of the 2D model. We use the same two layers, averaged from the detailed profile, as in model II, together with a much simpler geometry [15]

We analyzed all available microtremor measurements [19, 27]. In this stage, we used the amplitude of the microtremor H/V ratios to assign credibility to the dominant frequency values. It is generally accepted that large amplitude of H/V is correlated with reliable dominant frequency values. Even if the amplitude of HVSR of noise records is still a matter of debate, the correlation of the dominant frequencies with the known subsoil structure at the location of the 2D NS cross section and two indicative Vs profiles at BRG and DAB along the EW direction, shows that they reflect the vibration characteristics of the complete sediment column (Premygdonian plus Mygdonian sediments).

Additionally, we computed transfer functions from earthquake records at 14 sites, using data from a temporary seismic network [19, 27]. We first compared the estimates that are obtained from different techniques of analysis (SSR, HVSR and GIS), using different time windows of the recorded seismograms (noise before P-wave arrival, S-waves, and coda) (Fig. 11.14). We verified that each time we obtained similar transfer functions. At most sites close to the centre of the basin, transfer functions do not show subsequent peaks, which could be interpreted as corresponding multiples of a single layer over a half space. The observed peaks are closely spaced in frequency to be interpreted as overtones suggesting that, at the centre of the basin, lateral constrains are significant or that different layers contribute to ground motion at different frequencies. On the other hand, close to the edges of the basin, at the stations along the boundary between sediments and bedrock, few transfer functions show rather simple shapes. The comparison between transfer functions estimated from earthquake records with those determined from microtremor measurements showed similar results in terms of dominant frequencies, even if (as usually observed) transfer functions from microtremors have lower amplitudes.

One significant point concerning the seismic response of the 3D basin is the large vertical amplifications observed at some locations, similar to that observed in the horizontal components. This non-negligible vertical amplification occurs at the same frequencies at which the corresponding horizontal components are amplified, suggesting that, at least part of this amplification is due to locally generated Rayleigh waves, amplifying motion in both radial and vertical components [27].

11.3.4 Engineering Implications of Complex Site Effects

2D and 3D computations are usually performed to account for complex site effects and the engineering community is well aware of the need to take them into account. However, the dominant feature of building codes is based on 1D concept that applies to both soil classification, which is exclusively based on the vertical Vs profile, in particular only to that of the first 30 m, and the definition of the design response spectrum. Because of the lack of well documented subsoil structures and high quality earthquake records, there is a long way to go until the complex site effects will be treated systematically and taken into account in future seismic codes. Nevertheless, the well known soil model in Euroseistest and the large amount of data sets, allowed us to propose some practical rules in order to take into account the complex site effects in the seismic design. Our aim has been mainly to stimulate the discussion

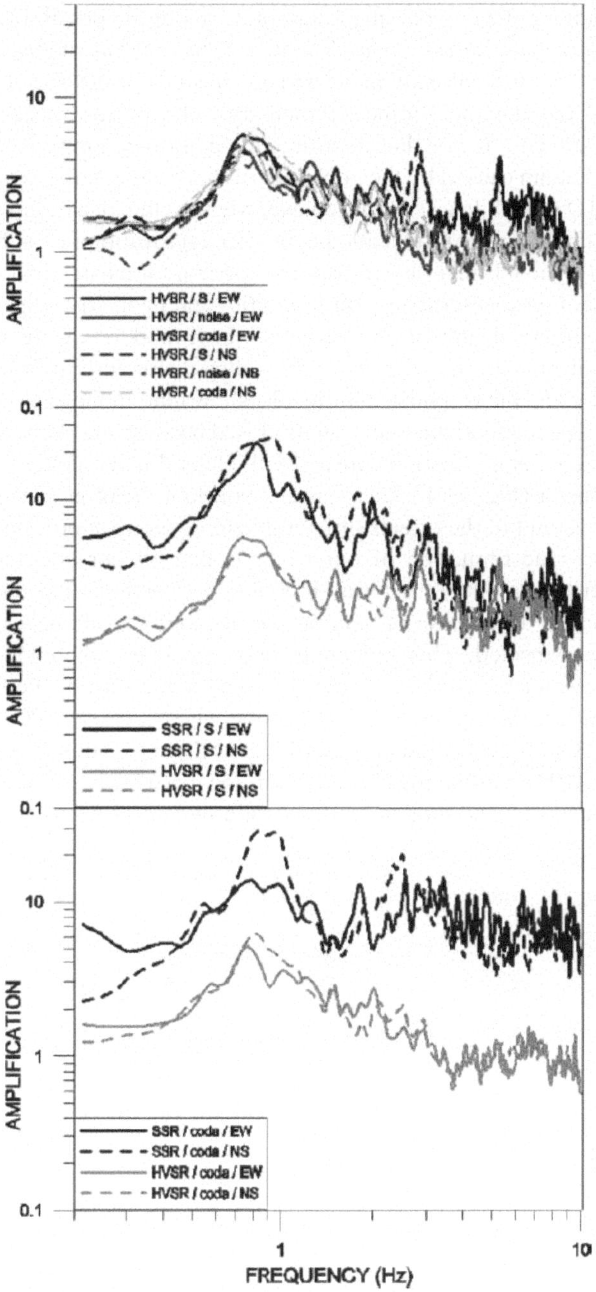

Fig. 11.14 Comparisons of different transfer functions estimated at site RSAR using earthquake records. Each curve identified in the legend corresponds to spectral ratios relative to the reference station (SSR), or spectral ratios of the horizontal components relative to the vertical component at the site (HVSR). Results are shown for three different time windows: noise before P-wave arrival (noise), S-wave (S), and coda (coda). *EW* and *NS* indicate the direction of corresponding horizontal component [27]

among the engineering community about the possibility of including some kind of provision for such complex site effects in terms of a next generation seismic codes.

We have already presented conclusions of 2D and 1D numerical modeling of the detailed model for the sedimentary valley of Euroseistest (Fig. 11.2). Then, we have defined a factor that measures the additional amplification in the response spectra domain caused by the geometry of this basin relative to site effects measured using a 1D model. In a second step, we extended the idea of this factor and we computed the amplification ratio, called here after aggravation factor, between 2D and comparable 1D models in terms of response spectra, for the detailed soil model as well as simplified ones in terms of soil properties and geometry (Fig. 11.12 [15]). Comparing the numerical simulations among all these models, we are able to suggest the minimum information needed about the valley's structure in order to capture the essence of its seismic response. Our results show that, in the case of Euroseistest, site response depends fundamentally on its closed basin shape. Moreover, if we quantify the additional amplification caused by the lateral heterogeneity in terms of the aggravation factor (Fig. 11.15 left), a rough constant factor between 2 and 3 seems to take into account of the effects of lateral heterogeneity appropriately.

One limitation of our study is that all our computations are based on linear behavior of soil material. It could be argued that our results are probably not applicable to the case of very strong damaging earthquakes where strong non-linearity is expected. This objection, however, does not hold in the regions of moderate seismicity and in any case it is a useful starting point and especially for many future damaging events in Europe where expected PGA on rock is less than 0.2 g. Moreover, it could be argued that as the extra amplification expressed as the ratio of 2D/1D motion, the possible non-linear effects is probably similar in both 1D and 2D analysis, and thus the ratio is valid in case of a strong ground motion as well. This

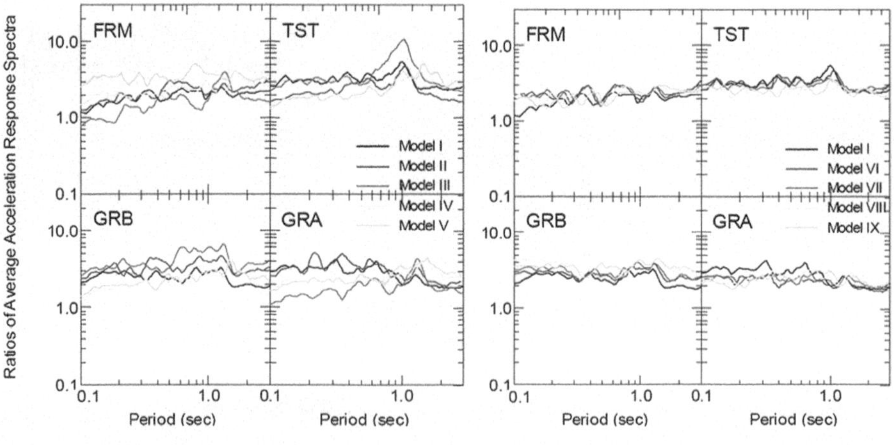

Fig. 11.15 2D/1D ratios of response spectra for 5% damping (*aggravation factor*) for the different 2D models I, II, III, IV and V (*left*) and I, VI, VII, VIII and IX (*right*), relative to the 1D response spectra computed using the 1D profile at the location of each receiver of the corresponding 2D soil model [15, 16]

does not intend to minimize the research that is still required to understand non-linear soil behavior phenomena and their consequences on the ground motion. We would rather like to call attention to the importance of the *aggravation factor* and the need to continue our research effort in the fields of major practical importance for example in bridge engineering.

In order to account, although crudely, for possible non-linear effects, we assumed an incremental decrease of Vs values and attenuation for the soil formations of the detailed 2D soil model while the shape of the interfaces of this model is kept the same. We, then, modeled seismic response with the same numerical code that accounts for the equivalent linear inelasticity with constant damping behavior of soil materials [16]. Namely, we have computed site responses for 95, 90, 85 and 80% of the initial Vs and respective decrease of Qs (S-wave quality factor) values (models VI – IX), as well as the corresponding aggravation factor (Fig. 11.15, right panel). It is worth noticing that the aggravation factor is rather similar (around 2 and 3) for the models with simplified dynamic properties compared to that of the detailed 2D soil model, while aggravation factor for models with different geometrical characteristics (models IV and V) is more discrepant compared to the detailed one. Aggravation factor for models of simplified dynamic properties with depth (models II and III) seems to stand in between the aforementioned cases. The variations of aggravation factor, although depending on frequency, are not very large and could be disregarded in favor of simplicity. These figures show that, at least as a first step, we could consider the *aggravation factor* to be constant throughout the basin. Trying to implement the location of each site relative to the significant lateral boundary in this concept decreases the general usefulness of the aggravation factor in engineering practice.

11.4 Conclusions – Discussion

Most of the recent observational site effect studies deal with the evaluation of empirical transfer functions. In practice, this has encouraged the continuous use of 1D models, because of the simplicity of computing a theoretical transfer function easily compared with the observations, despite the fact that it has been almost 20 years since the demonstration that in shallow alluvial valleys, such as Euroseistest, locally generated surface waves should be the predominant phenomenon.

In the particular case of Euroseistest, frequency domain analysis of weak motion data showed that the sediments filling the basin amplify ground motion by factors larger than 10 between 0.6 and 3 Hz. In time domain, numerical modeling of a very detailed cross section (NS) shows that site response in this valley is dominated by lateral propagation of locally generated surface waves. Looking the data in time domain, both seismograph and accelerograph records confirm the theoretical results. We observe that surface waves are excited at the same frequencies in which we expect the "resonance frequency" to occur. It is not surprising that the predominant frequencies for body and surface waves coincide because both are two manifestations of the same phenomena.

Discrepancies appear when we want to interpret the empirical transfer functions. We have shown that, if we want to account for the physics governing site effects, ground motion at this valley cannot be modeled using 1D models. The reason is clearly that the 1D "resonance" peak includes the contribution of locally generated surface waves, which occur at different times in the seismogram. This phase information is lost in the determination of the empirical transfer functions. An additional problem is that amplitude transfer functions tend to make us forget the physics of site effects. We suggest that the determination of empirical transfer functions is not enough to differentiate between 1D and 2D site effects, and that a successful modeling of a partial aspect of site response (i.e. amplitude) cannot guarantee that we can make successful predictions of ground motion at a site.

Our results from Euroseistest indicate also that, in order to improve current schemes accounting for site effects in the building codes, the more to be gained so far comes from consideration of lateral heterogeneity, at least in the case of shallow alluvial valleys, where locally generated surface waves are likely to be important. Having in mind that in terms of engineering applications and code specifications, it is not the detailed response of each valley that is of interest, we would rather like to study in a simple way the effects of each structure in terms of response spectra, and to be able to quantify the differences between different 2D models. The main usefulness of Euroseistest experimental site is the contribution to the effort required to better characterize a given site and to evaluate the expected amplitudes and frequency content of the ground motion. For these reasons, it seems that our proposal to use an aggravation factor that quantifies the additional and rather constant amplification of ordinary shaped alluvial valleys, stemming from the consideration of lateral heterogeneity is practical.

Acknowledgments Major part of the studies as well as the instrumentation of the EUROSEISTEST strong motion array was financially supported by EU under EUROSEISTEST (EV5V-CT.93-0281), EUROSEISMOD (ENV4-CT.96-0255) and EUROSEISRISK (EVG1-CT-2001-00040) Projects.

References

1. Aoi S. (2002) Boundary shape waveform inversion for estimating the depth of three dimensional basin structures. Bull Seismol Soc Am 92: 2410–2418, Doi: 10.1785/0120010245
2. Apostolidis P (2002) Determination of the soil structure using microtremor. Application on the estimation of the dynamic properties and the geometry of the soil formations at Thessaloniki city, Greece, Ph D Thesis AUTH. LSMFE
3. Bard PY, Bouchon M (1985) The two-dimensional resonance of sediment filled valleys. Bull Seismol Soc Am 75:519–541
4. Chávez-García FJ, Faccioli E (2000) Complex site effects and building codes: making the leap. J Seismol 4:23–40
5. Chávez-García FJ, Pedotti G, Hatzfeld D, Bard PY (1990) An experimental study of site effects near Thessaloniki (Northern Greece). Bull Seismol Soc Am 80:784–806
6. Chávez-García FJ, Raptakis D (2008) Inversion of soil structure and analysis of the seismic wavefield from a vertical array. Proc. 14th World Conference on Earthquake Engineering, October 12–17, 2008, Beijing, China

7. Chávez-García FJ, Raptakis DG, Makra KA, Pitilakis KD (2000) Site effects at Euroseistest-II. Results from 2D numerical modeling and comparison with observations. Soil Dyn Earthq Eng 19(1):23–39
8. He Z, Ye T, Su W (2005) 3-D velocity structure of the middle and upper crust in the Yunnan region, China. Pure Appl Geophys. doi: 10.1007/s00024-005-2780-x
9. IGME (2001) Inventory–recording of water boreholes in the graben of Koronia, Thessaloniki (coord.: I Milopoulos, editors: Veranis and Katirgioglou). Geophysical surface survey in the graben of Koronia, Thessaloniki (coord.: I Milopoulos, editor: Atzemoglou et al). Reports of the project entitled "Investigation of the exploitation possibilities of the deepest water table in the graben of Koronia, Thessaloniki", Water Supply Company of Thessaloniki (in Greek)
10. Japanese National Working Group on the Effects of Surface Geology on Seismic Motion (JESG) (1991) Ashigara valley blind prediction test
11. Jongmans D, Pitilakis K, Demanet D, Raptakis D, Riepl J, Horrent C, Tsokas G, Lontzetidis K, Bard PY (1998) EURO-SEISTEST: determination of the geological structure of the Volvi graben and validation of the basin response. Bull Seismol Soc Am 88:473–487
12. Kudo K (1995) Topics of effects of surface geology on strong-ground motion from the recer Earthquakes in Japan and the activity of Japanese Working Group on Effects of Surface Geology. Proc. l0th Conf. on Earthquake Engineering, 2634–2641
13. Kudo K, Kanno T, Okada H, Sasatani T, Morikawa N, Apostollidis P, Pitilakis K, Raptakis D, Takahashi M, Ling S, Nagumo H, Irikura K, Higashi S, Yoshida K (2002) S-Wave velocity Structure at EURO-SEISTEST, Volvi, Greece determined by the Spatial Auto-Correlation method applied for array records of microtremors, Proc. 11th Japan Earthquake Engineering Symposium, Paper No. 62
14. Lin G, Shearer PM, Hauksson E, Thurber CH (2007) A three dimensional crustal seismic velocity model for southern California from a composite event method. J Geophys Res 112(B11306). doi:10.1029/2007JB004977
15. Magistrale H, Day S, Clayton RW, Graves R (2000) The SCEC southern California reference three-dimensional seismic velocity model version 2. Bull Seismol Soc Am 90: S65–S76
16. Makra K, Chávez-García FJ, Raptakis D, Pitilakis K (2005) Parametric analysis of the seismic response of a 2D sedimentary valley: implications for code implementations of complex site effects. Soil Dyn Earthq Eng 25:303–315
17. Makra K, Raptakis D (2007) How sensitive the effects of lateral heterogeneity on seismic ground motion are? Proc. 4th International Conference of Earthquake Geotechnical Engineering, June 25–28, Thessaloniki, Greece, paper No 1687
18. Makra K, Raptakis D (2008) Analysis of strong ground motion recordings at a downhole array and correlation with simulation results. Proc 3rd Hellenic Conference on Earthquake Engineering, November 5–7, Athens, Greece, paper No 1965 (in Greek)
19. Makra K, Raptakis D, Chávez-García FJ, Pitilakis K (2001) Site effects and design provisions: the case of EUROSEISTEST. Pure Appl Geophys (PAGEOPH) Special issue on microzoning 158(11):2349–2367
20. Manakou M (2007) Contribution to the determination of a 3D soil model for site response analysis. The case of the Mygdonian basin. Ph D Thesis (in Greek with English abstract), Department of Civil Engineering, Aristotle University of Thessaloniki. http://invenio.lib.auth.gr/
21. Manakou M (2009) Determination of the geometrical and dynamic properties of the geological formations located close to the edges of the Mygdonian basin. PostDoc Thesis, Aristotle University of Thesalloniki
22. Manakou M, Raptakis DG, Apostolidis PI, Chávez-García FJ, Pitilakis KD (2010) 3D soil structure of the Mygdonian basin for site response analysis. Submitted for publication to Soil Dynamics and Earthquake Engineering Soil Dyn Earthq Eng 30: 1198–1211. doi: 10.1016/j.soildyn.2010.04.027
23. Mayne PW, Rix GJ (1993) G_{max}-q_c relationships for clays. ASTM Geotechnic Testing J 16(1):54–60

24. Parolai S, Bard PY (2003) Evaluation of amplification and frequency-dependent lengthening of duration of seismic ground motion due to local effects by means joint analysis of sonogram and standard spectral ratio. J Seismol 7(4):479–492
25. Pitilakis K, Raptakis D, Lontzetidis K, Tika-Vassilikou T, Jongmans D (1999) Geotechnical and geophysical description of EURO-SEISTEST, using field, laboratory tests and moderate strong motion recordings. J Earthq Eng 3:381–409
26. Raptakis D (1995) Contribution to the determination of the geometry and dynamic properties of soil formations, and their site response. Ph D thesis (in Greek with English abstract), Department of Civil Engineering, Aristotle University of Thessaloniki
27. Raptakis DG, Chavez-Garcia FJ, Makra KA, Pitilakis KD (2000) Site effects at Euroseistest-I. 2D determination of the valley structure and confrontation of the observations with 1D analysis. Soil Dyn Earthq Eng 19(1):1–22
28. Raptakis D, Lontzetidis K, Pitilakis K (1996) Surface Waves Inversion Method: a Reliable Method for the In Situ Measurements of Shear Wave Velocity. Proc. 4eme Colloque Nationale de Genie Parasismique et Aspects Vibratoires dans le Genie Civil, A.F.P.S, Avril 10–12, Paris, France. Vol. I:160–169
29. Raptakis D, Manakou M, Chávez-García FJ, Makra K, Pitilakis K (2005) 3D configuration of mygdonian basin and preliminary estimate of its site response. Soil Dyn Earthq Eng 25: 871–887
30. Raptakis D, Theodulidis N, Pitilakis K (1998) Data analysis of the Euroseistest strong motion array in Volvi (Greece): standard and horizontal to vertical ratio techniques. Earthq Spectra 14:203–224
31. Real CE, Tucker BE (1988) Turkey Flat, USA site effects test area site characteristics, Proc. of IASPEI/IAEE Joint Working Group on Effects of Surface Geology on Seismic Motion, 2nd Workshop, Tokyo
32. Satoh T, Kawase H, Mathsusima S (2001, Apr) Estimation of S-wave velocity structures in and around the Sendai basin, Japan, using array records of microtremors. Bull Seismol Soc Am 91(2):206–218
33. Semblat JF, Kham M, Parara E, Bard PY, Pitilakis K, Makra K, Raptakis D (2005) Seismic wave amplification: basin geometry vs soil layering. Soil Dyn Earthq Eng 25:529–538
34. Tournas D (2005) Study for the geometry of the Mygdonian basin, in the area of Euroseistest site with the use of geophysical measurements, MSc Thesis (in Greek), Department of Geology, Aristotle University of Thessaloniki

Chapter 12
Deployment of New Strong Motion Seismographs of K-NET and KiK-net

S. Aoi, T. Kunugi, H. Nakamura, and H. Fujiwara

Abstract Following the occurrence of the 1995 Hyogoken-Nanbu (Kobe) earthquake, National Research Institute for Earth Science and Disaster Prevention (NIED) has constructed two strong motion seismograph networks, K-NET and KiK-net. These networks cover uniformly the country with an inter-station interval of about 20–25 km, and a total number of the stations is about 1,700. To grasp the hazard (ground motion) of urban or downtown areas, most stations of K-NET are located in public offices, schools and parks, and a three-component accelerometer is installed on the free-surface. On the other hand, KiK-net stations are located in quiet places to avoid the artificial noise. Each KiK-net station has a borehole of 100 m or more in depth and strong motion seismographs have been installed both on the ground surface (uphole) and at the bottom of the boreholes (downhole). Recently, based on the request for quicker hazard information, all instruments of the K-NET and KiK-net have been renewed by including the change of accelerometers at the surface, new recorders, and also a new data collection system between stations and the Data Management Center (DMC) in Tsukuba. The newly developed instruments, which are state of the art in strong motion instrumentation, have several advantages such as real-time capability, larger measurable range and lower noise. This paper explains the new generation system of K-NET and KiK-net.

12.1 Introduction

Strong motion observation in Japan dates back as far as 1950s. Since then, many organizations have put a lot of effort to construct and maintain strong motion networks. The Japan Meteorological Agency (JMA; www.jma.go.jp/jma/en/Activities/earthquake.html) has operated an earthquake observation network for the monitoring of earthquakes. The Building Research Institute (BRI; http://smo.kenken.go.jp) has installed strong motion instruments in major cities throughout Japan to enhance

S. Aoi (✉)
National Research Institute for Earth Science and Disaster Prevention, 3-1 Tennodai, Tsukuba, Ibaraki 305-0006, Japan
e-mail: aoi@bosai.go.jp

S. Akkar et al. (eds.), *Earthquake Data in Engineering Seismology*, Geotechnical, Geological, and Earthquake Engineering 14, DOI 10.1007/978-94-007-0152-6_12,

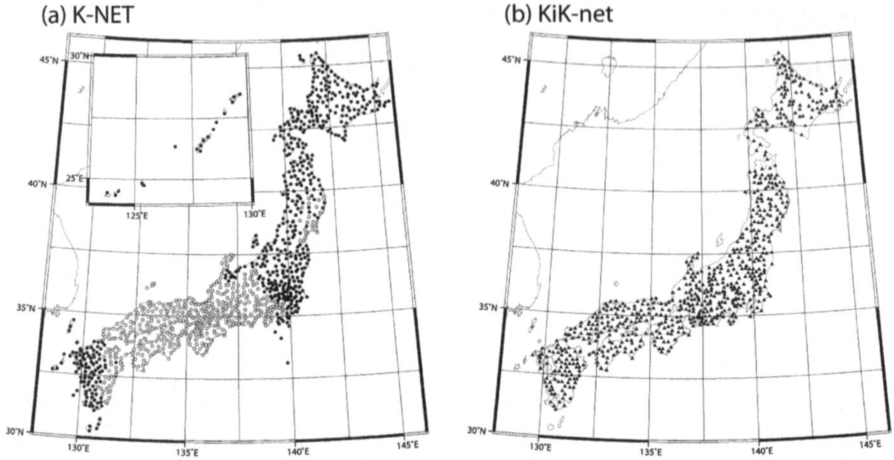

Fig. 12.1 Distribution of **a** K-NET and **b** KiK-net stations. *Open* and *closed circles* in (**a**) show stations where K-NET02 and K-NET02A are installed

the seismic safety of buildings. Strong motion observations at port facilities have been performed since 1962 under the Ministry of Land, Infrastructure, Transport and Tourism (MLIT; www.mlit.go.jp/kowan/kyosin/eq.htm). Some universities and research institutes including NIED have had local area strong motion networks.

After the 1995 Hyogoken-Nanbu earthquake, the development of the Japanese strong motion observation was drastically stimulated. Prompted by the fact that an insufficient coverage of the strong motion observation networks in Japan did not facilitate to grasp the overall characteristics of earthquakes, the Japanese government decided to improve the quality and quantity of the strong motion observations in Japan. JMA have increased the number of strong motion stations and the central government subsidized the local governments to install several thousands of observatories for seismic intensity observation. Since 1996, NIED has been in charge of constructing of two strong motion networks, K-NET and KiK-net (Fig. 12.1). These dense seismograph networks have successfully recorded many near source ground motions, and nearly 290,000 digital records from more than seven thousands events are now available through a public access web-site (Table 12.1 and Appendixes 1 and 2). These data are significantly contributing to disaster mitigation programs, earthquake resistant designs and scientific research in Japan and abroad.

12.1.1 K-NET (Kyoshin NETwork)

Just after the 1995 Hyogoken-Nanbu earthquake, construction of K-NET began and was completed in just 1 year [6]. The term K-NET stands for "Kyoshin network"

Table 12.1 Number of events and records released in the K-NET and KiK-net web sites

	K-NET		KiK-net		K-NET and KiK-net	
Year	Events	Records	Events	Records	Events	Records
1996	154	3,790	0	0	154	3,790
1997	312	7,908	7	64	312	7,972
1998	277	6,524	155	898	291	7,422
1999	227	5,315	169	2,118	262	7,433
2000	1,048	11,153	477	7,690	1,147	18,843
2001	339	8,630	325	7,875	375	16,505
2002	364	7,331	295	7,776	368	15,107
2003	664	12,950	624	17,557	690	30,507
2004	776	16,703	794	19,795	855	36,498
2005	571	14,114	540	13,500	591	27,614
2006	448	12,004	375	8,507	449	20,511
2007	719	15,825	634	12,115	725	27,940
2008	663	17,429	629	21,206	668	38,635
2009	550	13,494	499	15,216	554	28,710
Total	7,112	153,170	5,523	134,317	7,441	287,487

where “Kyoshin” means “strong motion” in Japanese. At its inception, K-NET consisted of 1,000 stations whose average station-to-station distance was about 20 km. All the stations were outfitted with the same type of strong motion accelerographs, K-NET95, which were installed on the free surface (Fig. 12.2). Afterwards, some stations were added to K-NET, such as the existing stations within the Kanto-Tokai area or the cable type ocean-bottom strong motion accelerometers at the Sagami Bay [3]. The number of stations currently is 1,032.

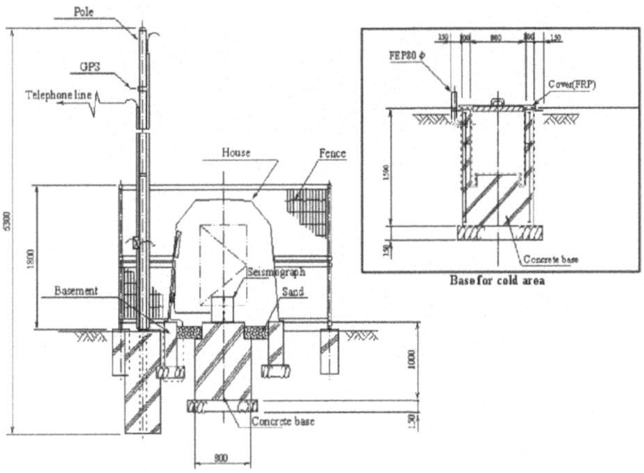

Fig. 12.2 Observation facility for K-NET. A basement for cold areas is shown at the inset

12.1.2 KiK-net (Kiban Kyoshin network)

The Japanese government established "the Headquarters for Earthquake Research Promotion" on July 18, 1995, and the "Fundamental Survey and Observation for Earthquake Research" plan was developed under their direction. In short, this plan is called KIBAN, which is a Japanese word meaning fundamental or infrastructure. The main part of the "Kiban" project encompasses many types of observation networks [12], such as the high sensitivity seismic network (Hi-net [9]), the strong motion observation network (KiK-net; Kiban Kyoshin network [1]), the broadband seismic observation network (F-net [4]), and the continuous GPS observation network (GEONET [5]). Each KiK-net station (Figs. 12.3 and 12.4) has an observation borehole of more than 100 m deep (Table 12.2), and a pair of tri-axial accelerometers (V404 manufactured by Akashi Corporation) is installed on the ground surface and also at the bottom of the observation boreholes together with high sensitivity velocity seismometers of Hi-net. The acquisition system, SMAC-MDK, was also manufactured by Akashi Corporation.

K-NET and KiK-net are pioneer networks in Japan to freely releasing all digital data through the Internet immediately after an earthquake, and nowadays this open-data policy is becoming a common practice. Although our initial policy was to release all data within 1 week after the occurrence of an earthquake, the request for a quicker hazard information from many local governments who are responsible for

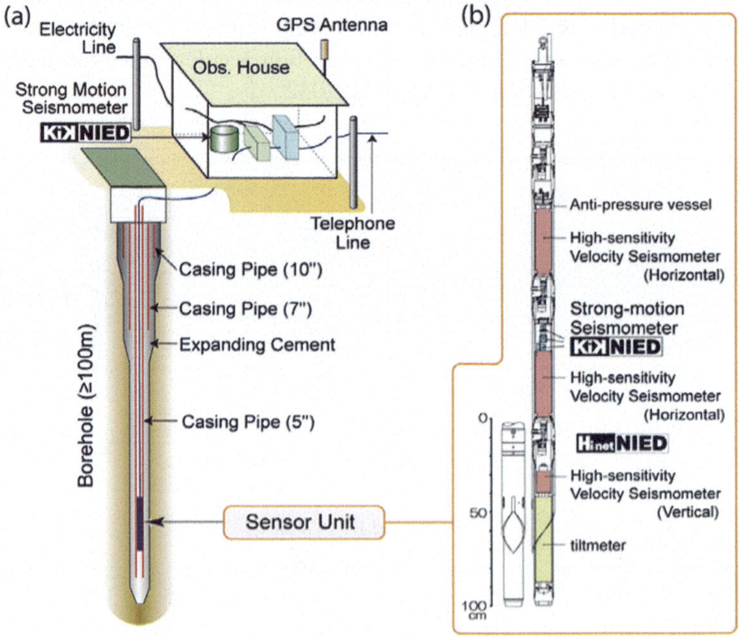

Fig. 12.3 **a** Observation facility and **b** a downhole instrument for KiK-net

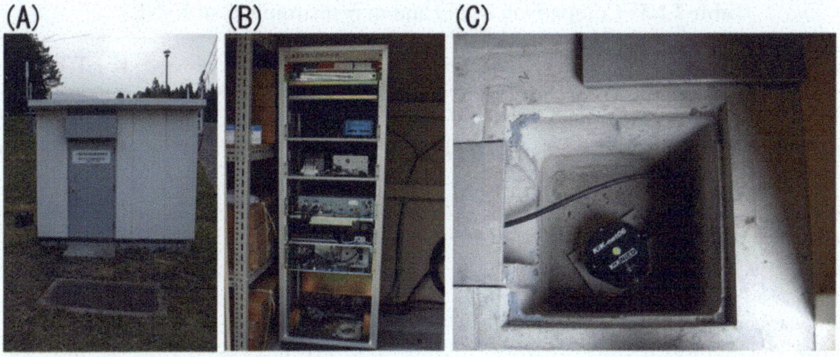

Fig. 12.4 **a** Observatory building of KiK-net. **b** Acquisition system of KiK-net and Hi-net installed in a rack **c** Accelerograph of KiK-net installed in the pit

Table 12.2 Distribution of the KiK-net borehole depths

Depth (m)	Number of stations
100–149	420
150–249	182
250–499	43
500–999	17
1,000–1,999	13
2,000–	13
Total	688

the initial response after an earthquake, prompted us to update our network. Based on these requests, we conducted the replacement of all the instruments at about 1,700 K-NET and KiK-net stations (excluding the downhole sensors) between 2003 and 2008.

12.2 New Instruments of K-NET and KiK-net

New instruments were developed by introducing new technologies including information technologies and high-performance sensors. The networks are equipped with three types of instruments (K-NET02, K-NET02A and KiK-net06) which have similar function and performance. K-NET02 and K-NET02A systems are installed in the stations indicated by open and closed circles in Fig. 12.1a and KiK-net06 in the all KiK-net stations (Fig. 12.1b). Summary of the new K-NET and KiK-net systems are shown in Tables 12.3 and 12.4, respectively, by comparing them with the old systems, K-NET95 and SMAC-MDK.

Table 12.3 Comparison of old and new instruments of K-NET

	K-NET95	K-NET02	K-NET02A
Maximum measurable acceleration	2,000 gal	4,000 gal	4,000 gal
Dynamic range (RMS noise/full scale)	114 dB (19 bit)	132 dB (22 bit)	132 dB (22 bit)
Accelerometer	V403[a]	FBA–ES[b]	JA40GA[c]
Calculation of JMA seismic intensity	×	o	o
Calculation of response spectrum	×	o	o
Recording capacity	8 MB	512 MB	768 MB
User's programmability	×	o(Linux OS)	o(Linux OS)
Continuous data recording	×	o	o
Data communication	RS232C	TCP/IP	TCP/IP

[a]Akashi Corporation.
[b]Kinemetrics Inc.
[c]Japan Aviation Electronics Industry, Ltd.

Table 12.4 Comparison of old and new instruments of KiK-net

	SMAC-MDK	KiK-net06
Maximum measurable acceleration surface/borehole	2,000/2,000 gal	4,000/2,000 gal
Dynamic range (RMS noise/full scale)	114 dB (19 bit)	132 dB (22 bit)
Accelerometer surface	V404[a]	JA40GA[b]
Borehole	V404	V404
Calculation of JMA seismic intensity	×	o
Processing for EEW (B-delta method)	×	o
Calculation of strong motion indexes[c]	×	o
Recording capacity	80 MB	768 MB
User's programmability	×	o(Linux OS)
Continuous data recording	×	o
Data communication	RS232C	TCP/IP

[a]Akashi Corporation.
[b]Japan Aviation Electronics Industry, Ltd.
[c]Continuous calculation of PGA, PGV, PGD, real-time intensity and response spectrum.

12.2.1 Sensor Module

At each station, a tri-axial accelerometer whose maximum measureable range is ±4,000 gal, is installed on the free surface. K-NET02 employs Episensor FBA-ES-DECK (Kinemetrics Inc.), and K-NET02A and KiK-net06 use the JA-40GA sensor [16] manufactured by Japan Aviation Electronics Industry Ltd. Though the Episensor is a high performance accelerometer for precise strong motion observation, sometimes step-wise noise, which is a rather common phenomenon, appears. This kind of noise is brought by the release of the stress accumulated in the spring (flexure). Though the amplitude of the stepwise noise is very small (much less than 1 gal in most cases), they have harmful influence when one integrates the accelerogram into velocity or displacement and it is very difficult to remove as it overlaps the data (e.g. [2]). To avoid this type of noise, the spring of the new sensor, JA-40GA, is made of quartz and therefore the occurrence of stepwise noise in the data is significantly reduced as compared to the old sensor.

Regarding KiK-net stations, in addition to installation of new free surface sensors, a tri-axial accelerometer (V404) whose maximum measureable range is ±2,000 gal, was also installed at the bottom of the borehole, stored in a pressure-resistant tube made of stainless steel together with the Hi-net high-sensitivity seismometers. The downhole sensors which were originally installed at the time of construction of the stations have not been replaced at this time. The orientation of horizontal components of downhole seismometers used in the KiK-net has some uncertainty due to difficulties during installation. Orientations are therefore estimated by evaluating correlation of teleseismic waveform data [15]. Sensor orientations may change every time the sensor is re-installed for maintenance or some other reason. The updated orientations are also available through the Internet.

12.2.2 Acquisition System

The acquisition system consists of a measurement module and a communication module (Fig. 12.5), and each of them is controlled by individual built-in Linux

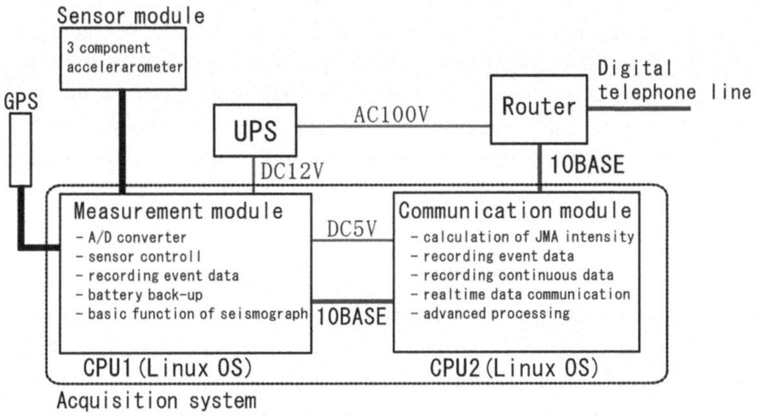

Fig. 12.5 Block diagram of K-NET02 which consists of a sensor module and acquisition system (measurement module and communication module)

boards which allow an easy update of the system program through the network. The measurement module functions as a conventional strong motion seismograph with high precision of observation. The communication module processes advanced tasks, such as the calculation of the JMA seismic intensity, the recording of continuous data and a real-time data transmission. When a power shortage continues more than an hour, the measurement module stops power supply from the backup battery to the router and the communication module to save battery life for recording data. Internal clock is synchronized with GPS (Global Positioning System) and the accuracy of absolute timing is within 0.1 ms. as long as the GPS signal is received. To protect the acquisition system from the damages owning to lightning, arresters are installed between the system and outside modules such as power supply, telephone line and GPS antenna. Arresters are also installed inside the sensor module.

Time history of the ground motion is recoded by an event triggering system. When the ground motion exceeds the threshold, the system begins to record ground acceleration with 100 Hz sampling including a 15 s pre-trigger data. To avoid aliasing, a high-cut filter is applied to the data before recording. Thus the total response characteristic is almost flat from 0 (direct current component) to 30 Hz. At the same time of triggering, the communication module begins to make connection to the DMC through a digital telephone line (ISDN; Integrated Services Digital Network) using a dial-up router. It takes typically 5–7 s to establish the connection and a few seconds to send the pre-trigger data. Typically after 7–10 s the system is triggered, real-time data can be obtained. This data includes a short 1.5 s delay, which corresponded to the time spent for making 1 s packet data plus 0.5 s system delay.

The capability of the new system of making the connection from the station to the DMC has two advantages compared to the old DMC-to-station dialing-up system. First, the new system substantially reduces the time to obtain the data because it eliminates the need to wait for information of the earthquake parameters (location and magnitude), which was a requirement of the old system for deciding a station priority dial-up list. Second, it helps avoiding any over-crowding of telephone lines because the connection can be established before an eventual congestion, which may extend for several hours or days following an earthquake.

12.2.3 Performances of New Instruments

To examine the performance of old and new instruments, we installed K-NET95 and K-NET02 at Tsukubane strong motion observatory where noise level is much lower than the instrumental noise and obtained the record of instrumental noise by observing horizontal ground motion. Figure 12.6 shows the noise records of the whole system including sensor and acquisition systems in time domain and spectral domain, respectively. Spectra of noise were obtained by the standard method

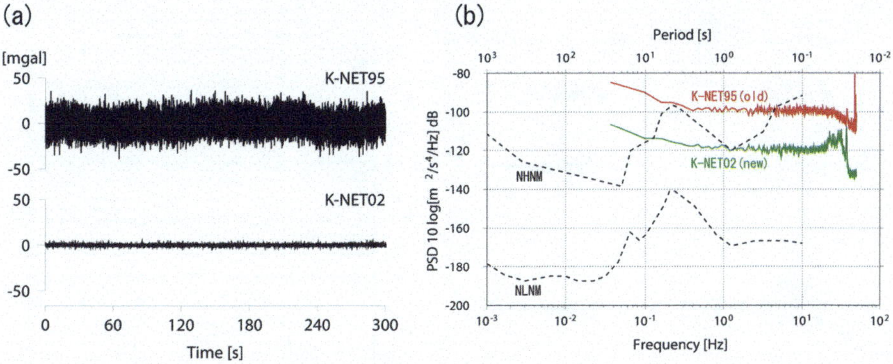

Fig. 12.6 **a** Records of instrumental noise in time domain obtained by K-NET95 and K-NET02 by observing vertical ground motion at Tsukubane strong motion observatory where noise level is much lower than the instrumental noise. **b** Records of instrumental noise in spectral domain obtained by K-NET95 (*upper*), and K-NET02 (*lower*). *Broken lines* show the quietest (NLNM) and noisiest (NHNM) of IRIS broadband stations

of evaluating the noise level of broadband stations [13]. For comparison we show the noise level of NHNM and NLNM, the noisiest and quietest broadband stations of IRIS all over the world. Comparison of the noise in time and spectral domains shows that the level of noise of the new instruments is roughly ten times lower than old one.

The noise level of the new instruments is mainly due not to the sensor but to the acquisition system because an effective dynamic range of the acquisition system is 132 dB and the dynamic range of the sensor is larger than this value. The practical advantage of the new sensors, especially the JA-40GA, is a larger maximum measureable-range and the occurrence of few stepwise noise. Maximum measurable range of the accelerometer for the surface was increased from 2,000 to 4,000 gal because recent dense strong motion seismograph network revealed that extremely large shaking exceeding gravity is not very rare. Figure 12.7a shows ground acceleration recorded just above the reverse fault during the 2008 Iwate-Miyagi Nairiku earthquake. The maximum acceleration of UD component on the surface was 3,866 gal which would not have been possible to record without saturation by the old sensor. Permanent displacements as well as time history of displacements (Fig. 12.7b) are easily obtained by double integration of these accelerograms because they are practically stepwise noise-free.

12.3 Continuous Observations

The acquisition system of the new instruments also has a continuous observation capability. Continuous data are stored in the pre-reserved area of a compact flash memory on the communication modules. The recording capacity of continuous data

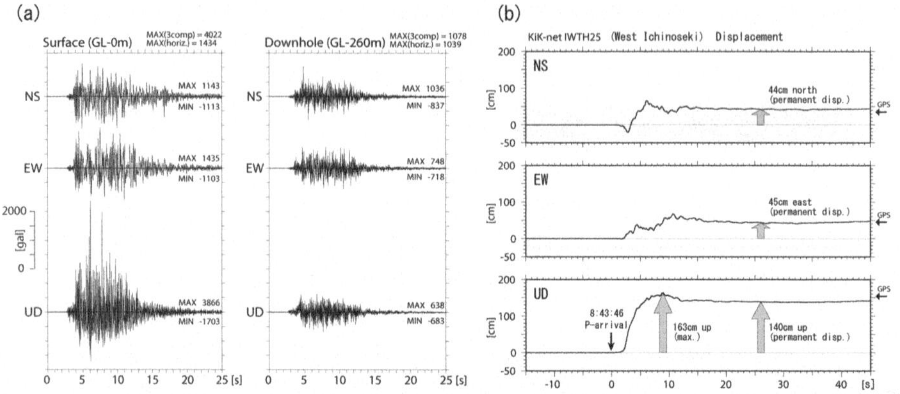

Fig. 12.7 Time history of **a** acceleration and **b** displacement recorded by the surface and borehole sensors during the 2008 Iwate-Miyagi Nairiku earthquake at KiK-net IWTH25 station (West Ichinoseki). Displacement waveforms are derived by the double integration of the original acceleration records with baseline correction. *Rightmost arrows* indicate the static displacement observed by the GPS station (ICNS) of Tohoku University, which is located a few hundred meters from IWTH25 station

is of several days, and when the pre-reserved area reaches to full the oldest data are automatically over-written by new data. Several kinds of strong motion indexes such as real-time intensity [8], peak values of acceleration, velocity and displacement, and response spectra are continuously calculated every 1 s and stored in the memory. Those data and indexes are easily downloaded from DMC through the network of the data collection system by specifying the required time window by the user.

If a continuous communication line is available, continuous telemetric observation is also possible using the new instrument. As a feasibility study for continuous strong motion observation, and taking advantage of the fact that each KiK-net station shares the same seismic observatory with Hi-net, we set up the continuous telemetric observation of the downhole UD component of KiK-net through a shared use of the EarthLAN. The EarthLAN, by which Hi-net continuous data is transmitted, is a continuous telephone line service provided by a communication company [10]. The reason for limiting the continuous transmission of KiK-net to only one component is because most of the bandwidth of the EarthLAN is occupied by Hi-net data, and the extra capacity is not enough for all six components, three components both for surface and downhole accelerometers. The continuously calculated strong motion indexes above explained, as well as the UD component of the downhole data, are sent every 1 s using the surplus bandwidth of the EarthLAN. And also, the epicentral distance and magnitude estimated by the B-Delta method [11] and *P*-wave arrival time are promptly estimated at each station and sent to the DMC at optional timing through the EarthLAN. These indexes have the potential to be implemented in earthquake early warning systems.

From the viewpoint of disaster mitigation, it will become necessary in the near future to provide the data that is required to estimate the spatial distribution of not only the hazard (ground motion) but also the risk (damage) of the earthquake in real-time. To meet this need, more rapid data transmission may be required. Because of the infrequent occurrence of large earthquakes, strong motion has been commonly observed by event triggering system which requires connection of the telephone-line only during the data collection. To achieve more rapidity, continuous observation is one of the most likely options. Data recorded by an event triggering system provides important information of past earthquakes and helps to estimate the hazard and risk of a future earthquake. With a continuous observation system, owning to the rapid progress of information technologies, soon we would be able to fully monitor in real-time and thus directly contribute to the mitigation of ongoing seismic disasters. Our strong motion network capability currently lies somewhere in between these two observation systems.

12.4 Data Flow and Data Release

Real-time ground motion maps (acceleration, real-time intensity and response spectrum) of Japan are always generated and uploaded to the web site (www.kyoshin.bosai.go.jp/kyoshin) every 5 s, by using the strong motion indexes which are continuously calculated and sent from every KiK-net stations through the EarthLAN (Fig. 12.8).

When an earthquake occurs and the data receiving system of DMC in Tsukuba receives an event data from the earliest station, an event number is assigned for making a group of data which is expected to belong to same earthquake. Event data receiving within 600 s are considered to be the same group of data. Each data are sent to the web site one by one. This procedure does not require to wait until the transmission of data from all the station have finalized, and make it possible to release the data from the near source of an earthquake in a few minutes for most cases. Since this prompt release of data do not go through any manual check, some inappropriate records such as noise data (ground notion of artificial noise or data due to observational trouble), or the data from a different earthquake occurring at almost the same time. Figures such as waveforms and response spectrum at each station, and maps of peak ground motion (PGA, PGV, response spectrum and seismic intensity) distributions are uploaded at almost the same time. These maps are created using interpolations by an optimal Delaunay triangulation [17] from all data observed by K-NET and KiK-net.

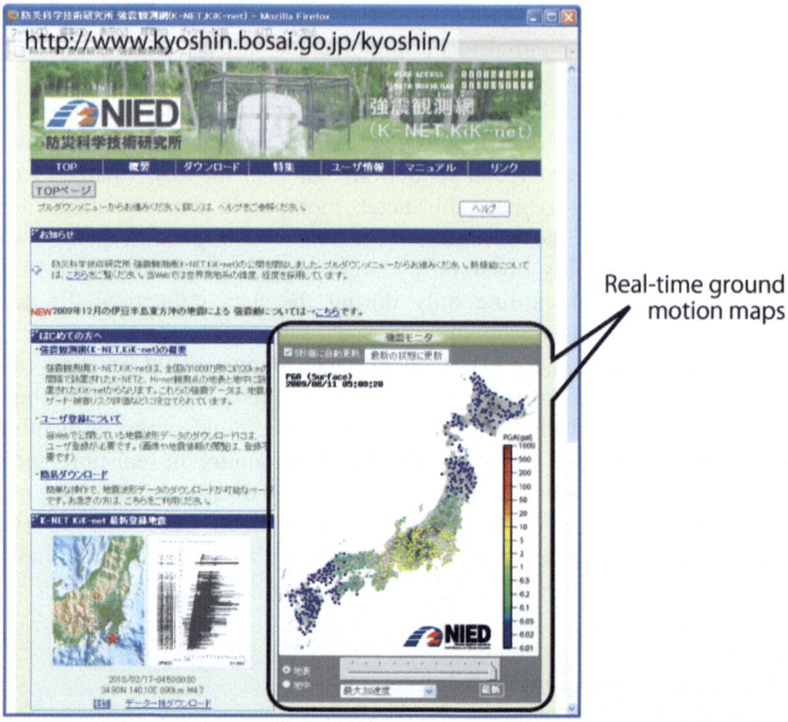

Fig. 12.8 A real-time ground motion map, which is generated from real-time strong motion indexes (PGA, real-time intensity, response spectrum) continuously calculated and telemated from every KiK-net stations every 1 s. The map is routinely shown on web site and automatically re-loaded every 5 s

Seismic intensities (e.g. [7, 14]) of an event observed by K-NET are promptly transmitted to DMC (within 2 min), as soon as their values are calculated by using 60 s of ground motion data counted from arrival time by the communication module of the acquisition system at each station. The seismic intensities are sent to the local governments and the mass media through JMA and this information is immediately run as a telop on television. The seismic intensities are also directly sent to the local governments by request.

All data are manually checked in office hours. The operators take away noise data, and records are related to source information (occurrence time, location and magnitude) and formally upload the archives, digital data and a set of related figures, to the web site.

12.5 Summary

All the instruments (excluding downhole sensors) of the Japanese nationwide strong motion networks, K-NET and KiK-net, have been replaced recently, to answer to the request for a quicker hazard information immediately after an earthquake. The instruments have been newly developed by introducing new technologies including information technologies and high-performance sensors, which make them state of the art in strong motion instrumentation.

The new instruments have the capability of automatically calling the DMC at NIED in Tsukuba, several seconds after being triggered and are able to transmit waveform data even while recording. This function not only significantly reduces the time for data collection but also helps avoiding any overcrowding of telephone lines. K-NET02 and K-NET02A have been officially approved as a seismic intensity meter by the JMA. Measured seismic intensity, which is the most popular and important index for national and local government during an earthquake, is automatically transmitted to them through JMA within 2 min after being triggered. This information is greatly contributing to the decisions and actions by the administration during an earthquake, and is widely broadcasted through television and radio.

The basic performance of the strong motion instruments has been greatly improved. Measureable range of accelerometers on the ground surface has been extended from ±2,000 to ±4,000 gal and the dynamic range has been improved by about a factor of 10 as compared to the old system. The newly introduced sensor, the JA-40GA, which has a quartz spring (flexure) significantly reduces the occurrence of stepwise noise which very frequently appears in data observed by a sensor with a metal spring. This will improve the accuracy of data analysis especially in the long-period range such as waveform inversion analysis of source processes and real-time seismology.

New instruments also have a continuous observation capability. Continuous data are stored in the memory. Several kinds of strong motion indexes such as real-time intensity, peak values of acceleration, velocity and displacement, and response spectra are continuously calculated by the communication modules. Those data and indexes are easily downloaded from DMC through the network by specifying the required time window by the user. Continuous telemetric observation is also possible using the new instrument if a continuous communication line is available. The communication module sends the data by packets of 1-s. duration. Indeed, UD components of the downhole sensor of KiK-net06 are continuously sent to DMC by using leeway bandwidth of continuous line prepared for Hi-net, as a feasibility study for continuous observation of strong motion. From the viewpoint of disaster mitigation, data from continuous strong motion observation will become important both for the earthquake early warning, and for grasping the spatial distribution of the hazard and the risk (damage) immediately after an earthquake.

Digital data and related figures and movies are available soon after the earthquake (typically within 15 min) without any manual check. This data is checked by an operator in the next business hour and the new archive is re-uploaded as the formal data.

One of the new capabilities of the NIED strong motion web site are the maps of real-time ground motion all over Japan. One can always see the intensity, PGA and response spectrum which are continuously sent from all KiK-net station every 5 s and observe the wave propagation in real-time.

Appendix 1 Remarkable records obtained by K-NET and KiK-net (sorted by peak ground acceleration on the surface)

No.	Site code	Site name	PGA (3 comp.) [gal]	PGA (NS) [gal]	PGA (EW) [cm/s]	PGV (UD) [cm/s]	Earthquake Name	Origin time	M_{JMA}	Depth [km]	Hypocentral distance [km]	Epicentral distance [km]
1	IWTH25	ICHINOSEKI-W	4022	1143	1433	3866	2008 Iwate-Miyagi Nairiku Earthquake	2008/06/14-08:43	7.2	8	9	3
2	AKTH04	HGASHINARUSE	2600	1318	2449	1094	2008 Iwate-Miyagi Nairiku Earthquake	2008/06/14-08:43	7.2	8	23	22
3	NIG021	TOHKAMACHI	1750	1716	850	564	2004 Mid Niigata prefecture Earthquake	2004/10/23-17:56	6.8	13	25	21
4	MYG011	OSHIKA	1571	1103	1112	825		2003/05/26-18:24	7.0	71	92	59
5	NIG019	OJIYA	1502	1147	1308	820	2004 Mid Niigata prefecture Earthquake	2004/10/23-17:56	6.8	13	15	7
6	IWTH26	IGHINOSEKI-E	1372	888	1056	927	2008 Iwate-Miyagi Nairiku Earthquake	2008/06/14-08:43	7.2	8	15	12
7	IWTH04	SUMITA	1305	730	723	1280		2003/05/26-18:24	7.0	71	86	48
8	IWTH02	TAMAYAMA	1186	1019	684	593		2008/07/24-00:26	6.8	108	111	24
9	HKD020	MINATOMACHI	1177	536	1127	368		2004/12/14-14:56	6.1	9	12	9
10	TTRH02	HINO	1142	927	753	776	2000 Western Tottori prefecture Earthquake	2000/10/06-13:30	7.3	11	13	7
11	IWTH27	RIKUZENTAKATA	1098	888	556	637		2003/05/26-18:24	7.0	71	76	28
12	IWT007	KAMAISHI	1056	594	1039	592		2003/05/26-18:24	7.0	71	89	54
13	IWTH25	ICHINOSEKI-W	1004	345	939	378		2008/06/14-23:42	4.8	10	11	3
14	NIG021	TOHKAMACHI	990	816	811	220		2004/10/23-18:34	6.5	14	29	26
15	HKD100	HIROO	988	810	973	461	2003 Tokachi-oki Earthquake	2003/09/26-04:50	8.0	42	94	84
16	KGS005	MIYANOJOH	977	902	901	288		1997/05/13-14:38	6.2	8	18	16
17	MYG002	UTATSU	949	593	858	339		2003/05/26-18:24	7.0	71	73	17
18	ISK006	TOGI	945	717	849	462	2007 Noto Hanto Earthquake	2007/03/25-09:42	6.9	11	13	7
19	IWTH12	KUNOHE	944	716	670	318		2008/07/24-00:26	6.8	108	119	50
20	NIG028	NAGAOKA-SHISHO	921	870	706	436	2004 Mid Niigata prefecture Earthquake	2004/10/23-17:56	6.8	13	–	15
21	KGS002	IZUMI	903	727	542	246		1997/03/26-17:31	6.3	8	14	12

Appendix 1 (continued)

No.	Site code	Site name	PGA (3 comp.) [gal]	PGA (NS) [gal]	PGA (EW) [cm/s]	PGV (UD) [cm/s]	Earthquake Name	Origin time	M_{JMA}	Depth [km]	Hypocentral distance [km]	Epicentral distance [km]
22	ISK005	ANAMIZU	903	473	782	556	2007 Noto Hanto Earthquake	2007/03/25-09:42	6.9	11	22	19
23	KSRH03	SHIBECHA-N	884	806	300	588	2003 Tokachi-oki Earthquake	2003/09/26-04:50	8.0	42	189	184
24	KSRH06	TSURUI-E	879	756	563	629		2004/11/29-03:32	7.1	48	89	75
25	OKY004	NIIMI	878	528	817	171	2000 Western Tottori prefecture Earthquake	2000/10/06-13:30	7.3	11	40	38
26	MYGH03	KARAKUWA	872	809	651	460		2003/05/26-18:24	7.0	71	72	13
27	IWT001	TANEICHI	870	826	649	275		2008/07/24-00:26	6.8	108	132	76
28	MIEH10	GEINOU	863	850	374	227		2007/04/15-12:19	5.4	16	17	4
29	IWTH21	YAMADA	856	550	586	711		2003/05/26-18:24	7.0	71	105	77
30	IWTH02	TAMAYAMA	854	796	675	500		2003/05/26-18:24	7.0	71	136	116
31	HRS009	YUKI	853	415	832	218	2001 Geiyo Earthquake	2001/03/24-15:28	6.4	51	76	57
32	SMNH01	HAKUTA	849	720	607	631	2000 Western Tottori prefecture Earthquake	2000/10/06-13:30	7.3	11	14	8
33	NIGH01	NAGAOKA	841	818	655	375	2004 Mid Niigata prefecture Earthquake	2004/10/23-17:56	6.8	13	20	15
34	NIGH01	NAGAOKA	827	742	502	384		2004/10/23-18:03	6.3	9	15	12
35	IWTH25	ICHINOSEKI-W	824	688	547	670		2008/06/16-23:14	5.3	11	12	2
36	NIG019	OJIYA	820	794	637	355		2004/10/23-18:34	6.5	14	19	12
37	NIG018	KASHIWAZAKI	813	667	514	369	2007 Niigataken Chuetsu-oki Earthquake	2007/07/16-10:13	6.8	17	27	21
38	MYG004	TSUKIDATE	812	740	678	224	2008 Iwate-Miyagi Nairiku Earthquake	2008/06/14-08:43	7.2	8	36	35
39	TTR007	KOHFU	803	725	573	404	2000 Western Tottori prefecture Earthquake	2000/10/06-13:30	7.3	11	17	13
40	IWTH25	ICHINOSEKI-W	785	782	210	461		2008/06/14-09:20	5.7	6	23	22
41	MIE004	KAMEYAMA	771	716	356	323		2007/04/15-12:19	5.4	16	18	8
42	KGS002	IZUMI	756	728	443	189		1997/05/13-14:38	6.2	8	18	16
43	NIGH11	KAWANISHI	754	741	534	300		2004/10/23-18:34	6.5	14	26	22
44	IWTH03	IWAIZUMI	736	475	549	575		2008/07/24-00:26	6.8	108	109	8

Appendix 1 (continued)

No.	Site code	Site name	PGA (3 comp.) [gal]	PGA (NS) [gal]	PGA (EW) [cm/s]	PGV (UD) [cm/s]	Earthquake Name	Origin time	M_{JMA}	Depth [km]	Hypocentral distance [km]	Epicentral distance [km]
45	TTR009	NICHINAN	729	629	595	289	2000 Western Tottori prefecture Earthquake	2000/10/06-13:30	7.3	11	17	13
46	HRSH01	MIHARA	723	461	493	615	2001 Geiyo Earthquake	2001/03/24-15:28	6.4	51	65	40
47	MYG005	NARUKO	722	324	708	292		1996/08/13-11:13	5.0	10	11	5
48	NIG020	KOIDE	719	527	524	329		2004/10/23-18:34	6.5	14	17	9
49	SZO002	ITOH	703	313	639	229		2009/12/18-08:45	5.3	0	4	4
50	HRS017	MIHARA	696	652	397	177	2001 Geiyo Earthquake	2001/03/24-15:28	6.4	51	69	47

Source: The earthquake information (location, magnitude and origin time) are based on the monthly report on earthquakes and volcanoes published by Japan Meteorological Agency and the hypocentral and epicentral distances are calculated using these locations.

Appendix 2 Remarkable records obtained by K-NET and KiT-net (sorted by peak ground velocity on the surface)

No.	Site code	Site name	PGV (3 comp.) [cm/s]	PGV (NS) [cm/s]	PGV (EW) [cm/s]	PGV (UD) [cm/s]	Earthquake Name	Origin time	M_{JMA}	Depth [km]	Hypocentral distance [km]	Epicentral distance [km]
1	TTRH02	HINO	146	124	87	56	2000 Western Tottori prefecture Earthquake	2000/10/06-13:30	7.3	11	13	7
2	NIG019	OJIYA	135	98	128	30	2004 Mid Niigata prefecture Earthquake	2004/10/23-17:56	6.8	13	15	7
3	NIG018	KASHIWAZAKI	127	110	84	27	2007 Niigataken Chuetsu-oki Earthquake	2007/07/16-10:13	6.8	17	27	21
4	ISK005	ANAMIZU	104	35	99	19	2007 Noto Hanto Earthquake	2007/03/25-09:42	6.9	11	22	19
5	HKD098	TAIKI	102	74	74	35	2003 Tokachi-oki Earthquake	2003/09/26-04:50	8.0	42	112	103
6	IWTH25	ICHINOSEKI-W	100	71	62	85	2008 Iwate-Miyagi Nairiku Earthquake	2008/06/14-08:43	7.2	8	9	3
7	TKCH07	TOYOKORO	98	91	75	15	2003 Tokachi-oki Earthquake	2003/09/26-04:50	8.0	42	130	123
8	IBUH03	ATSUMA	91	77	90	22	2003 Tokachi-oki Earthquake	2003/09/26-04:50	8.0	42	211	206
9	KSRH09	SHIRANUKA-S	85	41	84	18	2003 Tokachi-oki Earthquake	2003/09/26-04:50	8.0	42	140	134
10	AKTH04	HIGASHINARUSE	80	54	74	33	2008 Iwate-Miyagi Nairiku Earthquake	2008/06/14-08:43	7.2	8	23	22
11	HKD091	URAHORO	79	57	70	22	2003 Tokachi-oki Earthquake	2003/09/26-04:50	8.0	42	126	119
12	HKD020	MINATOMACHI	78	40	74	13		2004/12/14-14:56	6.1	9	12	9
13	NIGH01	NAGAOKA	75	60	62	28	2004 Mid Niigata prefecture Earthquake	2004/10/23-17:56	6.8	13	20	15
14	NIG028	NAGAOKA-SHISHO	71	64	65	25	2004 Mid Niigata prefecture Earthquake	2004/10/23-17:56	6.8	13	–	15
15	MYG005	NARUKO	71	70	33	20	2008 Iwate-Miyagi Nairiku Earthquake	2008/06/14-08:43	7.2	8	33	32
16	NIGH11	KAWANISHI	68	36	58	12	2004 Mid Niigata prefecture Earthquake	2004/10/23-17:56	6.8	13	22	17
17	NIG021	TOHKAMACHI	67	54	50	13	2004 Mid Niigata prefecture Earthquake	2004/10/23-17:56	6.8	13	25	21
18	HKD066	SHIBETSU	66	43	52	10	2003 Tokachi-oki Earthquake	2003/09/26-04:50	8.0	42	230	226
19	NIG019	OJIYA	64	63	54	11		2004/10/23-18:34	6.5	14	19	12
20	FKO006	FUKUOKA	63	59	32	10		2005/03/20-10:53	7.0	9	27	26
21	NIG021	TOHKAMACHI	62	58	49	10		2004/10/23-18:34	6.5	14	29	26

Appendix 2 (continued)

No.	Site code	Site name	PGV (3 comp.) [cm/s]	PGV (NS) [cm/s]	PGV (EW) [cm/s]	PGV (UD) [cm/s]	Earthquake Name	Origin time	M_{JMA}	Depth [km]	Hypocentral distance [km]	Epicentral distance [km]
22	IWTH26	ICHINOSEKI-E	61	58	44	28	2008 Iwate-Miyagi Nairiku Earthquake	2008/06/14-08:43	7.2	8	15	12
23	ISK006	TOGI	60	38	51	21	2007 Noto Hanto Earthquake	2007/03/25-09:42	6.9	11	13	7
24	MYG011	OSHIKA	60	51	37	20		2003/05/26-18:24	7.0	71	92	59
25	HKD092	IKEDA	60	55	57	17	2003 Tokachi-oki Earthquake	2003/09/26-04:50	8.0	42	144	138
26	HKD110	SAMANI	57	56	34	18	2003 Tokachi-oki Earthquake	2003/09/26-04:50	8.0	42	110	102
27	HKD085	SHIRANUKA	56	38	53	13	2003 Tokachi-oki Earthquake	2003/09/26-04:50	8.0	42	137	131
28	HDKH04	MONBETSU-W	55	28	55	16	2003 Tokachi-oki Earthquake	2003/09/26-04:50	8.0	42	192	187
29	TTR008	YONAGO	54	36	54	10	2000 Western Tottori prefecture Earthquake	2000/10/06-13:30	7.3	11	20	16
30	KSRH02	AKAN-S	52	44	47	16	2003 Tokachi-oki Earthquake	2003/09/26-04:50	8.0	42	154	148
31	NIG017	NAGAOKA	51	48	23	15	2004 Mid Niigata prefecture Earthquake	2004/10/23-17:56	6.8	13	21	17
32	HKD109	URAKAWA	50	41	47	13		2003/09/26-06:08	7.1	21	94	92
33	MYG004	TSUKIDATE	50	45	39	15	2008 Iwate-Miyagi Nairiku Earthquake	2008/06/14-08:43	7.2	8	36	35
34	HKD100	HIROO	49	44	48	19	2003 Tokachi-oki Earthquake	2003/09/26-04:50	8.0	42	94	84
35	HKD067	NAKASHIBETSU	48	33	42	8	2003 Tokachi-oki Earthquake	2003/09/26-04:50	8.0	42	215	210
36	NIG019	OJIYA	48	21	46	9	2007 Niigataken Chuetsu-oki Earthquake	2007/07/16-10:13	6.8	17	36	32
37	SZO002	ITOH	48	20	44	8		2009/12/18-08:45	5.3	0	4	4
38	HKD084	AKAN	47	43	46	16	2003 Tokachi-oki Earthquake	2003/09/26-04:50	8.0	42	154	148
39	HKD105	MOMBETSU	47	38	37	12	2003 Tokachi-oki Earthquake	2003/09/26-04:50	8.0	42	189	184
40	HKD109	URAKAWA	47	38	37	20	2003 Tokachi-oki Earthquake	2003/09/26-04:50	8.0	42	124	117
41	KGS005	MIYANOJOH	47	35	46	9		1997/05/13-14:38	6.2	8	18	16
42	HKD126	MUKAWA	47	46	35	16	2003 Tokachi-oki Earthquake	2003/09/26-04:50	8.0	42	202	198
43	HDKH07	SAMANI	47	28	42	23	2003 Tokachi-oki Earthquake	2003/09/26-04:50	8.0	42	112	104
44	HKD095	OBIHIRO	46	33	38	17	2003 Tokachi-oki Earthquake	2003/09/26-04:50	8.0	42	152	146
45	NIGH11	KAWANISHI	46	41	32	10		2004/10/23-18:34	6.5	14	26	22
46	TTR009	NICHINAN	45	38	26	13	2000 Western Tottori prefecture Earthquake	2000/10/06-13:30	7.3	11	17	13

Appendix 2 (continued)

No.	Site code	Site name	PGV (3 comp.) [cm/s]	PGV (NS) [cm/s]	PGV (EW) [cm/s]	PGV (UD) [cm/s]	Earthquake Name	Origin time	M_{JMA}	Depth [km]	Hypocentral distance [km]	Epicentral distance [km]
47	TTR007	KOHFU	45	42	26	11	2000 Western Tottori prefecture Earthquake	2000/10/06-13:30	7.3	11	17	13
48	TKCH08	TAIKI	45	23	43	16	2003 Tokachi-oki Earthquake	2003/09/26-04:50	8.0	42	117	109
49	HKD078	TOHRO	44	35	37	13	2003 Tokachi-oki Earthquake	2003/09/26-04:50	8.0	42	161	156
50	NIG019	OJIYA	44	11	44	4		2004/10/27-10:40	6.1	12	25	22

Source: The earthquake information (location, magnitude and origin time) are based on the monthly report on earthquakes and volcanoes published by Japan Meteorological Agency and the hypocentral and epicentral distances are calculated using these locations.

References

1. Aoi S, Obara K, Hori S, Kasahara K, Okada Y (2000) New strong-motion observation network: KiK-net. EOS trans. Am Geophys Union 81:F863
2. Boore DM (2001) Effect of baseline corrections on displacements and response spectra for several recordings of the 1999 Chi-Chi, Taiwan, earthquake. Bull Seismol Soc Am 91: 1199–11211
3. Eguchi T, Fujinawa Y, Fujita E, Iwasaki S, Watabe I, Fujiwara H (1998) A real-time observation network of ocean-bottom-seismometers deployed at the Sagami trough subduction zone, central Japan. Mar Geophys Res 20:73–94
4. Fukuyama E, Ishida M, Hori S, Sekiguchi S, Watada S (1996) Broadband seismic observation conducted under the FREESIA Project. Rep Nat'l Res Inst Earth Sci Dsas Prev 57:23–31
5. Geographical Survey Institute (1998) Crustal deformation of Japan detected GEONET. http://mekira.gsi.go.jp/ENGLISH/index.html . Accessed 29 Jan 2010
6. Kinoshita S (1998) Kyoshin net (K-NET). Seismol Res Lett 69:309–332
7. Kunugi T (2000) Relationship between Japan meteorological agency instrumental intensity and instrumental modified mercalli intensity obtained from K-NET strong-motion data. Zisin2 53:89–93
8. Kunugi T, Aoi S, Nakamura H, Fujiwara H, Morikawa N (2008) A real-time processing of seismic intensity. Zisin2 60:243–252
9. Obara K (2002) Hi-net: high sensitivity seismograph network, Japan. Lect Notes Earth Sci 98:79–87
10. Obara K, Shiomi Y, Haryu Y, Matsumura M, Shimanuki T (2008) Development of network platform for data transmission and practical use for NIED Hi-net system. Japan geoscience union meeting 2008:S144–P012
11. Odaka T, Ashiya S, Tsukada S, Sato S, Ohtake K, Nozaka D (2003) A new method of quickly estimating epicentral distance and magnitude from a single seismic record. Bull Seismol Soc Am 93:526–532
12. Okada Y, Kasahara K, Hori S, Obara K, Sekiguchi S, Fujiwara H, Yamamoto A (2004) Recent progress of seismic observation network in Japan – Hi-net, F-net, K-NET and KiK-net – . Earth Planets Space 56:15–28
13. Peterson J (1993) Observations and modeling of seismic background noise. USGS Open File Rep. 93-322
14. Shabestari KT, Yamazaki F (2001) A proposal of instrumental seismic intensity scale compatible with MMI evaluated from three-component acceleration records. Earthq Spectra 17:711–723
15. Shiomi K, Obara K, Aoi S, Kasahara K (2003) Estimation on the azimuth of the Hi-net and KiK-net borehole seismometers. Zisin2 56:99–110
16. Tomioka T, Yamamoto S (2006) Development of Low Noise Accelerometer (JA-40GA), Japan Aviation Electronics technical report 29:14. www.jae.co.jp/gihou/gihou29/pdf/g_14.pdf. Accessed 29 Jan 2010
17. Wessel P, Smith W (1995) New version of the generic mapping tools released. EOS trans. Am Geophys Union 76:329

Chapter 13
Integrating the European Observational Seismology Infrastructure: NERIES Developments

Torild van Eck, D. Giardini, R. Sleeman, and B. Dost

Abstract Europe and its surroundings operate an impressive number of seismometers and accelerometers, accommodating high and lower quality sensors, permanent and mobile stations, land-based and ocean bottom systems. The equipment and observatories are largely funded by national resources and priorities. Recently a number of initiatives, among them NERIES, have accomplished a significant step towards integrating this diverse network into a homogeneous real-time network, data archival and data access facility, thus providing a powerful research tool for future earth science research and hazard assessment.

13.1 Introduction

An increased exploitation of the Earth's surface and its crust and an increased vulnerability of the society's infrastructure to geological and geomorphological processes, together with a steadily lower acceptance of risks due to these processes, demand an increasing knowledge of the structure and understanding of plate dynamics and its surface effects. This demand for earth science research requires a high-quality in-situ earth observation infrastructure providing continuous observations from a spatially distributed network of sensors. Together with the build-up of a secured, homogeneous data archive and robust data access facilities these are the basic elements of such a high-quality observation system, that is preferably complemented by advanced, multidisciplinary analysis facilities. Precisely this approach has been one of the main goals of the EC project NERIES (Network of Research Infrastructures for European Seismology).

T. van Eck (✉)
ORFEUS c/o Seismology Division, Royal Netherlands Meteorological Institute, 3730 AE De Bilt, The Netherlands
e-mail: torild.van.eck@knmi.nl

S. Akkar et al. (eds.), *Earthquake Data in Engineering Seismology*, Geotechnical, Geological, and Earthquake Engineering 14, DOI 10.1007/978-94-007-0152-6_13,

Since 2000 European seismological observatories and research institutes are successfully building an integrated and efficient research infrastructure. The accomplishments show that such integration is possible in a diversified social and political region such as Europe and its surroundings. Two consecutive EC Research Infrastructure projects, MEREDIAN (2000–2005) and NERIES (2006–2010) had a significant impact on coordinating the different national and regional initiatives within and around Europe. In this paper we present these recent developments towards a comprehensive observation and analysis research infrastructure, specifically with regard to seismic broadband and accelerometric networks.

13.2 Current Seismological Monitoring Networks

In-situ monitoring of ground motion in and around Europe is well developed. Earthquakes are considered a serious threat in many southern and central European countries, but with a decreasing risk acceptance also European countries with low seismicity need to consider seismic hazard seriously.

The political and geographical realities in Europe favoured a patchwork coverage of widely-distributed operational responsibilities and a large variety of equipment types. In spite of this, European networks represent some of the densest and best-equipped land-based networks in the world (Fig. 13.1).

Complementary coverage is provided by large-scale temporary deployments of dense networks and borehole measurements, mainly motivated by diverse research interests. Examples are the MIDSEA [19] and EIFEL [4] projects. A coordinated European approach of these temporary deployments has recently been launched with the Euro-Array concept [5]. The Topo-Iberia project [6] marked the first step. Although initially only being a national Spanish initiative, it recently has been extended to Morocco as a joint US, Moroccan and Spanish cooperation, as well as to France as a joint French and Spanish initiative.

A less satisfactory coverage is found in the seas and oceans, notably the north Atlantic and the Mediterranean Sea, where many earthquakes occur. This is not surprising as sea and ocean bottom monitoring costs are significantly higher than those for land-based observations. Ocean-bottom seismometers (OBS) have been installed notably in France and Italy. Regular temporary OBS experiments have taken place around Spain, France, Iceland, Greece and Turkey showing their potential for significant additional data [10]. Within the NERIES and ESONET projects a closer coordination with land-based mobile deployments is being initiated. The Arctic area, increasingly relevant with regard to possible future exploitation, still lacks proper in-situ observatories.

Around Europe, Greenland is currently being monitored more intensively thanks to the multinational GLISN project [11]. Most countries in northern Africa and the Middle East are rapidly building modern observational networks. Unfortunately data exchange between northern African countries and Europe are currently still limited.

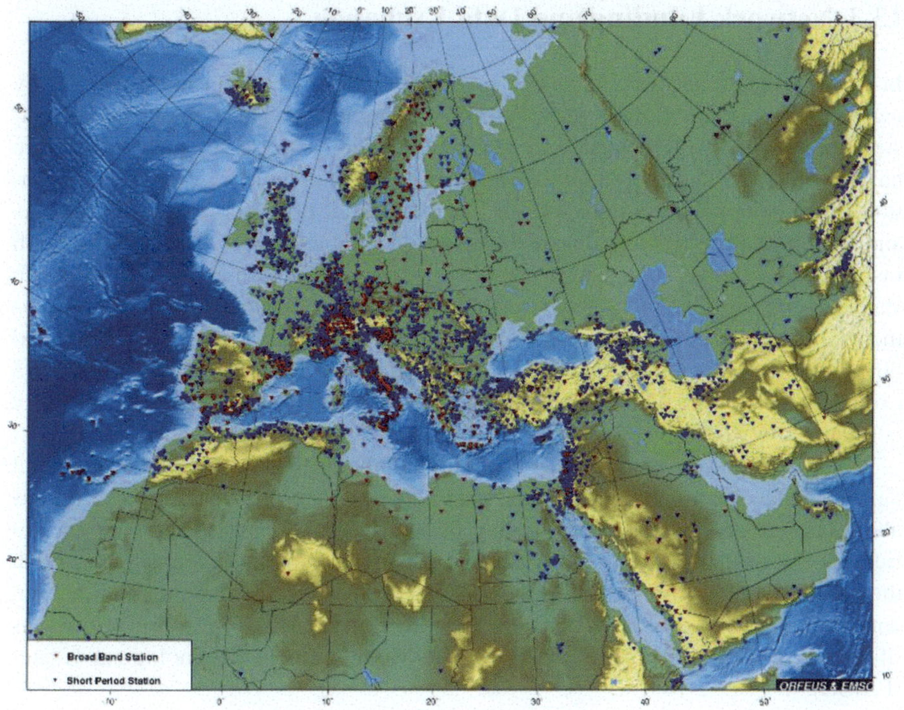

Fig. 13.1 About 4,000 seismic stations operated by about 180 networks are currently installed in and around Europe [2008 overview provided by EMSC and ORFEUS]. This network provides the seismic phase arrival time data for the operations of the EMSC

However, clear and steady progress is being made with countries like Morocco, Tunisia and Algeria. Joint effective monitoring of, and innovative research on, earthquakes and complicated plate boundary structures in the Mediterranean area will certainly develop once data exchange on a broad scale across the Mediterranean area is realised.

Along the eastern border of the European Union data exchange with Turkey, deploying two new modern seismological networks, and Georgia has strongly improved. In countries at the eastern border of the European Union with low seismicity, the coverage with modern broadband seismometers is significantly less dense. However, countries around the Caucasus (e.g. Azerbaijan) are rapidly developing their networks. Many countries in south-west Asia do operate advanced networks, but as yet only a few do exchange data between each other.

In summary the current seismological network in and around Europe is generally dense and modern and it is developing rapidly towards a large virtual seismological observatory. Significant and challenging initiatives are still necessary to create an integrated, homogeneous coverage of in-situ ground-motion sensors on the European plate and its boundaries.

13.3 Network Funding and Data Exchange

In-situ permanent networks in and around Europe are currently primarily funded by national organisations often with well-defined mandates. These mandates emphasize: monitoring current local and regional seismicity, the assessment of seismic hazard and risk, and the supply of national and local rapid earthquake information. Most seismological networks in Europe are not primarily research networks and therefore they are not funded by research organisations. Academic research institutes, like universities and academies of sciences in former eastern European states, operating permanent networks are exceptions in Europe. Temporary deployments, often targeting specific research goals, are on the other hand often supported through research funding. One consequence of this funding structure is that data exchange for global seismological research is not considered a first priority for many permanent networks.

Historically, however, the research community has often been driving open global data exchange. Facilitating open rapid parameter, i.e. interpreted, data exchange has always been a generally-accepted policy among seismological (research) observatories since around 1895 [17]. Current networks often maintain this practice through the International Seismological Center (ISC). However, rapid regional, cross-border, data exchange is apparently considered more important for the observatories as illustrated by the success of the European-Mediterranean Seismological Center (EMSC).

Open waveform data exchange took off at the same time as broadband and high-dynamic range, digital recording became the standard goal in the research community at the end of the 1980s. Technical and political difficulties had to be overcome in order to enable efficient waveform data exchange. With the creation of a number of research networks the seismological communities in the US and Europe coordinated and facilitated data exchange early on. They established non-governmental organisations, IRIS in 1984 and ORFEUS in 1987 [13, 20] in the US and Europe respectively. A broader global collaboration followed with the creation of the International Federation of Digital Seismograph Networks (FDSN) in 1986 [16]. The FDSN was crucial in establishing in 1986 the Standard for the Exchange of Earthquake Data (SEED), which is currently the standard digital format agreed on by seismological observatories and data centers for the exchange and archive of waveform data. At present most major data acquisition manufacturers provide their equipment with options for SEED output.

This preparatory work of the early global/international research networks at the end of the 1980s paved the way to continue the traditions of open data exchange for digital waveform data. European networks, although mostly not funded by the research community, still aim to maintain or establish this open data exchange practice, provided it does not interfere with their current operations. Implicitly, the huge advantage of opening this data for fundamental earthquake research is recognized by the same observatories. Unfortunately, mobile research deployments in Europe aiming at lithospheric studies did not follow a coherent archiving and formatting strategy, except more recently in, among others, the UK, Germany, France and Switzerland.

13.4 European Scale Open Rapid Data Exchange

A European-scale observatory research infrastructure implies that an open data exchange policy of waveform data is implemented and broadly accepted as the standard. As explained above, due to individual funding priorities, open exchange of waveform data in and around Europe is not self evident. Real-time or near real-time open data exchange, which became technical possible through the Internet in the 1990s, is even less self-evident. Therefore clear agreements between the parties are necessary and being made. Although not always easy, the last decade has shown a significant transformation towards this concept of open and free exchange of real-time data. A number of developments have been crucial in this process.

The initial framework was created by long-term international research networks in Europe, like, NARS (University of Utrecht) in The Netherlands [14], Geoscope (IPGP) in France [15], MEDNET (INGV) in Italy [3] and GEOFON (GFZ) in Germany [8]. These networks pioneered the installation of significantly improved sensors (broadband), data acquisition systems (dataloggers) and data transmission equipment and protocols (Fig. 13.2). After the middle of the 1990s these modern seismic stations were installed in most local and regional permanent networks on a large scale (Fig. 13.2). The research networks also pioneered open real-time waveform data access to facilitate long-time archiving.

Since about 2000 Europe's Internet network became sufficiently established as to enable free, or very low cost, reliable data transmission. Pursuing the real-time data exchange concept pioneered by GEOFON, the EC-project MEREDIAN started the Virtual European Broadband Seismic Network (VEBSN [21]). Consequently, the VEBSN concept of real-time Internet data exchange between networks in different countries in and around Europe became possible by three coinciding crucial developments.

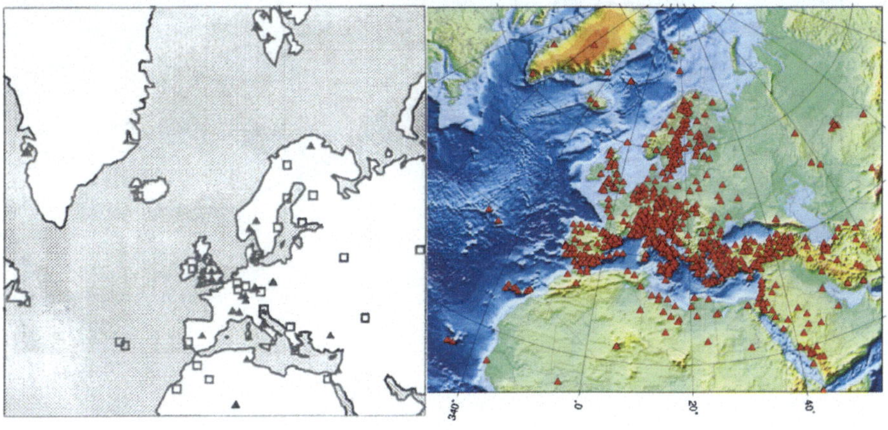

Fig. 13.2 **a** Broadband stations installed and in operation in 1987. **b** Broadband stations installed and in operation in 2008. About 50% of the broadband waveform data currently being produced is archived in standard SEED format within the European Distributed Waveform Data Archive (EIDA) coordinated within ORFEUS

1. The reliable and fast Internet network created within Dante [9]. Currently the National Research & Education Networks (NRENs) coordinated within Terena provide low-cost, reliable and fast data exchange facilities for research data in Europe.
2. The robust and effective data exchange protocol SeedLink [7], initially based on the ComServ software package developed by Quanterra®. This was later developed into an autonomous software package by the GEOFON group at GFZ and made publicly available. This protocol, together with incorporated software modules ("plugins") to communicate with standard dataloggers, provided a simple data exchange tool easy to be implemented by a network.
3. The EC project MEREDIAN (2000–2005) providing the seed money for the networks to implement data exchange in Europe, including Eastern Europe, and to develop additional SeedLink plugins to interface a variety of data acquisition systems. Within this project SeedLink became a de facto standard in Europe and surrounding area.

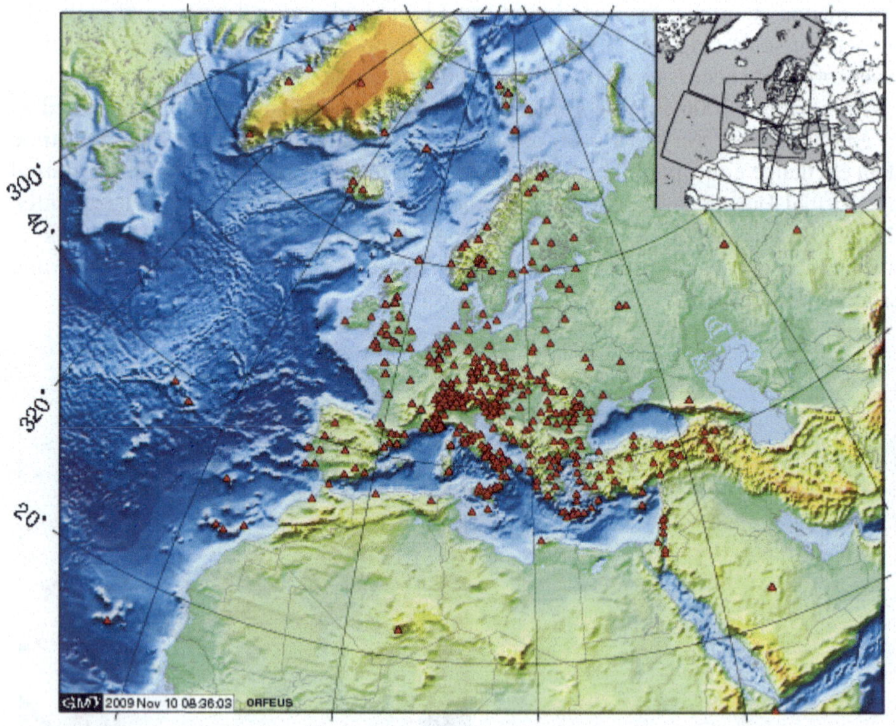

Fig. 13.3 The European Virtual Broadband Seismic Network (VEBSN), a collaborative observatory network initiative, providing access to near real-time continuous waveform data. This is the situation as of January 2010

The current VEBSN concept enables each network in Europe to operate a regional, or even, a global (virtual) network of stations, relying on real-time data streams supplied by many different providers. Obviously from the viewpoint of many current observatories this concept can have both advantages and disadvantages depending on their service agreements with funding providers. Clear agreements between ORFEUS and the data providers are therefore being established.

In 2010 the European-Mediterranean observatories and the NERIES project realised a VEBSN that exchanges and archives in near real-time digital waveform data from around 500 permanent stations from 55 networks (Fig. 13.3). This is a significant step towards an integrated European observational seismology infrastructure.

13.5 Observational Networks and Research Tools

The EC infrastructure project NERIES (2006–2010) involved national observatories, research networks, coordinating organisations like ORFEUS and EMSC and academia. This provided a fertile ground for cross collaboration.

One of the aims of the NERIES project has been to provide integrated access to a significant larger set of data from observatories than has been able before. For this, NERIES took advantage of already existing and recently upgraded European infrastructure of national or local seismological networks. Many of these observatories record high-quality data for local or regional monitoring and hazard research. These same data are, however, also core data for research on large-scale plate observations or global seismological research.

Therefore, another aim of the NERIES project has been to provide research tools for these data and to make these openly available for the whole seismological community, both for basic research and applied studies more relevant for the observatories. Tools developed within NERIES include, for example, ShakeMaps, loss estimation software, site response software, data analysis software modules and time varying hazard estimation tools. Many of those tools are certainly relevant to improve operational services of the involved networks. Consequently, the NERIES project has been promoting closer collaboration between the research community, developing advanced analysis tools, and the observatories, providing the data and implementing the research tools.

To facilitate this collaboration ORFEUS organises, since 2002, annual observatory coordination workshops and meetings at which recent developments are discussed among observatories and, recently, also academia. At these meetings, the technical and operational teams of the observatories initially exchanged common practice and tools. Currently, the observatory teams are regularly discussing and exchanging their experience with the research communities on newly developed or improved tools. One illustrative consequence of this closer collaboration between observatories and research institutes can be found in the software development.

Observatories in and around Europe have since the beginning of the 1990s been moving from software developed in house to commercial or community-developed software packages. For example Antelope® and Apollo® (commercial packages) and EarthWorm, Seismic Handler, and SeisComP3 (open software packages) are rapidly replacing in-house software in observatories. Locally developed software has become increasingly too expensive to maintain, and adapting complete local packages for increasing automatic and real-time operations too cumbersome.

Seismological observatories are increasingly required to provide very rapid information. To meet this goal automatic location tools and rapid interpretation tools like automatic locations, magnitudes and ShakeMaps are currently being implemented within the observatories. Modular software packages for Antelope®, EarthWorm and SeisComP3 are often developed in research environments and these provide observatories with new or improved tools facilitating routine observatory practices. Moreover, observatories can implement suitable community developed modules within their standard packages.

European observatories and research institutes are thus returning to their earlier symbiotic relation, by creating a collaborative observatory research infrastructure facility for data driven research. Observatory and research communities seem, however, to remain funded through different resources.

13.6 Digital Waveform Data Archives

From the 1980s waveform data was primarily archived through exchange of tapes, disks and CD-ROMs between network operators and the archiving facilities. Unfortunately, this required a significant and increasing effort. As shown above real-time data transfer, or even delayed automatic data transfer by Internet proved since about 2,000 to be a far more efficient and sustainable option in Europe. Taking advantage of this development, the VEBSN is currently the major practical tool for archiving waveform data. Delayed data retrieval mechanisms, however, may still be required to fill data gaps in the archives due to incomplete real-time data transfers.

Given the current European funding structure for research infrastructures, it is at this stage not feasible to create and maintain a single European waveform data archive like IRIS-DMS in the US. Consequently, the NERIES project implemented the European Integrated Distributed Waveform Data Archive (EIDA) coordinated within ORFEUS. The EIDA consists of a number of existing large-scale national facilities that share the task of archiving the exponentially increasing amount of digital waveform data becoming available from the European-Mediterranean observatories. The data exchange communication protocol for this distributed archive, ArcLink, has been developed on the basis of SeedLink by the GEOFON group at GFZ.

A major task still remains ahead in recovering older data and to turn large off-line archives into on-line archives. Systematic scanning of old seismograms at SISMOS (http://sismos.rm.ingv.it) is one of the many different initiatives in this direction.

13.7 Waveform Data for Research

Seismological digital waveform data is the major observational material for research in global, and earthquake seismology. Easy access to all available data and the high quality of the data are major requirements of the research community. However, the above described integration procedure of the widely diverse observational structure and limited resources poses a serious challenge in fulfilling these requirements.

Quality control and user oriented access tools have therefore become important together with data provenance and integrity. Automated quality control procedures [18] are playing an increasingly important role in the ORFEUS community, providing feedback to both data providers and data users. PQLX [12] and [2] is an excellent example of such a tool, but more tools are rapidly being implemented. User feedback or "crowdsourcing" is one of the tools being considered, facilitating data users to participate in data quality control.

A broad range of data access facilities are being implemented. IRIS and ORFEUS aim at coordinate traditional access tools like Wilber, BreQFast and NetDC, enabling users to access different archives using the same tools. NERIES started implementing a more versatile approach with webservices, facilitating the integration of different services. The Seismic Data Portal (www.seismicportal.eu) is one implementation of the integration of earthquake parametric data, broadband seismic waveform data and accelerometric waveform data enabling users to browse through seismic data. Another application allows users to retrieve large volumes of data from the EIDA through a simple command line procedure using web services.

Integrating the observational seismology infrastructure, in itself a challenge, has also revealed quality control challenges as well as new options for efficient QC feedback involving data providers and data users. The current infrastructure also enables us to build a sustainable one-stop-shop data access for the research community. In both aspects we will benefit in the future from an increased collaboration with the information technology community.

13.8 Accelerometer Networks

Europe and its surroundings also operate an impressive number of accelerometers (Fig. 13.4). Strong-motion networks are usually motivated by geotechnical and earthquake engineering purposes and are often operated by engineering-related organisations or public civil protection authorities. However, a significant number of mobile equipment pools are maintained by research organisations and they are often being deployed after important earthquakes, for example in the recent l'Aquila earthquake in Italy in 2009. Seismological networks are increasingly incorporating accelerometric networks within their observational infrastructure. Accelerometric data are becoming increasingly of interest to seismologist as well.

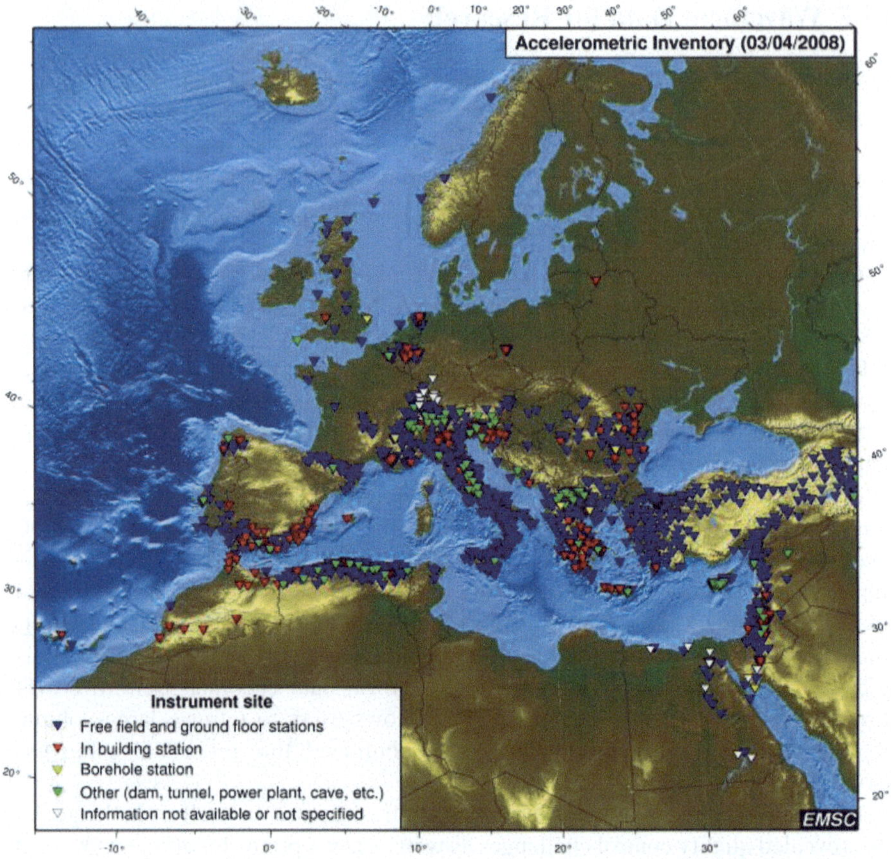

Fig. 13.4 An overview of nearly 4,000 accelerometers in operation in and around Europe in 2008 (courtesy EMSC). Within the NERIES project more homogeneous data access has been created as a first step to facilitate open access to accelerometric data

13.9 Acceleration Data Exchange

Open exchange of acceleration data has currently not reached the same level as seismological broadband data, although significant projects have been undertaken [1]. Standardized digital metadata formats, more complicated than for broadband seismic data, are still under discussion hampering efficient open data exchange. Also since strong ground motion observations were considered most important for local and regional hazard and risk investigations, standardized open data exchange was not given a high priority.

Within the NERIES project the situation has significantly improved in the last few years. Several larger acceleration networks are currently providing open access to their data and the NERIES seismic data portal, mentioned before, provides integrated access to accelerometric data from a number of networks as discussed in

Chapter 9 by Roca et al. and Chapter 10 by Pequegnat et al., this volume. This data exchange and integrated access initiatives require further developments in which strong-motion data users are served according to their requirements.

13.10 Discussion and Conclusions

An exciting European observational seismology infrastructure is being created in and around Europe in spite of the political and technical challenges. A will to coordinate and collaborate together with available EC funds have been crucial elements in the realisation.

The coordination and collaboration between observatories across borders, and between observatories, funded through local mandates, and research institutes, using data for basic research, are necessary. Clear agreements on data exchange policies and a mutual beneficial symbiotic relation between observatories and research institutes facilitate this.

The EC infrastructure projects MEREDIAN and NERIES have provided timely glue to integrate the recently modernized national observational seismology infrastructures. The EC provided small investments as compared with the recent total investment in national infrastructures, which is difficult to estimate but easily reaches beyond the 150 M€ level. Most likely, there are only a few better investments with such a high cost/benefit ratio then integrating this infrastructure into one effective European – Mediterranean observational seismology Infrastructure with open data access for the research community. The impact on the understanding of: earthquake hazard and risk, the dynamics and structure of the earth, lithosphere, crust and near-surface structure will be significant. Related initiatives in the US (e.g. Earthscope) and Japan provide a hint to what the benefits could be.

The European observational seismology infrastructure currently created within NERIES is well in-line with related developments in GEO(SS) and other distributed environmental observational infrastructures. The current infrastructure will enable further integration of seismological data with other geological and geophysical data.

Acknowledgements The described infrastructure is a joint initiative of many observatories and research institutes, among others cited in ORFEUS VEBSN contributors (www.orfeus-eu.org/Data-info/vebsn-contributors.html) and the European-Mediterranean observatories (www.orfeus-eu.org/Links/euromed.html). We thank John Douglas for critical remarks improving the paper.

NERIES is an EC funded project under contract RII3-CT-2006-026130.

References

1. Ambraseys N, Smit P, Douglas J, Margaris B, Sigbjörnsson R, Olafsson S, Suhadolc P, Costa G (2004) Internet-site for European strong-motion data. Boll Geofis Teor Appl 45(3):113–129
2. Boaz RI, McNamara DE (2008) PQLX: a data quality control system: uses and applications. ORFEUS Newsl 8(1). www.orfeus-eu.org/Organization/Newsletter/vol8no1/PQLX/PQLX.htm

3. Boschi E, Morelli A (1994) The MedNet program. Ann Geofis 37(5):1066–1070
4. Budweg M, Weber M, Bock G, Ritter J, Christensen U Eifel Plume Team (1999) A 400 km long broadband antenna in the Eifel-region. ORFEUS Electron Newsl 1(3). www.orfeus-eu.org/Organization/Newsletter/vol1no3/eifel.html
5. Cloetingh S et al (2006) Neotectonics and intraplate continental topography of the northern Alpine Foreland. Earth Sci Rev 74:127–196
6. Díaz J, Villaseñor A, Gallart J, Morales J, Pazos A, Córdoba D, Pulgar J, García Lobón JL, Harnafi M TopoIberia Seismic Working Group (2009) The IBERARRAY broadband seismic network: a new tool to investigate the deep structure beneath Iberia. ORFEUS Electron Newsl 8(2). www.orfeus-eu.org/Organization/Newsletter/vol8no2/vol8no2_article1.pdf
7. Hanka W, Heinloo A, Jaeckel K-H (2000) Networked seismographs: GEOFON real-time data distribution. ORFEUS Electron Newsl 2(3). www.orfeus-eu.org/Organization/Newsletter/vol2no3/geofon.html
8. Hanka W, Kind R (1994) The GEOFON program. Ann Geofis 37(5):1039–1065
9. Kirstein PT (2004) European International Academic Networking. Selected papers TERENA Networking Conference (www.terena.org/publications/tnc2004-proceeding)
10. Krüger F et al (2010) Ocean Upper Mantle Initiative: the oceanic lithosphere and its interaction with mantle convection. Konzept und Planungspapier, Koordination Support Schiffsanträge Sonne, Polarstern, Meteor/Merian, Mittelgrosse Forschungsschiffe. 32 pp
11. Larsen TB, Anderson KR, Beaudoin BC, Butler R, Clinton JF, Dahl-Jensen T, Ekström G, Giardini D, Hanka W, Kanao M, McCormack D, Mykkelveit S, Nettles M, Agostinetti NP, Tsuboi S, Voss P (2010) The GreenLand Ice Sheet monitoring Network (GLISN). Geophysical Research Abstracts, 12, EGU2010-4836–2
12. McNamara DE, Hutt CR, Gee LS, Benz HM, Buland RP (2009) A method to establish seismic noise baselines for automated station assessment. Seismol Res Lett 80(4):628–637
13. Nolet G, Romanowicz B, Kind R, Wielandt E (1986) Observatories and Research Facilities for European Seismology. ORFEUS Science Plan, 45 pp
14. Nolet G, Vlaar NJ (1982) The NARS project: probing the Earth's interior with a large seismic antenna. Terra Cognita 2:17–25
15. Romanowicz B, Cara M, Fels JF, Rouland D (1984) GEOSCOPE: a French initiative in long period three component global seismic networks. EOS Trans AGU 65:753–754
16. Romanowicz B, Dziewonski AM (1987) Global digital seismographic network: research opportunities and recent initiatives. In: C Fuchs, C Froidevaux (eds) Composition, structure and dynamics of the lithosphere-asthenosphere system. A.G.U., Public, Geodynamics series, 16, pp 99–110
17. Rothé J-P (1981) Fifty years of history of the international association of seismology (1901–1951). Bull Seismol Soc Am 71(3):905–923
18. Sleeman R, Vila J (2007) Towards an automatic quality control manager for the VEBSN. ORFEUS Electron Newsl 7(1). www.orfeus-eu.org/Organization/Newsletter/vol7no1/QCM/QCM.html
19. Van der Lee S, Giardini D, Estabrook C, Deschamps A, Chiarabba C (1999) New temporary broadband stations in the larger Mediterranean region. ORFEUS Electron Newsl 1(1). www.orfeus-eu.org/Organization/Newsletter/vol1no1/midsea.html
20. van Eck T, Dost B (1999) ORFEUS, a European initiative in broadband seismology: status and future plans. Phys Earth Plan Int 113:45–55
21. van Eck T, Trabant C, Dost B, Hanka W, Giardini D (2004) Setting up a virtual broadband seismograph network across Europe. EOS Trans AGU 85(13):125–129

Chapter 14
The Turkish National Accelerometric Network: 1973–2010

Polat Gülkan

Abstract For a country with the size and seismicity of Turkey, the number of strong motion recording instruments has been, and still remains, meager. The national accelerometric network dates its genesis to 1973, some 40 years after the first instruments had been deployed in California as the world's first. The article traces the procurement and emplacement of the first batch of analog sensors from 1973, and the stages of technological and numerical improvements that have been achieved since.

14.1 Setting the Stage: The 1939–1973 Period

Turkey has been one of the few countries in the world to create a ministry solely for the purpose of re-housing its homeless citizens following natural disasters, and reconstructing the affected urban areas. This is not to place a claim for exemplary communal compassion, and may even be interpreted as a tacit acknowledgment of the fact the public policies for disaster mitigation were largely impossible to enforce [4]. From 1939 when a disastrous M7.9 earthquake struck Erzincan through the 1940s not much could be done to counter earthquake losses except to provide initially tents and later permanent homes to survivors. Academics in earth sciences determined the earthquake hazard zones of the country and were also involved in the preparation of building construction requirements.

Until 1958 the Ministry of Public Works (MPW) had been entrusted the duty of tidying up after natural disasters (mostly earthquakes), preparing the earthquake zones map for the country and the design requirements for buildings to be constructed in seismic zones. The Turkish Grand National Assembly, the country's parliament, passed a law in 1958 that created the Ministry for Reconstruction and Resettlement (MRR), and these responsibilities were passed to the new government branch.

P. Gülkan (✉)
Department of Civil Engineering, Earthquake Engineering Research Center, Middle East Technical University, 06531 Ankara, Turkey
e-mail: a03516@metu.edu.tr

S. Akkar et al. (eds.), *Earthquake Data in Engineering Seismology*, Geotechnical, Geological, and Earthquake Engineering 14, DOI 10.1007/978-94-007-0152-6_14,

Turkish engineers active in the earthquake field were aware that ground acceleration records from strong earthquakes had been successfully acquired by analog instruments deployed in critical locations in the US, Japan and New Zealand. Their participation in the early World Conferences on Earthquake Engineering and the increasing number of publications on the interpretation and use of the response spectrum aroused interest for the few digitized accelerograms that were in use in those days. The sparse collection included the ubiquitous El Centro trace from the May 18, 1940 earthquake in Imperial Valley. With a mixture of curiosity and wistfulness the earthquake professionals in Turkey hoped to record the first all home-made ground acceleration wave form.

In 1971, MRR created the Earthquake Research Institute (ERI) from what had been a small unit within its General Directorate of Disaster Affairs (GDRA), the principal branch of the Ministry that dealt with building regulation development, earthquake hazard zones mapping and post-earthquake damage assessment and planning. In spite of its academically inspired name ERI was really an elite unit within the Ministry reporting directly to the minister and directed its work to applied research objectives. What was unique about ERI was the fact that by law, it had been accorded powers to draw from a pool of money called the Disasters Fund that was diverted to it from profits of State Economic Enterprises. When earthquakes caused unexpectedly costly damages during 1966–1971, a second resource to support ERI activities was introduced by the creation of the Earthquake Fund to which money was diverted from sale of tobacco and alcoholic beverages. The financial freedom to spend from the "Earthquake Fund" was unique in the Turkish bureaucracy that normally is tightly under the control of the Court of Accounts and the Ministry of Finance. This freedom was abrogated in 1983 when MRR was re-united with MPW for a single ministerial entity along with most types of extra-budgetary fund pools.

Until well into the 1970s the accelerometer design and production industry worldwide relied on analog technology [2]. The difficulties and uncertainties associated with converting analog traces on photographic paper into digital format were recognized, but these were not considered to be insurmountable. In spite of the availability of the funds for the acquisition of strong motion accelerographs, these instruments fell within the broad definition of capital equipment, so it was necessary for ERI to submit a proposal for a project to the State Planning Organization (SPO) for their procurement. The request was approved for the 1973 budget, and the first 15 sensors of the Kinemetrics SMA-1 vintage were procured for deployment as they were delivered over the next year.

The policy for the deployment of the sensors was determined with consideration of ease of access for regular maintenance and following earthquakes, safety from unwanted interference, availability of power and a phone interrogation point with a designated person. The cities were all selected with reference to the perceived likelihood of recording a future earthquake. These criteria were mostly fulfilled by the provincial representation branches of MRR, so the early generation of sensors was emplaced in these buildings that are architecturally and structurally identical. A few of them were also placed inside rural health clinics and buildings belonging to other ministerial functions. In hindsight it is unfortunate that little awareness was displayed for the possible contamination of the records by soil–structure interaction. All of the early sensors were stand-alone stations so that none had a twin station nearby to study wave propagation or site effects. No consideration was given to achievement of site geology variability (to this day nearly all stations in Turkey are on Class C or D sites, and stations on rock are very few). This is not surprising, given the very few number of stations in the first place. Double stations in the same city were a luxury that simply could not be afforded. For the same reason, instrumented buildings did not exist.

14.2 The First Vernacular Accelerogram: The 1976 Denizli Record

By 1975, a paltry total of 34 stations had been deployed in the country that has a surface area of 800,000 km^2. I am showing their locations in Fig. 14.1. The table in the Appendix lists them in aggregated form until 2010. The coordinates of the stations in the table are current as of this writing even though it may have been moved following its first emplacement. In the interest of brevity a single table is provided where the year of reporting of that station is listed.

Fig. 14.1 Status in 1976 (Denizli is marked with the *star*)

The station in Denizli (a city in western Turkey with 500,000 residents now) was shaken by an M4.9 shallow earthquake on August 19, 1976, and produced the eagerly awaited record with a PGA of about 0.34 g. This record is shown in Fig. 14.2 for its historic relevance.

14.3 Expansion of the National Network Until 1996

The Division of Earthquake Research of GDDA produced reports at approximately 5 year intervals from 1985 to inform about the state of the Turkish National Network under its management. In 1985 there were 65 SMA-1 instruments deployed as

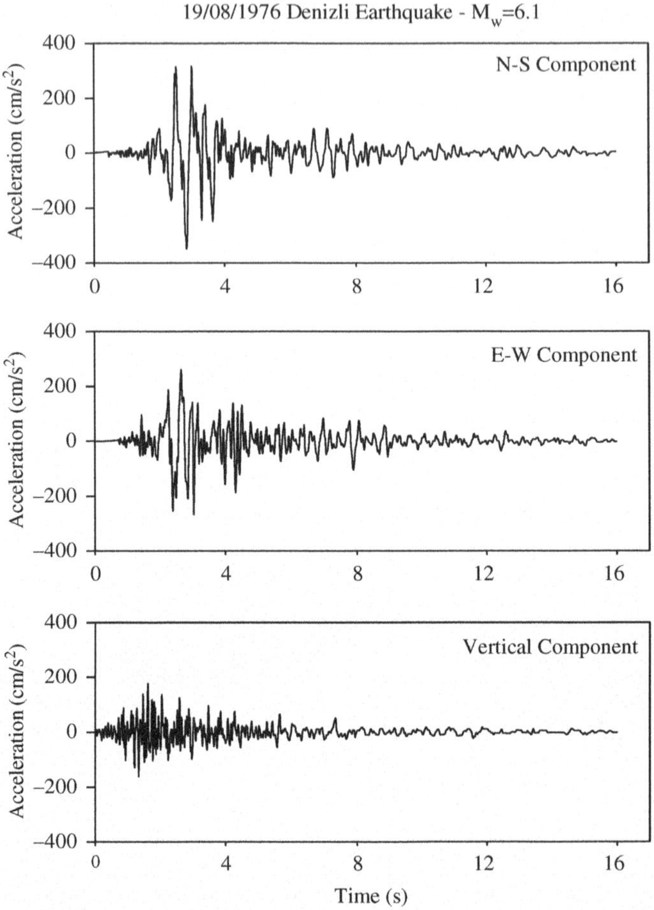

Fig. 14.2 The 1976 Denizli record

shown in Fig. 14.3 [3]. By 1990 [1] the number had increased only marginally to 69, and the location of a few stations had been displaced. These differences that had occurred by 1990 are marked in Fig. 14.3.

Changes in strong motion sensor technology were reflected in the instrument inventory of the national network. While no major procurements had been made by 1996 [6] the breakdown of the instruments was listed as 73 SMA-1 s, 19 digital SIG-SA (Swiss produced) and one Kinemetrics SSA-1 sensors for a total of 93. The status as of [6] is shown in Fig. 14.4.

14.4 Post-1999 Status

The two major events that occurred in the Sea of Marmara region in Turkey during 1999 rank among the largest earthquakes to have occurred in the eastern Mediterranean Basin during the last 100 years. In terms of their visible effects (such

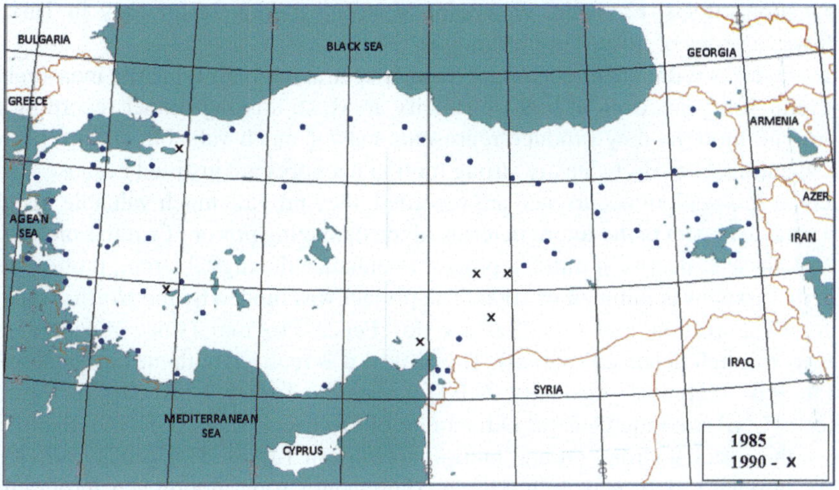

Fig. 14.3 Status in 1985 and 1990

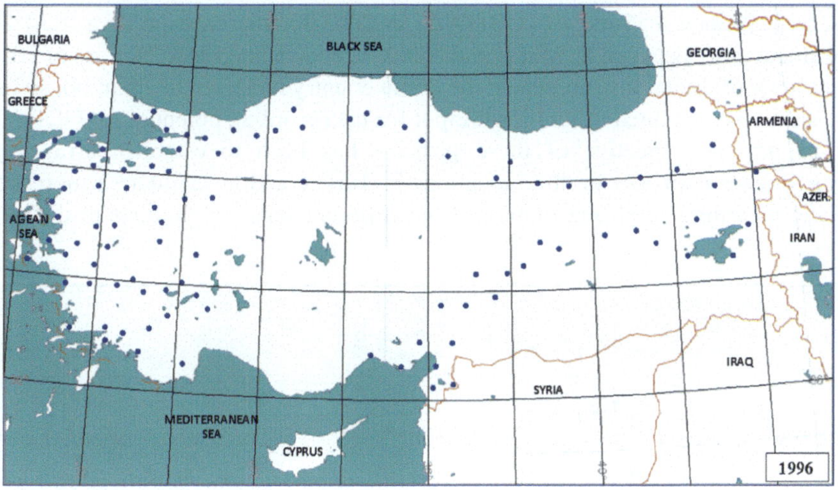

Fig. 14.4 Status in 1996

as ground deformation patterns, widespread damage and economic losses) these earthquakes established new thresholds. The first one with $\mathbf{M}_W 7.4$ struck on August 17, 1999, and the second $\mathbf{M}_W 7.2$ on November 12, 1999, and yet the number and quality of strong motion records recovered from the main shocks was very disappointing. With the few and functional instruments located inside institutional buildings there was interference in the ground acceleration traces caused by the buildings. The instruments were stand-alone devices, a combination of analog and digital types, and were triggered independently, making it difficult to study wave propagation effects. From the viewpoint of strong motion seismology in Turkey 1999 was a series of missed opportunities.

Earthquakes with magnitude larger than 7 occur at intervals typically measurable in several 100 years even in high seismicity areas, so missing proper recording of the ground motions they produce represents loss of much valuable and irreplaceable scientific data. High density strong motion networks are justified because when effects of this seldom occurrence are recorded, they provide much valuable insight as to what to expect in the future in terms of the damaging power of similar motions.

With these concerns in mind, a project to enhance the digital strong motion network in Turkey was initiated in 2000. The project was funded by the North Atlantic Treaty Organization under its Science for Peace Program (since renamed the Security through Science Program). The project was realized with the partnership of Middle East Technical University (METU), Ankara, Turkey, University of Nevada at Reno (UNR) and the General Directorate of Disaster Affairs (GDDA), the operator of the national strong ground motion network in Turkey. This project provided for the purchase of 20 new digital three-component strong-motion accelerographs. The NATO project permitted the emplacement of two arrays, code-named BYTNet and DATNet, established between Bursa-Yalova and Denizli-Aydın [5], respectively (Fig. 14.5). An identical program was funded by the Scientific and Technological Research Agency of Turkey (TUBITAK) in 2003 for the creation of a third 20-instrument array between K. Maraş and Antakya, code named MATNet. As of 2003, the number of digital instruments in the entire country had risen to about 60, so that this addition constituted a significant input for the recording potential for GDDA.

The primary objective of these projects has been to establish array type accelerometer networks in three locations in Turkey and to facilitate a more systematic recording of strong motions from different types of faulting in different

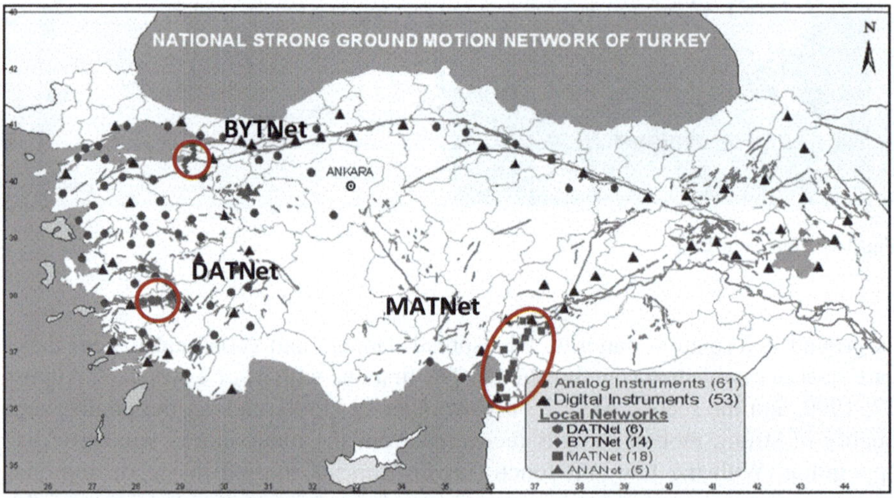

Fig. 14.5 BYTNet, DATNet and MATNet arrays

tectonic settings. Thus, the network geometries suggested by Iwan [7] were considered a relevant starting point. Although the number of instruments we had available did not allow that plan to be fully implemented, these instruments provided a good start. The site conditions are well-documented to make the results most useful for improvement of models of the amplitudes of ground motions in Turkey.

14.5 Current Status

The current number of triaxial accelerographs maintained by GDDA is in excess of 320, including earlier generation analog devices that have been converted to digital technology, instruments procured under different projects and recently acquired instruments. These instruments are shown as dots in Fig. 14.6, and are listed for future reference in the Appendix. Thanks to the recently completed TUBITAK-funded program for the "Compilation of Strong Motion Data from National Strong Motion Data According to International Standards" (www.daphne.deprem.gov.tr) all records that have been obtained to date (some 1,250 three-component traces) have been processed uniformly and their recording station site characteristics have been determined through geophysical surveys. A search engine facilitates downloading digital data conforming to prescribed requirements.

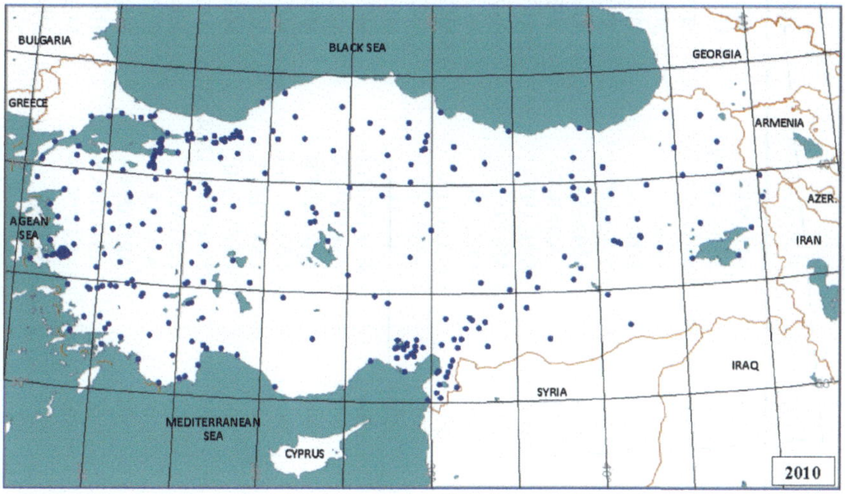

Fig. 14.6 Status in 2010

The latest figure is 327 instruments deployed as of 2010 as part of the national system. This number still needs to be increased to at least 1,000.

Acknowledgments I thank O. Ergünay for factual information concerning the funding sources at GDDA.

APPENDIX List of Stations

Station No	Station Abb.	Station Code	Station Province	Station (City/Town)	Latitude	Longitude	Date of Deployment
1	ADN	0118	ADANA	MERKEZ	37.021	35.191	2010
2	ADY	0201	ADIYAMAN	MERKEZ	37.761	38.267	2010
3	AFY	0301	AFYON	MERKEZ	38.776	30.534	2010
4	AGR	0401	AGRI	MERKEZ	39.720	43.016	1985
5	AJK	1202	BINGOL	MERKEZ	38.919	40.410	2010
6	AKDG	6602	YOZGAT	AKDAGMADENI	39.660	35.897	2010
7	AKS	4502	MANISA	AKHISAR	38.911	27.823	2010
8	AKSR	6801	AKSARAY	MERKEZ	38.353	34.027	2010
9	AKY	5402	SAKARYA	AKYAZI	40.670	30.623	2010
10	ALA	4503	MANISA	ALASEHIR	38.356	28.516	2010
11	ALC	1905	CORUM	ALACA	40.167	34.858	2010
12	AMS	0501	AMASYA	MERKEZ	40.644	35.804	1976
13	AMS	0502	AMASYA	MEPKEZ	40.668	35.853	2010
14	AMS	0503	AMASYA	MERKEZ	40.667	35.835	2010
15	ANA01	2601	ESKISEHIR	MERKEZ	39.814	30.528	2010
16	ANA02	2602	ESKISEHIR	MERKEZ	39.789	30.497	2010
17	ANA03	2603	ESKISEHIR	MERKEZ	39.880	30.453	2010
18	ANA04	2604	ESKISEHIR	MERKEZ	39.773	30.510	2010
19	ANA05	2605	ESKISEHIR	MERKEZ	39.723	30.533	2010
20	ANA05	2606	ESKISEHIR	MERKEZ	39.749	30.496	2010
21	AND	4604	KAHRAMANMARAS	ANDIRIN	37.570	36.357	1996
22	ANT	0701	ANTALYA	MERKEZ	36.894	30.667	2010
23	AOK	0106	ADANA	CEYHAN	36.908	35.567	2010
24	ARD	7501	ARDAHAN	MERKEZ	41.106	42.705	1985
25	ARD	7502	ARDAHAN	MERKEZ	41.112	42.701	2010
26	ART	0801	ARTVIN	MERKEZ	41.177	41.838	2010
27	AYD	0901	AYDIN	MERKEZ	37.837	27.838	2010
28	AYD	0902	AYDIN	MERKEZ	37.845	27.800	1985

APPENDIX (continued)

Station No	Station Abb.	Station Code	Station Province	Station (City/Town)	Latitude	Longitude	Date of Deployment
29	AYV	1005	BALIKESIR	AYVALIK	39.311	26.686	1976
30	BALA	0603	ANKARA	BALA	39.558	33.119	2010
31	BART	7401	BARTIN	MERKEZ	41.633	32.333	2010
32	BAYB	6901	BAYBURT	MERKEZ	40.262	40.210	2010
33	BCK	1503	BURDUR	BUCAK	37.460	30.588	1985
34	BCK	1504	BURDUR	BUCAK	37.458	30.588	2010
35	BDR	4802	MUGLA	BODRUM	37.033	27.440	1976
36	BEY	0601	ANKARA	BEYPAZARI	40.161	31.917	1976
37	BGA	1703	CANAKKALE	BIGA	40.232	27.263	2010
38	BGC	1008	BALIKESIR	BIGADIC	39.398	28.127	1976
39	BKL	0107	ADANA	CEYHAN	37.033	35.633	2010
40	BLC	1101	BILECIK	MERKEZ	40.141	29.977	2010
41	BLC	3510	IZMIR	BALCOVA	38.409	27.043	2010
42	BLD	2003	DENIZLI	BULDAN	38.045	28.834	1996
43	BLK	1001	BALIKESIR	MERKEZ	39.650	27.857	2010
44	BLK	1002	BALIKESIR	MERKEZ	39.660	27.860	2010
45	BLK	1004	BALIKESIR	MERKEZ	39.660	27.860	2010
46	BLKS	1003	BALIKESIR	MERKEZ	39.655	27.862	2010
47	BND	1006	BALIKESIR	BANDIRMA	40.332	27.997	2010
48	BND	1007	BALIKESIR	BANDIRMA	40.341	27.942	2010
49	BNG	1201	BINGOL	MERKEZ	38.897	40.503	1976
50	BNG4	1205	BINGOL	MERKEZ	38.897	40.503	2010
51	BOL	1401	BOLU	MERKEZ	40.746	31.607	1976
52	BOS	3515	IZMIR	BOSTANLI	38.465	27.094	2010
53	BOZ	1102	BILECIK	BOZOYUK	39.904	30.053	2010
54	BRD-1	1501	BURDUR	MERKEZ	37.722	30.294	1996
55	BRD-2	1502	BURDUR	MERKEZ	37.704	30.221	2010
56	BRN	3501	IZMIR	BORNOVA	38.459	27.167	2010

APPENDIX (continued)

Station No	Station Abb.	Station Code	Station Province	Station (City/Town)	Latitude	Longitude	Date of Deployment
57	BRN	3502	IZMIR	BORNOVA	38.455	27.227	2010
58	BRN	3511	IZMIR	BORNOVA	38.421	27.256	2010
59	BRS	1602	BURSA	MERKEZ	40.194	29.051	1976
60	BRS	1604	BURSA	MERKEZ	40.183	29.127	2010
61	BUC	3512	IZMIR	BUCA	38.401	27.152	2010
62	BYN	3513	IZMIR	MERKEZ	38.458	27.167	2010
63	BYR	3514	IZMIR	BAYRAKLI	38.476	27.158	2010
64	BYT01	1603	BURSA	MERKEZ	40.182	29.130	2010
65	BYT02	1601	BURSA	MERKEZ	40.226	29.075	2010
66	BYT03	1605	BURSA	DEMIRTAS	40.273	29.096	2010
67	BYT04	1606	BURSA	GEMLIK	40.363	29.122	2010
68	BYT05	1607	BURSA	GEMLIK	40.394	29.098	2010
69	BYT06	1608	BURSA	GEMLIK	40.410	29.179	2010
70	BYT07	1609	BURSA	GEMLIK	40.425	29.167	2010
71	BYT08	1615	BURSA	ORHANGAZI	40.422	29.291	2010
72	BYT09	1616	BURSA	ORHANGAZI	40.450	29.259	2010
73	BYT10	1617	BURSA	ORHANGAZI	40.494	29.299	2010
74	BYT11	7701	YALOVA	MERKEZ	40.564	29.306	2010
75	BYT12	7702	YALOVA	MERKEZ	40.590	29.267	2010
76	BYT13	7703	YALOVA	MERKEZ	40.651	29.279	2010
77	BYT14	7704	YALOVA	MERKEZ	40.658	29.247	2010
78	CAM	2004	DENIZLI	CAMELI	37.074	29.346	2010
79	CEK	3403	ISTANBUL	K.CEKMECE	41.026	28.759	2010
80	CER	1801	CANKIRI	CERKES	40.815	32.883	1976
81	CKK	0101	ADANA	MERKEZ	37.044	35.226	2010
82	CKR	6603	YOZGAT	CEKEREK	40.067	35.500	2010
83	CKV	0102	ADANA	MERKEZ	36.996	35.369	2010
84	CLK	1203	BINGOL	MERKEZ	38.858	40.577	2010

APPENDIX (continued)

Station No	Station Abb.	Station Code	Station Province	Station (City/Town)	Latitude	Longitude	Date of Deployment
85	CMD	3522	IZMIR	BORNOVA	38.436	27.199	2010
86	CMR	5101	NIGDE	CAMARDI	37.832	34.987	2010
87	CNK	1701	CANAKKALE	MERKEZ	40.141	26.399	1985
88	CNK	1702	CANAKKALE	MERKEZ	40.160	26.410	2010
89	CNKR	1802	CANKIRI	MERKEZ	40.608	33.610	2010
90	COR	1904	CORUM	MERKEZ	40.546	34.937	2010
91	COT	0112	ADANA	YUREGIR	36.872	35.475	2010
92	CRD	2005	DINIZLI	CARDAK	37.824	29.668	1996
93	CRD	2006	DINIZLI	CARDAK	37.825	29.668	2010
94	CYH	0104	ADANA	CEYHAN	37.024	35.809	1990
95	CYH	0105	ADANA	CEYHAN	37.027	35.816	2010
96	DAD	0605	ANKARA	MERKEZ	39.908	32.753	2010
97	DAT01	2007	DENIZLI	SARAYKOY	37.933	28.923	2010
98	DAT02	0903	AYDIN	BUHARKENT	37.974	28.746	2010
99	DAT03	0907	AYDIN	KUYUCAK	37.912	28.465	2010
100	DAT04	0908	AYDIN	NAZILLI	37.913	28.343	2010
101	DAT05	0909	AYDIN	SULTANHISAR	37.884	28.151	2010
102	DAT06	0904	AYDIN	KOSK	37.857	28.050	2010
103	DBY	0402	AGRI	DOGUBEYAZIT	39.549	44.091	2010
104	DCE	0303	AFYON	DINAR	38.075	30.161	1996
105	DDH	0304	AFYON	DINAR	38.067	30.171	1996
106	DEPR1	0606	ANKARA	BALA	39.288	33.096	2010
107	DEPR10	0615	ANKARA	BALA	39.556	33.120	2010
108	DEPR11	0616	ANKARA	BALA	39.315	33.206	2010
109	DEPR2	0607	ANKARA	BALA	39.463	33.217	2010
110	DEPR3	0608	ANKARA	BALA	39.556	33.120	2010
111	DEPR4	0609	ANKARA	BALA	39.315	33.206	2010
112	DEPR5	0610	ANKARA	BALA	39.453	33.719	2010

APPENDIX (continued)

Station No	Station Abb.	Station Code	Station Province	Station (City/Town)	Latitude	Longitude	Date of Deployment
113	DEPR6	0611	ANKARA	BALA	39.549	33.125	2010
114	DEPR7	0612	ANKARA	BALA	39.549	33.125	2010
115	DEPR8	0613	ANKARA	BALA	39.288	33.097	2010
116	DEPR9	0614	ANKARA	BALA	39.464	33.217	2010
117	DHS	4201	KONYA	DOGANHISAR	38.141	31.685	2010
118	DIN	0302	AFYON	DINAR	38.060	30.154	1985
119	DJK	0305	AFYON	DINAR	38.069	30.160	1996
120	DKH	0306	AFYON	DINAR	38.053	30.139	1996
121	DKL	3503	IZMIR	DIKILI	39.074	26.888	1996
122	DLC	7102	KIRIKKALE	DELICE	39.945	34.031	2010
123	DMR	4504	MANISA	DEMIRCI	39.035	28.648	1976
124	DNET1	8103	DUZCE	MERKEZ	40.786	31.282	2010
125	DNET2	8104	DUZCE	MERKEZ	40.862	31.178	2010
126	DNET3	8105	DUZCE	MERKEZ	40.903	31.152	2010
127	DNET4	8106	DUZCE	MERKEZ	40.812	31.116	2010
128	DNET5	8107	DUZCE	MERKEZ	40.840	31.142	2010
129	DNET6	8102	DUZCE	MERKEZ	40.834	31.164	2010
130	DNET7	8108	DUZCE	MERKEZ	40.860	31.230	2010
131	DNZ	2001	DENIZLI	MERKEZ	37.762	29.092	2010
132	DNZ	2002	DENIZLI	MERKEZ	37.812	29.111	1976
133	DRSB	1009	BALIKESIR	DURSUNBEY	39.578	28.632	2010
134	DSH	4403	MALATYA	DOGANSEHIR	38.096	37.887	1990
135	DSI	0307	AFYON	DINAR	38.076	30.178	1996
136	DUR	1010	BALIKESIR	DURSUNBEY	39.604	28.627	1985
137	DYB	2101	DIYARBAKIR	MERKEZ	37.931	40.203	2010
138	DZC	8101	DUZCE	MERKEZ	40.844	31.149	1985
139	EDN	1011	BALIKESIR	EDINCIK	40.336	27.861	2010
140	EDN	1012	BALIKESIR	EDINCIK	40.346	27.862	1976

APPENDIX (continued)

Station No	Station Abb.	Station Code	Station Province	Station (City/Town)	Latitude	Longitude	Date of Deployment
141	EDR	1013	BALIKESIR	EDREMIT	39.590	27.019	1985
142	ELB	4605	KAHRAMANMARAS	ELBISTAN	38.204	37.198	1990
143	ELM	0702	ANTALYA	ELMALI	36.737	29.921	2010
144	ELZ	2301	ELAZIG	MERKEZ	38.670	39.193	1985
145	EMT	4303	KUTAHYA	EMET	39.335	29.250	2010
146	ERB	6003	TOKAT	ERBAA	40.698	36.596	2010
147	ERC	2401	ERZINCAN	MERKEZ	39.742	39.512	1976
148	ERC	2402	ERZINCAN	MERKEZ	39.752	39.487	2010
149	ERG	5903	TEKIRDAG	M.EREGLISI	40.973	27.950	2010
150	ERZ	2501	ERZURUM	MERKEZ	39.903	41.262	2010
151	ERZ	2502	ERZURUM	MERKEZ	39.906	41.256	1976
152	EZN	1704	CANAKKALE	EZINE	39.774	26.346	1976
153	FNK	0703	ANTALYA	FINIKE	36.302	30.146	1985
154	FOC	3504	IZMIR	FOCA	38.662	26.759	2010
155	FOC	3505	IZMIR	FOCA	38.640	26.770	1985
156	FTH	4803	MUGLA	FETHIYE	36.626	29.124	1976
157	GBZ	4106	KOCAELI	GEBZE	40.786	29.450	1996
158	GDZ	4304	KUTAHYA	GEDIZ	38.995	29.400	1996
159	GEC	0113	ADANA	YUREGIR	36.956	35.627	2010
160	GEM	5804	SIVAS	GEMEREK	39.185	36.008	2010
161	GL-1	1705	CANAKKALE	GELIBOLE	40.423	26.667	1996
162	GNN	1014	BALIKESIR	GONEN	40.114	27.642	1985
163	GOL	0202	ADIYAMAN	GOLBASI	37.787	37.653	2010
164	GOL	0203	ADIYAMAN	GOLBASI	37.781	37.641	1990
165	GOR	4505	MANISA	GORDES	38.940	28.283	1996
166	GRD	1402	BOLU	GEREDE	40.792	32.206	1996
167	GUMU	2901	GUMUSHANE	MERKEZ	40.449	39 497	2010
168	GYN	1403	BOLU	GOYNUK	40 398	30 790	2010

APPENDIX (continued)

Station No	Station Abb.	Station Code	Station Province	Station (City/Town)	Latitude	Longitude	Date of Deployment
169	GYN	1404	BOLU	GOYNUK	40.397	30.783	2010
170	GYNC	0505	AMASYA	GOYNUCEK	40.393	35.522	2010
171	GZL	3506	IZMIR	GUZELYALI	38.394	27.082	2010
172	GZL	3516	IZMIR	GUZELBAHCE	38.371	26.891	2010
173	GZP	0704	ANTALYA	GAZIPASA	36.235	32.316	2010
174	GZT	2703	GAZIANTEP	MERKEZ	37 058	37 350	2010
175	HAY	0602	ANKARA	HAYMANA	39.440	32.506	2010
176	HBK	0114	ADANA	YUREGIR	37.170	35.686	2010
177	HRS	2503	ERZURUM	HORASAN	40.042	42.174	1985
178	HRS	2504	ERZURUM	HORASAN	40.043	42.173	2010
179	HTY	3102	HATAY	MERKEZ	36.213	36.159	1976
180	IGD	7601	IGDIR	MERKEZ	39.926	44.054	1985
181	ILI	3507	IZMIR	ILICA	38.310	26.310	1996
182	ING	1610	BURSA	INEGOL	40.067	29.509	2010
183	INO	2607	ESKISEHIR	INONU	39.817	30.146	2010
184	IST	3401	ISTANBUL	MERKEZ	41.058	29.010	1996
185	IST	3402	ISTANBUL	MERKEZ	41.080	29.090	1976
186	IZN	1611	BURSA	IZNIK	40.430	29.719	2010
187	IZN	1612	BURSA	IZNIK	40.442	29.717	1990
188	IZT	4101	KOCAELI	MERKEZ	40.767	29.917	1985
189	KAG	3602	KARS	KAGIZMAN	40.144	43.121	2010
190	KAS	0705	ANTALYA	KAS	36.195	29.647	2010
191	KAST	3702	KASTAMONU	MERKEZ	41.416	33.797	2010
192	KAYS	3801	KAYSERI	MERKEZ	38.690	35.500	2010
193	KBK	7801	KARABUK	MERKEZ	41.205	32.624	2010
194	KELK	2902	GUMUSHANE	KELKIT	40.124	39.436	2010
195	KEM	0706	ANTALYA	KEMER	36.603	30.560	2010
196	KHM	0103	ADANA	MERKEZ	37.058	35.367	2010

APPENDIX (continued)

Station No	Station Abb.	Station Code	Station Province	Station (City/Town)	Latitude	Longitude	Date of Deployment
197	KIG	1207	BINGOL	KIGI	39.340	40.280	1985
198	KIL	0115	ADANA	YUREGIR	37.081	35.455	2010
199	KIR	4001	KIRSEHIR	MERKEZ	39.160	34.162	2010
200	KIRK	7101	KIRIKKALE	MERKEZ	39.850	39.850	2010
201	KLS	1613	BURSA	KELES	39.915	29.232	2010
202	KMR	4603	KAHRAMANMARAS	MERKEZ	37.580	36.931	1976
203	KNK	3508	IZMIR	KINIK	39.088	27.375	2010
204	KOC01	4102	KOCAELI	MERKEZ	40.785	30.026	2010
205	KOC02	4103	KOCAELI	MERKEZ	40.786	30.025	2010
206	KOC03	4104	KOCAELI	MERKEZ	40.680	29.970	2010
207	KOC04	4105	KOCAELI	MERKEZ	40.674	29.969	2010
208	KOC05	4108	KOCAELI	MERKEZ	40.760	29.932	2010
209	KOC06	4109	KOCAELI	MERKEZ	40.760	29.933	1976
210	KON	3518	IZMIR	KONAK	38.431	27.144	2010
211	KONY	4202	KONYA	MERKEZ	37.904	32.516	2010
212	KORK	0707	ANTALYA	KORKUTELI	37.001	30.350	2010
213	KOY	4804	MUGLA	KOYCEGIZ	36.970	28.687	1985
214	KRB	1706	CANAKKALE	KARABIGA	40.404	27.306	1996
215	KRG	1901	CORUM	KARGI	41.139	34.484	2010
216	KRL	1206	BINGOL	KARLIOVA	39.293	41.009	2010
217	KRMN	7001	KARAMAN	MERKEZ	37.172	33.226	2010
218	KRS	3601	KARS	MERKEZ	40.603	43.082	1976
219	KRT	0110	ADANA	KARATAS	36.568	35.390	1996
220	KSK	3519	IZMIR	KARSIYAKA	38.453	27.111	2010
221	KUM	0708	ANTALYA	KUMLUCA	36.337	30.292	2010
222	KUS	0905	AYDIN	KUSADASI	37.860	27.265	1976
223	KUS	0906	AYDIN	KUSADASI	37.862	27.260	2010
224	KUT	4301	KUTAHYA	MERKEZ	39.428	29.992	1996

APPENDIX (continued)

Station No	Station Abb.	Station Code	Station Province	Station (City/Town)	Latitude	Longitude	Date of Deployment
225	KUT	4302	KUTAHYA	MERKEZ	39.419	29.997	2010
226	KYM	2608	ESKISEHIR	KAYMAZ	39.520	31.183	2010
227	KYN	3517	IZMIR	BUCA	38.376	27.194	2010
228	KZK	0116	ADANA	YUREGIR	37.105	35.464	2010
229	LAD	5502	SAMSUN	LADIK	40.910	35.903	2010
230	LARA	0709	ANTALYA	LARA	36.879	30.722	2010
231	MAD	2302	ELAZIG	MADEN	38.392	39.675	2010
232	MAR	4805	MUGLA	MARMARIS	36.839	28.245	1985
233	MAT01	3110	HATAY	SAMANDAG	36.082	35.950	1985
234	MAT02	3103	HATAY	ALTINOZU	36.116	36.247	2010
235	MAT03	3101	HATAY	MERKEZ	36.214	36.160	2010
236	MAT04	3111	HATAY	SERINYOL	36.373	36.220	2010
237	MAT05	3108	HATAY	KIRIKHAN	36.498	36.366	2010
238	MAT06	3106	HATAY	ISKENDERUN	36.592	36.158	1996
239	MAT06	3107	HATAY	ISKENDERUN	36.582	36.185	2010
240	MAT07	3109	HATAY	KIRIKHAN	36.584	36.414	1985
241	MAT08	3104	HATAY	HASSA	36.693	36.489	2010
242	MAT09	3105	HATAY	HASSA	36.803	36.511	2010
243	MAT10	4607	KAHRAMANMARAS	PAZARCIK	37.485	37.298	2010
244	MAT11	4601	KAHRAMANMARAS	MERKEZ	37.539	36.982	2010
245	MAT12	4602	KAHRAMANMARAS	MERKEZ	37.575	36.915	2010
246	MAT13	4608	KAHRAMANMARAS	TURKOGLU	37.375	36.838	2010
247	MAT14	4606	KAHRAMANMARAS	NARLI	37.387	37.138	2010
248	MAT15	2701	GAZIANTEP	ISLAHIYE	37.025	36.636	1976
249	MAT16	2702	GAZIANTEP	NURDAGI	37.184	36.733	2010
250	MAT17	8001	OSMANIYE	MERKEZ	37.084	36.269	2010
251	MAT18	8002	OSMANIYE	BAHCE	37.192	36.562	2010
252	MDR	1406	BOLU	MUDURNU	40.468	31.210	2010

APPENDIX (continued)

Station No	Station Abb.	Station Code	Station Province	Station (City/Town)	Latitude	Longitude	Date of Deployment
253	MEN	1405	BOLU	MENGEN	40.938	32.076	2010
254	MKP	1614	BURSA	M.KEMALPASA	40.035	28.394	1996
255	MLS	4806	MUGLA	MILAS	37.303	27.781	2010
256	MLT	4401	MALATYA	MERKEZ	38.350	38.340	1976
257	MLT	4402	MALATYA	MERKEZ	38.350	38.340	2010
258	MLZ	4902	MUS	MALAZGIRT	39.144	42.531	1985
259	MNG	0710	ANTALYA	MANAVGAT	36.787	31.432	2010
260	MNS	4501	MANISA	MERKEZ	38.613	27.381	1996
261	MNV	3520	IZMIR	BORNOVA	38.478	27.211	2010
262	MRD	4701	MARDIN	MERKEZ	37.326	40.724	2010
263	MRS	3301	MERSIN	MERKEZ	36.781	34.603	1985
264	MRZ	0504	AMASYA	MERZIFON	40.880	35.459	1985
265	MUG	4801	MUGLA	MERKEZ	37.214	28.356	1976
266	MUR	6502	VAN	MURADIYE	38.990	43.763	1985
267	MUR	6503	VAN	MURADIYE	38.990	43.768	2010
268	MUS	4901	MUS	MERKEZ	38.761	41.504	1985
269	MVS	3521	IZMIR	KARSIYAKA	38.468	27.077	1976
270	NAC	0108	ADANA	CEYHAN	36.876	35.617	2010
271	NIG	5102	NIGDE	MERKEZ	37.968	34.673	2010
272	ODM	3509	IZMIR	ODEMIS	38.216	27.965	1996
273	OLT	2505	ERZURUM	OLTU	40.550	41.995	2010
274	ORD	5201	ORDU	MERKEZ	40.975	37.917	2010
275	OSM	1902	CORUM	OSMANCIK	40.971	34.799	2010
276	OSM	1903	CORUM	OSMANCIK	40.976	34.800	1976
277	PAL	2303	ELAZIG	PALU	38.696	39.932	2010
278	POT	4404	MALATYA	POTURGE	38.117	38.524	2010
279	REF	2403	ERZINCAN	REFAHIYE	39.899	38.769	1985
280	REF	2404	ERZINCAN	REFAHIYE	39.905	38.771	2010

APPENDIX (continued)

Station No	Station Abb.	Station Code	Station Province	Station (City/Town)	Latitude	Longitude	Date of Deployment
281	RES	6004	TOKAT	RESADIYE	40.394	37.329	1985
282	SAL	4506	MANISA	SALIHLI	38.483	28.123	1985
283	SAN	6301	SANLIURFA	MERKEZ	37.168	38.801	2010
284	SDL	0308	AFYON	SANDIKLI	38.437	30.252	2010
285	SERK	0711	ANTALYA	SERIK	36.918	31.088	2010
286	SGK	0109	ADANA	CEYHAN	37.178	35.689	2010
287	SHG	0111	ADANA	SEYHAN	37.166	35.226	2010
288	SKOC	0604	ANKARA	S.KOCHISAR	38.962	33.524	2010
289	SKR	5401	SAKARYA	MERKEZ	40.737	30.380	1985
290	SLH	1208	BINGOL	SOLHAN	38.968	41.054	1985
291	SLH	1209	BINGOL	SOLHAN	38.966	41.050	2010
292	SLT	0309	AFYON	SULTANDAGI	38.526	31.238	2010
293	SLV	3404	ISTANBUL	SILIVRI	41.073	28.256	2010
294	SMN	5501	SAMSUN	MERKEZ	41.344	36.256	1976
295	SMV	4305	KUTAHYA	SIMAV	39.093	28.978	2010
296	SNG	1015	BALIKESIR	SINDIRGI	39.255	28.164	2010
297	SNK	3201	ISPARTA	SENIRKENT	38.105	30.558	2010
298	SRK	5904	TEKIRDAG	SARKOY	40.615	27.123	1996
299	SSH	5801	SIVAS	SUSEHRI	40.169	38.106	1985
300	STG	2609	ESKISEHIR	SEYITGAZI	39.446	30.697	2010
301	SVS	5803	SIVAS	MERKEZ	39.750	37.160	2010
302	TAT	1301	BITLIS	TATVAN	38.503	42.281	1985
303	TCK	1204	BINGOL	MERKEZ	38.885	40.504	2010
304	TER	2405	ERZINCAN	TERCAN	39.777	40.391	1985
305	TER	2406	ERZINCAN	TERCAN	39.777	40.384	2010
306	TFN	1505	BURDUR	TEFENNI	37.316	29.779	2010
307	TKR	5901	TEKIRDAG	MERKEZ	40.958	27.497	1985
308	TKR	5902	TEKIRDAG	MERKEZ	40.979	27.515	2010

APPENDIX (continued)

Station No	Station Abb.	Station Code	Station Province	Station (City/Town)	Latitude	Longitude	Date of Deployment
309	TKT	6001	TOKAT	MERKEZ	40.329	36.555	2010
310	TKT	6002	TOKAT	MERKEZ	40.300	36.570	1976
311	TOS	3701	KASTAMONU	TOSYA	41.013	34.037	1985
312	TPT	4107	KOCAELI	TEPETARLA	40.720	30.079	2010
313	TRBZ	6101	TRABZON	MERKEZ	40.985	39.670	2010
314	URL	3523	IZMIR	URLA	38.328	26.771	2010
315	USK	6401	USAK	MERKEZ	38.671	29.404	1996
316	VAN	6501	VAN	MERKEZ	38.503	43.402	1976
317	VEZ	5503	SAMSUN	VEZIRKOPRU	41.138	35.466	2010
318	YER	4808	MUGLA	YERKESIK	37.134	28.284	1985
319	YLV	7705	YALOVA	MERKEZ	40.659	29.280	2010
320	YMN	3524	IZMIR	CIGLI	38.497	27.107	2010
321	YNC	1707	CANAKKALE	YENICE	39.929	27.259	1996
322	YOZG	6601	YOZGAT	MERKEZ	39.816	34.810	2010
323	YRD	0117	ADANA	YUREGIR	36.838	35.238	2010
324	YSL	3525	IZMIR	KONAK	38.372	27.108	2010
325	YTG	4807	MUGLA	YATAGAN	37.340	28.137	2010
326	ZAR	5802	SIVAS	ZARA	39.893	37.748	1996
327	ZONG	6701	ZONGULDAK	MERKEZ	41.451	31.779	2010

References

1 Bayülke N, İnan E, Güler HH (1990) "Strong Motion Accelerograph Records of Turkey: volume II," Turkish Ministry of Public Works and Resettlement, General Directorate of Disaster Affairs, Division of Earthquake Research, July, 255 pp.
2 Brady AG (2009) Strong-motion accelerographs: early history. Earthq Eng Struct Dyn 38: 1121–1134
3 Ergünay O, İnan E, Bayülke N, Koşan U (1985): "Strong Motion Accelerograph Records of Turkey," Turkish Ministry of Public Works and Housing, General Directorate of Disaster Affairs, Division of Earthquake Research, July, 62 pp.
4 Gülkan P (2000): "Building Code Enforcement Prospects: failure of Public Policy," Chapter 15 of 1999 Kocaeli, Turkey, Earthquake Reconnaissance Report, Supplement A to Volume 16, Earthquake Spectra, Dec, pp 351–367.
5 Gülkan P, Çeken U, Çolakoğlu Z, Uğraş T, Kuru T, Apak A, Anderson JG, Sucuoğlu H, Çelebi M, Akkar DS, Yazgan U, Denizlioğlu AZ (2007) Enhancement of the national strong motion network in Turkey. Sesimol Res Lett 78(4):429–438
6 Inan E, Çolakoğlu Z, KoçN, Bayülke N, Coruh E (1996) Catalog of Earthquakes with Acceleration Records (1975–1996), Turkish Ministry of Public Works and Resettlement, General Directorate of Disaster Affairs, Division of Earthquake Research, July, 98 pp.
7 Iwan WD (1978) Strong-Motion Earthquake Arrays, Proceedings of the International Workshop on Strong-Motion Arrays, May 2–5, 1978, at Honolulu, Hawaii. Pasadena: California Institute of Technology.

Disclaimer Reference in this article to any commercial products by commercial name, trademark or its manufacturers does not constitute or imply their endorsement or recommendation.

Chapter 15
The Current State of Strong Motion Monitoring in Switzerland

J. Clinton, C. Cauzzi, D. Fäh, C. Michel, P. Zweifel, M. Olivieri, G. Cua, F. Haslinger, and D. Giardini

Abstract The next generation Swiss Strong Motion Network has recently been funded by the Swiss Government: in the next 8 years the Swiss Seismological Service expects to install 100 new 24-bit broadband freefield stations in predominantly urban locations across the country with realtime, continuous data transmission at high sampling rates. This infrastructure will compliment the existing 30 comparable realtime stations installed over the last 3 years, and replace the original ~70 strong triggered dial-up network installed in the early 1990s. The introduction of these new stations provides an opportunity to reassess how strong motion data is used in Switzerland, for routine network operations, emergency response and scientific purposes. The strong motion data will be acquired in parallel with the existing broadband network, and will be processed together with broadband data for earthquake early warning, triggering and locations, near real time ShakeMaps, and moment tensor inversion. Challenges arise on how to archive and provide this type of data to the scientific and engineering communities. Metadata maintenance needs to parallel efforts for the broadband network. Although permanent online archival of large volumes of data is rapidly becoming more affordable, this size of the new dataset dwarfs the existing broadband data currently being generated. We discuss the optimal strategies to permanently archive the continuous data, both within Switzerland, and via the existing European data infrastructures.

15.1 Introduction

When compared to seismically more active regions, damaging earthquakes are rare in Switzerland. Nevertheless, 10–15 earthquakes are felt (on average) each year by the population and damaging events are expected every 5–10 years [36]. Over the past 800 years, Switzerland has suffered at least 28 events with attributed moment magnitude $M_W \geq 5.5$, 12 of these resulted in severe damage to buildings, with macroseismic intensity values reaching or exceeding VIII [16]. The historical

J. Clinton (✉)
Swiss Seismological Service (SED – ETHZ), Sonneggstrasse 5, 8092 Zürich, Switzerland
e-mail: j.clinton@sed.ethz.ch

S. Akkar et al. (eds.), *Earthquake Data in Engineering Seismology*, Geotechnical, Geological, and Earthquake Engineering 14, DOI 10.1007/978-94-007-0152-6_15,

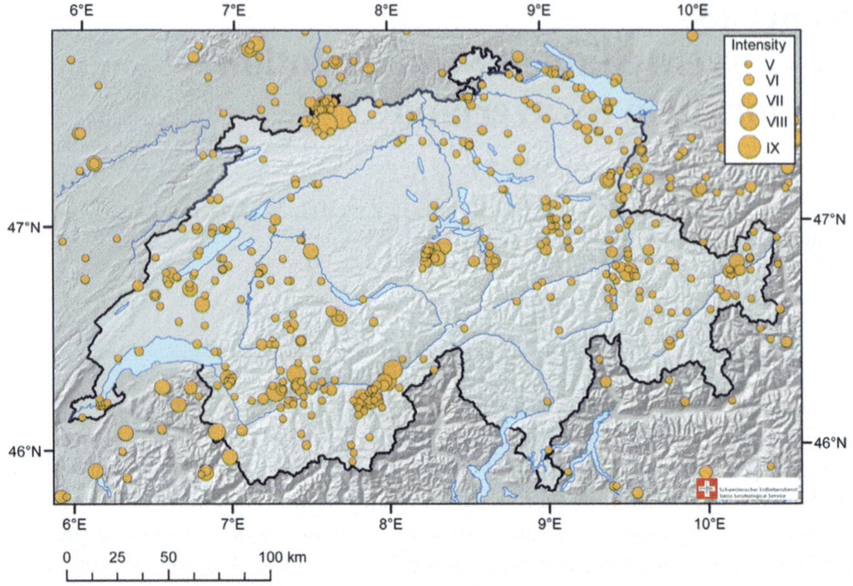

Fig. 15.1 Historical seismicity in Switzerland between 250 and 1974. Only events with attributed intensity $\geq$ V are shown. *Symbols* are proportional to macroseismic intensity. Note the high damaging historical earthquakes in the region of Basel (NW) and in the Wallis (SW)

seismicity distribution in Switzerland is depicted in Fig. 15.1. Seismic activity is highest in the Wallis (SW) and in the Basel region (NW). Central Switzerland, the Graubünden (SE), and the Rhine Valley of St. Gallen (NE) have also experienced significant shaking in the recorded history.

In spite of the moderate hazard level [36], a relatively high seismic risk exists in Switzerland [21], resulting from the (a) high population density, (b) high degree of industrialization, with critical infrastructures including dams, nuclear power and chemical plants, and (c) generally low level of preparedness due to the relatively long return periods of strong ground shaking.

Instrumental observations now complement the macroseismic observations and provide the basis for a homogeneous record of the seismicity in Switzerland since 1975. The instrumentally recorded earthquake catalogue in Fig. 15.2 shows close correlation with the macroseimic catalogue.

The Swiss Seismological Service (SED) at the ETH Zürich (ETH) is the Swiss federal agency responsible for monitoring the seismicity of Switzerland and surrounding areas, providing rapid notification of earthquakes to the local authorities and the public, and archiving and providing reliable data for seismological and earthquake engineering research studies. The SED manages both a high-gain broadband/short period seismometer network and a low-gain accelerograph network. The former is intended to monitor earthquake activity at magnitudes well below the human perception threshold [9], whereas the latter is principally aimed at engineering purposes and thus mainly records strong motion. Nevertheless, the

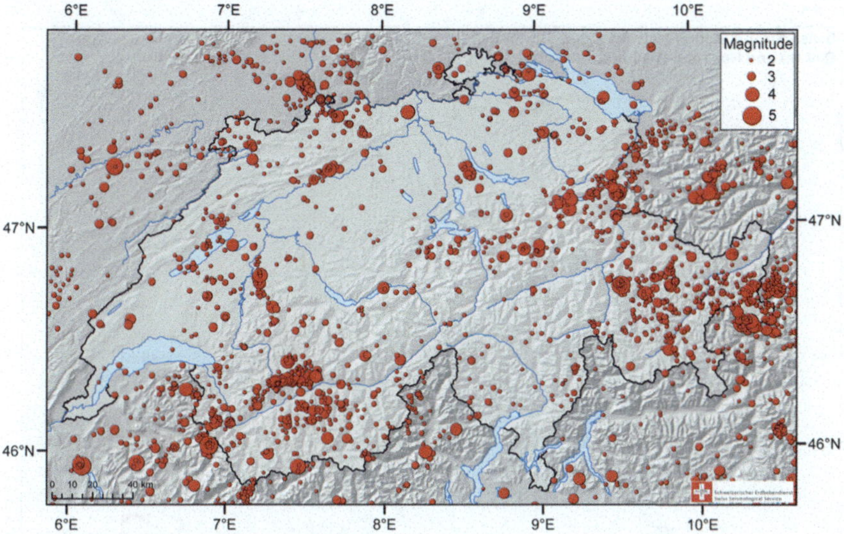

Fig. 15.2 Instrumental Seismicity (1977–2009) in Switzerland observed by the SED. Over 25,000 events were located by the SED in this period

progressive introduction of broadband low gain accelerometric sensors with very large dynamic range alongside high-dynamic range dataloggers is rapidly blurring the boundary between weak and strong motion instruments – leading to digital ground velocity records of engineering significance and acceleration records of seismic observation quality.

15.2 Seismic Monitoring in Switzerland

Instrumental monitoring of earthquake activity in Switzerland began at the onset of the twentieth century. Since then Switzerland has been at the forefront of seismic monitoring in Europe. By the 1970s, a telemetered high-gain short-period seismic network was installed, with data recorded on microfilm support. By the mid 1980s, the network consisted of a dense network of high gain sensors with radio communications digitized on a central computer system and automatically processed. In the early 1990s, the initial National Strong Motion Network was established, with dial-up triggered instrumentation.

The modern Swiss Seismic Network was created in the late 1990s, when broadband sensors alongside 24-bit dataloggers replaced the short period sensors, GPS timing was introduced, and communications were upgraded to a high bandwidth, secure internet system. This new network is known as the CHNet, and comprises the national broadband (SDSNet) and the national strong motion (SSMNet) networks, as well as some Special Networks (targeted local densifications based on temporary projects).

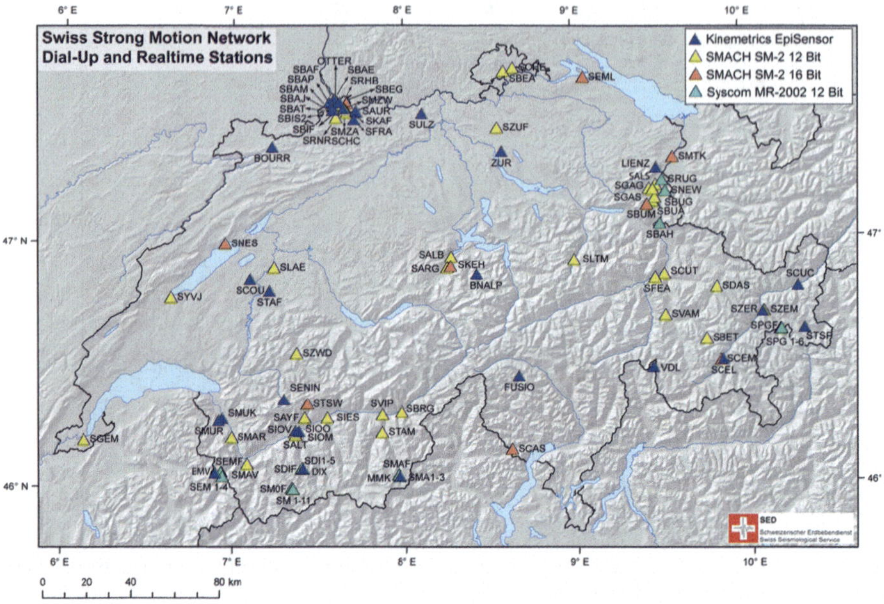

Fig. 15.3 Strong Motion Stations operated by the SED. Stations with EpiSensors are part of the continuous real-time network. The other sensors operate in dial-up mode and have limited dynamic range

15.2.1 Dial-Up Strong Motion Network

In the early 1990s, the Swiss National Strong Motion Network was installed, consisting of 59 free-field stations as well as of 5 dam-related arrays with a total of 34 strong motion instruments (see Fig. 15.3). This network, consisting of dial-up triggered 3-component sensors, was installed across Switzerland and would now be considered outdated due to the limited local storage, the narrow sensor frequency response and the digitiser dynamic range of only 12–16 bits [37]. The network goals were to (1) characterise strong ground motions at the different sites; (2) determine the attenuation of strong ground motion with distance; (3) investigate site effects due to different local geology; and (4) investigate the dynamic behavior of various key dams in Switzerland.

The dial-up network uses a mix of GeoSig SMACH SM-2 12/16 bit and SYSCOM MR-2002 12 bit sensors, with the trigger level varying according to local noise conditions. By 2000, all recording systems were upgraded with a time code receiver (DCF or GPS).

15.2.2 Broadband Seismic Network

From 1999 to 2002, the short-period network with radio based analogue telemetry was replaced by a fully digital, predominantly broadband, high dynamic

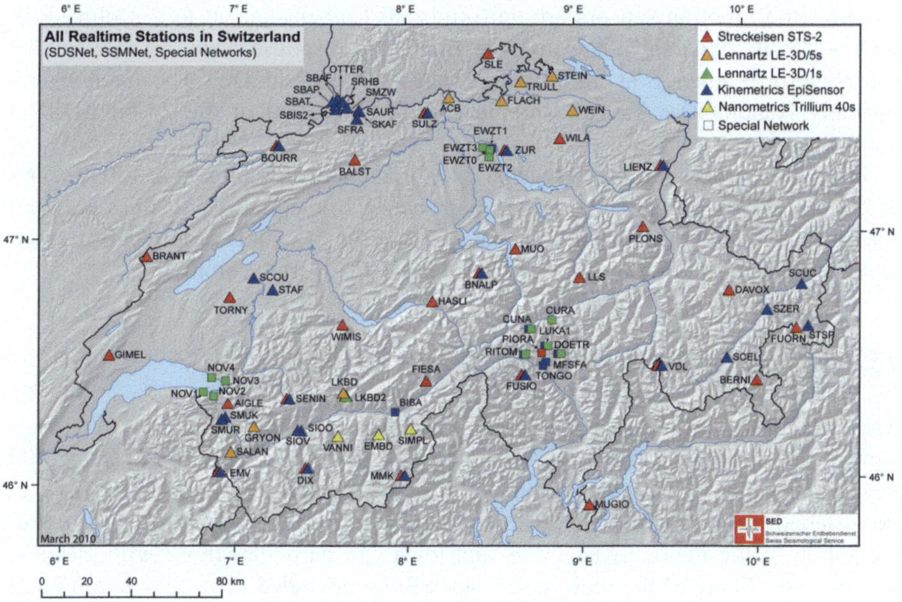

Fig. 15.4 All continuously monitored stations from the CHNet SED. The Swiss digital seismic network (SDSNet) mainly consists of Streckeisen STS-2 seismometers, complimented by short period velocity sensors (~35 stations). High dynamic range accelerometer stations from the SSMNet Swiss strong motion network comprise about 40 stations, some co-located with broadband sensors

range network. The new Swiss Digital Seismic Network (SDSNet) consists of 28 broadband Streckeisen STS-2 stations equally distributed across the country with inter-station spacing of about 30–50 km (Fig. 15.4). Data is digitized on 24-bit Nanometrics equipment (HRD-24, Trident and Taurus). Data is continuously telemetered via secure Government internet lines to Zürich in real-time for processing. Archives of both the continuous and event data at high sample rates (120 sps) provide the research community with state-of-the-art seismic data.

Currently, the Seismic Network also monitors some 30 sensors from outside Switzerland (Germany, Austria and Italy) to optimise earthquake detection and locations for the country. The system transmits data via SeedLink [23].

15.2.3 Modern Strong Motion Network

Since 2006, improvements to the SSMNet have followed the same structure as the SDSNet infrastructure – new strong motion stations are continuously monitored in real time with high dynamic range instruments at high sampling rates [1]. Currently, strong motion sensors are co-located with broadband sensors at 11 sites, and similar instrumentation is installed at ~30 free-field, mainly urban locations (Fig. 15.4). Note the densification of strong motion stations in the Basel Area, the Wallis and the Graubünden, where the Swiss earthquake hazard is highest.

The strong motion sensors are uniformly all EpiSensors, set to 2 g clip level, with ~155 dB dynamic range and flat frequency response from 200 Hz (above the Nyquist frequency at all sampling rates we use) to DC. The dataloggers, typically Nanometrics Taurus, are the current equivalent of the 24-bit SDSNet dataloggers.

Data from strong motion co-located with broadband are sampled at 120 sps, and at 250 sps at the standalone strong motion sites. Commercial ADSL lines are used for the standalone stations. Once the data reaches the SED, it is acquired, processed and archived in a similar manner to the broadband SDSNet stations.

15.2.4 Planned Strong Motion Network Upgrade

The renewal of the Swiss Strong Motion Network was funded in 2009 by a decision of the Swiss Federal Council. The project involves installation of some 100 new accelerometer stations over the next 8 years, with an initial 30 new stations to be installed in the first 4 years. The key goal of the network is to densify instrumentation in the major Swiss cities and towns, in particular those with elevated seismic risk. Many of the dial-up stations will be upgraded. Instrumentation quality and communications will also be state-of-the-art, very broadband accelerometers with 24 bit digitization and real time continuous communication of high sample rate data. The data will also be processed alongside the SDSNet data. Site selection will accommodate both scientific objectives and the level of cultural noise in order to maximise usage of the collected data.

As part of the project, a new housing solution has been designed (Fig. 15.5) for the standard new station. This is an improvement over existing urban strong motion station installations, and will minimize the effects of cultural and electrical noise. The vault consists of a concrete cylinder with a metallic cover, and hosts the sensor. A separate casing with the datalogger and communication systems can be placed either inside the vault, or several meters away from the station. The vault is anchored to the ground via steel bars in order to avoiding ground settlement, relative displacements and rotations [22]. Each station will have a detailed geophysical site assessment.

15.3 Strong Motion Data Archival, Dissemination and Storage Considerations

New strong motion stations at the SED are fully integrated into the Seismic Network. Data are collected at the SED in real-time (data latency is typically under 3 s) at high sampling rates, and passes through identical processing as broadband streams. This means if a strong motion station is considered quiet enough, it can be used for earthquake detection and automatic location. For manual event revisions, which are completed within tens of minutes following major events in Switzerland, all real-time strong motion streams are available, and often used, for both location

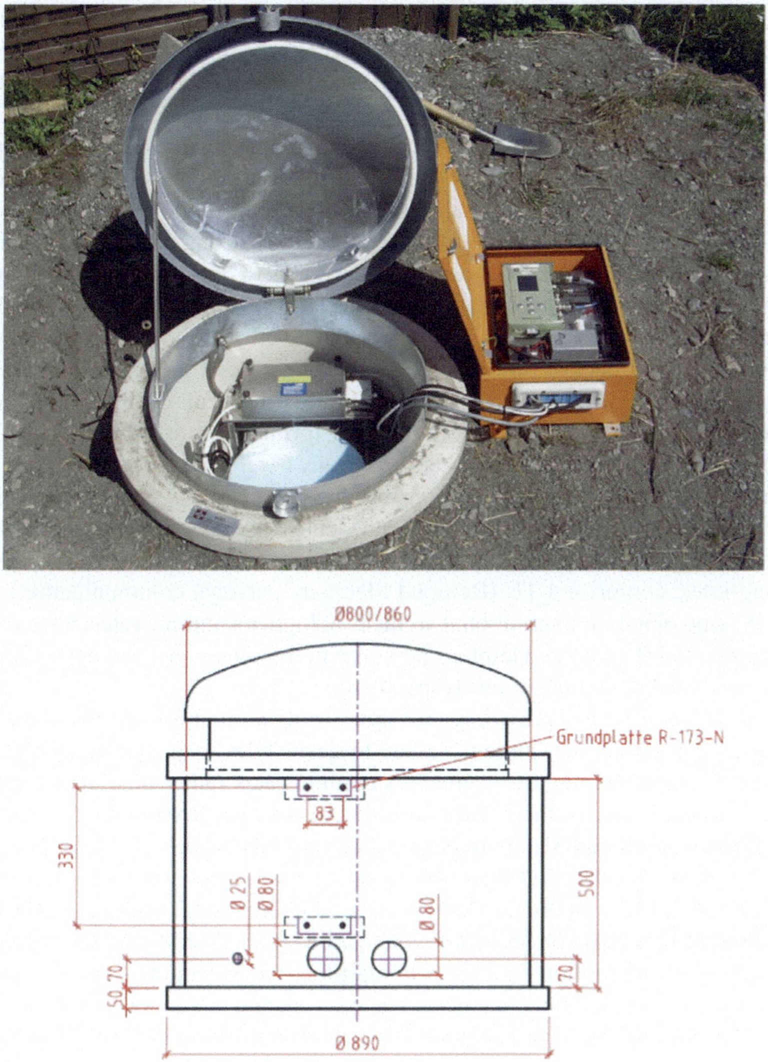

Fig. 15.5 Station housing concept for new SSMNet stations – photo from prototype installation at SSMNet station SCEL. The accelerometer is stored within the concrete cylinder. The datalogger and communication system (*orange* casing) can be hosted either in the same housing, or alternatively several meters away from the sensor

and magnitude estimates. Acquiring strong motion alongside our other datastreams also allows us to archive strong motion data identically to the broadband data, and to take advantage of the software created to optimise the quality and completeness of the waveform archives.

In nearly 20 years of operation so far, the SSMNet triggered dial-up network generated over 700 triggers that have been associated with earthquakes located by the SED Seismic Network. The total volume of data recorded in this archive is 16 MB. The SED has been producing waveform event files of ~100 s duration for all digital recordings of earthquakes since 1983, including short period, broadband, and strong motion data as available. The total volume of these over 25,000 event files is 48 GB. In contrast, a single continuous 250 sps channel recording requires approximately 30 Mb/day (~11 GB/year). Nevertheless, as data storage has become cheap, the SED now permanently archives all data collected by all stations within the CHNet, including the high sample rate continuous strong motion data. This has required a change in the approach the Network takes to data storage. The complete continuous waveform archives, dating back to late 1999 (the advent of broadband installations in the CHNet) are available on high availability rapid access disk that are maintained and backed up by professional IT group at the ETH. The archives are stored in both GSE2 format [24] and miniSEED format [33]. Currently, the complete archives of the CHNet in miniSEED format require 7 TB of disk space. The current growth of these archives is ~3 GB/day (~1.1 TB/year), of which 70% of the total is due to the ~40 strong motion stations.

For comparison, the entire ORFEUS Data Center (ODC) Archives, which archive broadband data from across Europe at until very recently at only lower 20 or 40 sps sampling rates, comprise 4 TB (Reinoud Sleeman, personal communication). GFZ and IRIS, who maintain international archives at higher sampling rates, have archive volumes of 30 TB (Andres Heinloo, personal communication) and 105 TB (Rick Benson, personal communication) respectively.

The SED has also been providing real-time strong motion waveforms to the ODC since mid-2009. These new streams, allied to a new ODC policy to begin archiving all seismic data at the highest available sampling rates rather than 20/40 sps, has lead to a dramatic rise in their daily archive growth rate, from under 3 GB/day to over 6 GB/day (Reinoud Sleeman, personal communication).

Beyond the ODC, the SED shares Swiss data with neighbouring seismic agencies (INGV, ZAMG, LED, BGR, GeoAzur Nice) and international agencies (IRIS/GFZ) over either an InterNaqs or SeedLink connection. Both ZAMG and the ODC have begun taking strong motion data feeds in realtime. We expect this sort of data sharing to grow as more and more agencies begin to operate continuous realtime strong motion stations, and become capable of processing this data in their own realtime seismic processing systems and event products (e.g. ShakeMaps).

In the past, the SSMNet triggered data has been made available on a dedicated web platform that was independent of the real-time continuous data streams. The continuous strong motion data was only available in event and continuous files on request. The SED now runs ArcLink [23], a software product from GFZ, which allows open access via scripting and web requests, to the entire continuous archives. The triggered data is added to this dataset.

The SED is an active contributor to the Accelerometic component of the European Earthquake Data Portal (http://www.seismicportal.eu/ – last accessed 1 April 2010). This project, involving a core group of 6 Accelerometric Networks as well as the European Mediterranean Seismological Centre (EMSC), has created a freely available distributed strong motion archive [31] that can be searched using

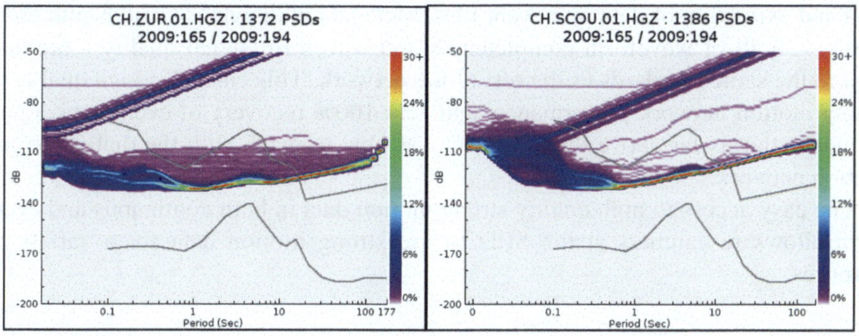

Fig. 15.6 Images from PQLX for 2 SSMNet stations at ZUR (collocated with STS2) and SCOU (standalone). The *grey lines* are the *high* and *low* noise models of Peterson [29]. The background signal at high frequencies is measuring the site noise. The microseismic peak is visible between 5 and 10 s. Between ~1 and 5 s, and beyond 10 s, the background signal typically is recording the sensor/datalogger noise

station, event and waveform parameterisations (e.g. PGA, PGV, Housner Intensity). As the SED now operates continuous stations, we were able to contribute all records from ML2.5 irrespective of the PGA, and thus our contribution does not have a sampling bias.

The continuous strong motion stations are now routinely processed via the health monitoring software tool PQLX [28]. This software, a standard tool for the broadband community, calculates power spectral densities from the continuous data illustrated as a probability density function. The software allows evaluation and temporal tracking of background station noise across the frequency spectrum. From this tool, it is clear the typical SED high quality continuous strong motion station stays below the Peterson [29] High Noise Model from high frequencies (~50 Hz) to just beyond the microseism (~15 s). The microseismic peaks are observed at all stations, as are occasional large teleseismic events at longer period. We observe high frequency cultural signals at f > 1 Hz, well above the sensor/datalogger noise at all sites except some of the best co-located sites in the SSMNet. Most stations record sensor noise rather than the background site noise only beyond the microseismic peak (T > 10 s), except for a short frequency window between 1 s and the emergence of the microseismic peak (~5 s) (Fig. 15.6). This suggests that strong motion data from the new SSMNet can be applicable for use in future noise study investigations.

15.4 Use of Strong Motion Data at the SED: Network Products and Research

The 40 strong-motion channels acquired continuously in real-time at the SED are processed identically to the broadband and short-period sensors also monitored by the Seismic Network. This allows use of the strong motion data for routine automatic network operations, manual locations and seamless archival of continuous data and extraction of data into event files. Metadata maintenance, and health monitoring – both of waveform completeness and waveform signal quality – are also kept to the same standards as the rest of the network. This ensures a high quality in strong motion network performance, and near 100% recovery of event data, a performance that is far above that the SED was able to reach with the dial-up strong motion network.

The easy access to high quality strong motion data in both continuous and event form allows researchers at the SED to use strong motion data for a variety of purposes.

15.4.1 Earthquake Detection

Most strong motion stations are located in urban sites and thus have high cultural noise, in particular in the 1–30 Hz frequency range suitable for local earthquake detection. Nonetheless, some stations are located at good sites on competent ground, and are included in the detection routine – this helps to locally improve the earthquake detection capacity of the network.

15.4.2 Earthquake Locations and Local Magnitudes

If a strong motion station contributes triggers to an event association, a P-pick will be searched for (using the Baer Picker [2]) and if found, the station will be included in the automatic location, though the magnitude will not be used. Once an automatic location is created, a single file is extracted for manual review containing all realtime waveforms monitored by the SED, including all the available strong motion stations. Manual locations, which are available within 10s of minutes following felt earthquakes in Switzerland (> ML2.5 inside the country), incorporate picks from strong motion stations wherever applicable. In 2009 about 10% of all reviewed earthquake picks were made on strong motion records, even though the great majority of events have magnitude below ML2.5. Although we can use the Wood Anderson amplitude for manual estimates of the Local Magnitude, due to poor signal-to-noise, this is typically only done when strong motion stations record events with ML > 3.5.

Dial-up strong motion data, if triggered, is manually added to the event files within hours/days of an event, and can also be used to refine manual locations and local magnitudes.

15.4.3 ShakeMaps

The SED produces USGS ShakeMaps [34] following all felt earthquakes [6, 35]. ShakeMaps are available within minutes of an event occurring, and use data recorded at all the continuous real-time strong motion stations. Once the dial-up stations have been added to the event files, this information is also added to the ShakeMap.

15.4.4 Moment Magnitude Estimates

Another automatic product available within minutes of an automatic earthquake location that is felt inside Switzerland is a Moment Tensor solution [5, 10] based on broadband time-domain waveform inversion, which provides the moment magnitude of the earthquake. High quality solutions are typically available for events with ML > 3.5. Strong motion data can be used in this solution, though this is not routine as events in Switzerland are typically too small to generate long period energy (10–40 s) above the strong motion sensor noise level.

A second estimate of the moment magnitude, based on the spectral fitting of large numbers of records [11] is also routinely estimated for all events. This method produces good quality solutions for all magnitudes recorded by the Seismic Network. Strong motion data is routinely used in these solutions.

15.4.5 Ground Motion Prediction Equations

Numerous researchers have analysed the high quality waveform and event catalogue datasets made available by the SED [3, 4, 6, 12]. Both the dial-up and continuous strong motion data contribute to the compiled datasets, and are particular important components as a large proportion of the high amplitude records measured under 5 km epicentral distance are recorded on accelerometers. Site investigations at seismic stations in Switzerland are essential to reduce uncertainties in attenuation models and microzonation studies, and therefore to improve seismic hazard products. Such studies were performed in two projects. The first project concerned sites in the Basel area for a microzonation study, in densely populated urban environment ([18, 26]; Havenith et al., 2007). During the second project [15, 30], mostly sites of permanent seismic stations on rock were investigated, often in mountain areas and therefore on rough topography. The products from site characterization are stored in dedicated database and will shortly be made available to the seismological community.

15.4.6 Site Effects

The SED operates strong motion stations on both hard rock for seismological monitoring, as well as on soft and very soft soils in complex geological configurations and at urban locations, with the aim of better understanding site effects and the ground motions that the populations are subjected to. In the past recorded ground motions were used to validate results from numerical simulations used for microzonation studies, covering among others the region of Basel [18, 19] and the Valais [17] with detailed studies in the Sion [32] and the Visp regions [13]. The microzonation for the cantons Basel Stadt and Basel Landschaft has been recently implemented and is replacing the code spectra in the Swiss building code SIA261 ([19]; http://www.geo.bl.ch/ and http://www.geo.bs.ch/erdbebenmikrozonierung/).

New permanent strong motion stations were placed in areas of observed damage during past earthquakes, e.g. the above regions with microzonation studies and Augusta Raurica [14, 17].

15.4.7 Monitoring of Geothermal Exploration

Fluid injections for geothermal energy production in Basel have triggered a series of earthquakes with magnitudes up to $ML = 3.4$ in December 2006, resulting in a large number of buildings with small damage and sizeable reimbursement claims which ultimately led to the halting of the project. The SED has operated a number of temporary strong-motion instruments in addition to the permanent network around Basel, resulting in a high-quality and high-density seismic record of these small events. This allowed a detailed analysis of the induced earthquake series and the related impact in the Basel region [27, 30]. The monitoring of geothermal exploration since then has been widely implemented at sites of geothermal exploration in Switzerland (e.g. Brigerbad in the Valais, Triemli Zürich) [20].

15.4.8 Early Warning

The SED participates in early warning efforts in California, and methods developed in that region are being applied within Switzerland [6, 7, 8]. Densely spaced stations providing high quality waveform data with minimal latency are critical for any system that attempts to determine earthquake locations and magnitude estimates as rapidly as possible. The large number and quality of strong motion stations currently available, and the planned densification, will make strong motion a core component of an Early Warning System for Switzerland.

15.5 Conclusions

The strong motion network in Switzerland is one of the densest not only in Europe has been in operation but also the world. A dial-up, low dynamic range network of 70 stations has been operation for nearly 20 years, and is in the process of being replaced by a large number of seismic observation quality strong motion instrumentation providing continuous real-time data with over 24 bit dynamic range, and frequency resolution from 200 Hz to DC. Currently, about 40 of these stations are operated by the SED, and we plan to install 30 more in the next 3 years, with an expectation to continue the expansion to a final total of 100 new stations.

The Swiss Seismological Service is the sole official source of earthquake information in Switzerland, and the SED Seismic Network group operates both the national strong motion and broadband networks. In practice, there is no separation in acquisition, processing and archival of the continuous real-time strong motion and broadband/short period data. The mature tools and software standard in the seismic network community can thus be applied to our strong motion datasets: these include monitoring of station uptime and data quality, rapid earthquake detection, location and source estimation, real-time data exchange, permanent archival and event extraction.

Acknowledgments The regional upgrades of the strong motion network during 2003–2009 were supported by the Cantons Basel Stadt, Basel Landschaft, Valais and Graubünden through different Interreg III projects. The upgrade of the Swiss national strong motion network is supported by the Federal Office for the Environment (FOEN/BAFU) Federal Roads Office (FEDRO/ASTRA), Swiss Federal Nuclear Safety Inspectorate (ENSI), Swiss Federal Railways (SBB), Vereinigung Kantonaler Feuerversicherungen (VKF), and ETH Zürich. CCES project COGEAR supports the installation of a dense seismic network in the region of Visp and Matter valley in the Valais. The network maps were created by Sabine Wöhlbier.

References

1. Baer M, Deichmann N, Braunmiller J et al (2007) Earthquakes in Switzerland and surrounding regions during 2006. Swiss J Geosci 100:517–528. doi: 10.1007/s00015-007-1242-0
2. Baer M, Kradolfer U (1987) An automatic phase picker for local and teleseismic events. Bull Seismol Soc Am 77:1437–1445
3. Bay F, Fäh D, Malagnini L, Giardini D (2003) Spectral shear-wave ground-motion scaling in Switzerland. Bull Seismol Soc Am 98(4):414–429. doi: 10.1785/0120010232
4. Bay F, Wiemer S, Fäh D, Giardini D (2005) Predictive ground motion relations for Switzerland: best estimates and uncertainties. J Seismol 9(2):223–240
5. Clinton JF, Hauksson E, Solanki K (2006) An evaluation of the SCSN moment tensor solutions: robustness of the M_W magnitude scale, style of faulting, and automation of the method. Bull Seismol Soc Am 96(5):1689–1705. doi: 10.1785/0120050241
6. Cua GB, Fischer M, Clinton JF, Wiemer S, Heaton TH, Giardini D (2008a) Calibrating and Implementing the Virtal Seismologist Approach for Earthquake Early Warning in Switzerland, American Geophysical Union, Fall Meeting 2007, abstract #S23E-03
7. Cua GB, Fischer M, Heaton TH et al (2008b) Real-time and off-line performance of the virtual seismologist earthquake early warning algorithm in California and Switzerland. American Geophysical Union, Fall Meeting 2008, #S11A-1724.
8. Cua GB, Fischer M, Heaton T, Wiemer S (2009) Real-time performance of the virtual seismologist earthquake early warning algorithm in Southern California. Sesimol Res Lett 80(5):740–747. doi: 10.1785/gssrl.80.5.740
9. Deichmann N, Baer M, Clinton J et al (2008) Earthquakes in Switzerland and surrounding regions during 2007. Swiss J Geosci 101:659–667. doi: 10.1007/s00015-008-1304-y
10. Dreger DS (2003) TDMT_INV: time domain seismic moment tensor inversion. In: Lee WHK, Kanamori H, Jennings P, Kisslinger C (eds) International handbook of earthquake and engineering seismology. Part B, Chapter 85–11. Academic Press, San Diego, CA
11. Edwards B, Allmann A, Fäh D, Clinton J (2010) Automatic computation of moment magnitudes for small earthquakes and the scaling of local to moment magnitude. Geophys J Int 183:407–420. doi: 10.1111/j.1365–246X.2010.04743.x
12. Edwards B, Fäh D, Allmann B, Poggi V (2009) Stochastic ground motion model for Switzerland. Pegasos Refinement Project. Report of the Swiss Seismological Service, ETH ZurichSED/PRP/R/006/20091130, 30. Nov 2009.
13. Fäh D COGEAR Working Group (2008). Coupled seismogenic Geohazards in Alpine Regions. Proceeding of the 14th World Conference on Earthquake Engineering October 12–17, 2008, Beijing, China. Paper Number 13-0004.

14. Fäh D (2009) Ein Erdbeben in Augusta Raurica? Jahresberichte aus Augst und Kaiseraugst 30:S291–S305
15. Fäh D, Fritsche S, Poggi V, Gassner-Stamm G, Kästli P, Burjanek J, Zweifel P, Barman S, Clinton J, Keller L, Renault P, Heuberger S (2009) Determination of Site Information for Seismic Stations in Switzerland. Work Package 4: pegasos Refinement Project. Swiss Seismological Service ETH, Zürich, Report SED/PRP/R/004/20090831.
16. Fäh D, Giardini D, Bay F et al (2003) Earthquake catalogue of Switzerland (ECOS) and the related macroseismic database. Eclogae Geol Helv 96(2):219–236
17. Fäh D, Havenith H, Roten D, Alvarez S, Giardini D (2007) Site Effects in the Rhone Valley, Switzerland: measurements, Observations, Modelling and Verification of the Building Code Spectra. Interreg Project SISMOVALP Seismic Hazard and Alpine Valley Response Analysis. Final Report Swiss Seismological Service ETH Zurich18.1.2007.
18. Fäh D, Steimen S, Oprsal I, Ripperger J, Wössner J, Schatzmann R, Kästli P, Spottke I, Huggenberger P (2006) The earthquake of 250 A.D. in Augusta Raurica, a real event with a 3D site-effect? J Seismol 10(4):459–477. doi: 10.1007/s10950-006-9031-1
19. Fäh D, Wenk T (2009) Mikrozonierung für die Kantone Basel Stadt und Basel Landschaft: Optimierung der Form der Antwortspektren und der Anzahl der Mikrozonen. Abschlussbericht: Teilbericht B Projekt "Umsetzung der Mikrozonierung in den Kantonen Basel Stadt und Basel Landschaft". Schweizerischer Erdbebendienst ETH Zürich.
20. Giardini D (2009) Geothermal quake risks must be faced. Nature 462:848–849
21. Giardini D, Wiemer S, Fäh D et al (2004) Seismic hazard assessment of Switzerland. Available at http://www.earthquake.ethz.ch/research/Swiss_Hazard/downloads/Hazard_report_2004.pdf/ (last accessed 5 Oct 2010).
22. Gorini A, Nicoletti M, Marsan P et al (2009) The Italian strong motion network. Bull Earthq Eng doi: 10.1007/s10518-009-9141-6
23. Hanka W, Saul J, Weber B, Becker J GITEWS Team (2008) Timely regional Tsunami warning and rapid global earthquake monitoring. ORFEUS Newsl. http://www.orfeus-eu.org/Organization/Newsletter/vol8no1/vol8no1.pdf last accessed 5 Oct 2010.
24. Havenith H-B, Fäh D, Polom U, Roullé A (2007) S-wave velocity measurements applied to the seismic microzonation of Basel, Upper Rhine Graben. Geophysical Journal International 170(1): 346–358. doi: 10.1111/j.1365-246X.2007.03422.x
25. IMS1.0 Manual (2000) International Data Center Documentation IDC3.4.1Rev2, http://ftp://ftp.isc.ac.uk/pub/isf/isf.pdf
26. Kind F, Fäh D, Giardini D (2005) Array measurements of S-wave velocities from ambient vibrations. Geophys J Int 160:114–126
27. Kraft T, Mai M, Wiemer S, Deichmann D, Ripperger J, Kästli P, Bachmann C, Fäh D, Wössner J, Giardini D (2009) Enhanced Geothermal Systems: mitigating Risk in Urban Areas. Eos 90(32):273–280
28. McNamara DE, Buland RP (2004) Ambient noise levels in the continental United States. Bull Seismol Soc Am 94(4):1517–1527. doi: 10.1785/012003001
29. Peterson J (1993) Observations and modelling of background seismic noise. Open-file report 93-322, U. S. Geological Survey, Albuquerque, New Mexico.
30. Poggi V, Edwards B, Fäh D (2010) Derivation of a reference shear wave velocity model from empirical site amplification. BSSA 101 (1). doi: 10.1785/0120100060
31. Ripperger J, Kästli P, Fäh D, Giardini D (2009) Ground motion and macro-seismic intensities of a seismic event related to geothermal reservoir stimulation below the city of Basel – observations and modelling. Geophys J Int 179:1757–1771
32. Roca A, Guéguen P, Godey S et al (2010), The European-Mediterranean distributed acceleometric data-base (this volume).
33. Roten D, Fäh D, Olsen KB, Giardini D (2008) A comparison of observed and simulated site response in the Rhone valley. Geophys J Int 173(3):958–978
34. SEED (Standard for the Exchange of Earthquake Data) (2004) V2.4 Reference Manual, http://www.iris.edu/manuals/SEEDManual_V2.4.pdf last accessed 5 Oct 2010

35. Wald DJ, Quitoriano V, Heaton TH et al (1999) TriNet "ShakeMaps": rapid generation of peak ground-motion and intensity maps for earthquakes in southern California. Earthq Spectra 15(3):537–556
36. Wiemer SB, Cua GB, Kästli P et al (2007) ShakeMaps at the Swiss Seismological Service: current status, innovations, and outlook. American Geophysical Union, Fall Meeting 2007, #S51A-0216.
37. Wiemer S, Giardini D, Fäh D et al (2009) Probabilistic seismic hazard assessment for Switzerland: bets estimates and uncertainties. J Seismol 13:449–478. doi: 10.1007/s10950-008-9138-7
38. Wyss A (2004) Swiss national strong motion network. Strong motion bulletin January 2004–December 2004. Publication Series of the Swiss Seismological Service, no 117, Swiss Federal Institute of Technology, Zurich, Switzerland, 2005 Available at http://seispc2.ethz.ch/strong_motion/download/Bulletin04.pdf/ (last accessed 5 Oct 2010).

Chapter 16
Real-Time Seismic Monitoring of Structures: Data Handling and Case Studies

M. Çelebi

Abstract Within the last decade, advances in the acquisition, processing and transmission of data from real-time seismic monitoring systems has contributed to the growth in the number structures instrumented with such systems. An equally important factor for such growth can be attributed to the demands by stakeholders to find rapid answers to important questions related to the functionality (or "state of health") of structures during and immediately following a seismic event. Hence, rapid and accurate assessment of the damage condition or performance of a building or a lifeline structure is of paramount importance to stakeholders, including owners, leasers, permanent and/or temporary occupants, users of infrastructures, city officials and rescue teams that are concerned with safety of those in the building, and those that may be affected in nearby buildings and infrastructures. In earlier papers, we described how observed data from sensors deployed in structures can be configured to establish seismic health monitoring of structures. In these configurations, drift ratios are the main parametric indicator of damage condition of a building. The process described for buildings can be applied directly for bridges as well. For bridges, the term, "drift ratio" is not generally used; however, relative displacements of critical elements of a bridge can be construed as such. While real-time data from structural arrays indicate that these methods are reliable and provide requisite information for owners and other parties to make informed decisions and to choose among pre-defined actions following significant events, there are several issues related to data ownership, transmission and archiving. This paper examines the real-time seismic monitoring systems deployed mainly in the United States, with particular attention to data issues – handling, dissemination, storage, and archiving. In most cases, due to the numerous channels involved, the deployments in each one of the real-time structures can be considered to be an individual array. Two detailed cases are described that demonstrate the variability in data ownership and dissemination.

M. Çelebi (✉)
U.S. Geological Survey (MS977), Menlo Park, CA 94025, USA
e-mail: celebi@usgs.gov

S. Akkar et al. (eds.), *Earthquake Data in Engineering Seismology*, Geotechnical, Geological, and Earthquake Engineering 14, DOI 10.1007/978-94-007-0152-6_16,

16.1 Introduction

16.1.1 The "Why" of Real-Time Seismic Monitoring of Structures

During and immediately following an earthquake, rapid and accurate assessment of the damage condition or performance of a building or of an important lifeline such as a long-span bridge is of paramount importance to stakeholders, including owners, leasers, permanent and/or temporary occupants of buildings, public using lifeline structures as well as city officials and rescue teams that are concerned with safety of those in the building, and those that may be affected in nearby buildings and infrastructures. These stakeholders will require answers to key questions such as: (a) is there visible or hidden damage? (b) if damage occurred, what is the extent? (c) does the damage threaten neighboring structures? (d) can the structure be occupied or functionality restored immediately without compromising life safety, or is life safety questionable? It is well known that, cost to rehabilitate a damaged property and consequently, the economic loss due to lack of permit to enter and/or re-occupy a building or use of an important lifeline like a long-span bridge may be significant.

16.1.2 Justification for Real-Time Data from Buildings or Lifelines

In the case of buildings, until recently, assessments of damage following an earthquake were essentially carried out through inspections by city-designated engineers following procedures similar to ATC-20 tagging requirements [1]. Tagging usually involves visual inspection only and is implemented by posting colored tags indicative of potential hazard to occupants: green indicating the building can be occupied – that is, the building does not pose a threat to life safety; yellow indicating limited occupation – that is, hazardous to life safety but not to prevent limited entrance to retrieve possessions; and red indicating entrance prohibited – that is, hazardous to life. However, one of the impediments to accurately assessing the damage level of structures by visual inspection is that some serious damage may be hidden by building finishes and fireproofing. In the absence of visible damage to the building frame, most steel or reinforced concrete moment-frame buildings will be tagged based on visual indications of building deformation, such as damage to partitions or glazing. Lack of certainty regarding the actual deformations that the building experienced may typically lead an inspector toward a relatively conservative tag. In such cases, expensive and time-consuming intrusive inspections may be recommended to building owners (e.g., it is known that, following the M_w=6.7 1994 Northridge, CA earthquake, approximately 300 buildings ranging in height from 1 to 26 stories were subjected to costly intrusive inspection of connections [7]). Thus enters the approach to real-time measurements of deformations of a structure for assessment of the structure's performance during an event. The rationale for such an approach is that, for example, a building owner and designated engineers are expected to use the response data acquired by a real-time health monitoring system to justify a reduced inspection program as compared to that which would otherwise be required by a city government for a similar non-instrumented building in the same area.[1]

This particular application to assess performance of structures in real-time has been the main catalyst for most of the buildings instrumented with real-time capability.

In the case of bridges, for example, several critical structural elements at various locations may be identified to define its performance (e.g. relative vertical displacement of the deck center, relative lateral displacement of the deck center, relative displacements of tower structures, if any). The real-time data can be automatically processed to monitor these critical locations to achieve the goals similar to those described for buildings.

16.1.3 Data, Hardware and Scope

Real-time measurement of displacements are acquired either by double integration of accelerometer time-series data, or by directly using GPS. Recorded sensor data are then related to the performance level of a building or another type of structure. For example, the performance-based design method stipulates that for a building the amplitude of relative displacement of the roof of a building with respect to its base indicates its performance.

Usually, drift ratio is computed using relative displacement between two consecutive floors. When accelerometers are used, determination of displacement is possible by strategically deploying them at a select number of pairs of consecutive floors – as shown in one of the cases presented later in this paper. For these determinations, software is used to compute displacements and drift ratios in real-time by double integration of accelerometer data. As later described for an actual case, several levels of threshold drift ratios can be postulated in order to make decisions for the need of inspections and/or re-occupancy.

For the purposes of this paper, discussions presented in this paper are limited to those systems utilizing accelerometers only. However, it should be mentioned that GPS can also be used to compute drift even though GPS-measured relative displacements are limited to being acquired only at the roof with respect to its reference base. Thus, drift ratio computed using GPS can only be the "average drift ratio" for the

---

[1]The City of San Francisco, California, has developed a "Building Occupancy Resumption Program" [2] whereby a pre-qualified Occupancy decision-making process, as described in this paper, may be proposed to the City as a reduced inspection program and in lieu of detailed inspections by city engineers following a serious earthquake.

whole building. In the United States, experience with real-time GPS monitoring for seismic purposes has been limited [5].[2,3]

16.2 Current Inventory of Real-Time Monitoring of Structures

Table 16.1 summarizes the current structures equipped with real-time seismic monitoring hardware and software. Most of these are in the United States. As noted in the table, there is great variability of how the data are handled, transmitted, owned and stored.

16.2.1 Data Streaming and Storage Issues

Data streaming and storage is accomplished in various ways for the structures listed in Table 16.1. Currently, there is no central organization that receives all or a substantial percentage of the real-time structural monitoring data, and there is uncertainty as to whether one or more organizations to receive, process and store real-time data will be created. Complete automatization of these processes has not been accomplished to date either. There are several reasons for this:

- Real-time monitoring of structures is rather new. Interest in real-time data from structures is just beginning to grow.
- Privately owned structures (e.g. banks) do not agree to releasing data. Hence, only the owner and their designated consultants have access to real-time data (or any data). Such privately owned data will probably never be disseminated for general engineering use.
- Publicly owned structures (e.g. Cape Girardeau Bridge, Atwood Building). Data from each one of these arrays listed in Table 16.1 is sizeable but not unmanageable. For various reasons, the data have not been channeled to a central location.

Table 16.2 summarizes bandwidth and storage (buffering) capacity required for only publicly-owned structures for which all data might be streamed and stored at one location. To date, limited arrangements have been made with different organizations (e.g. IRIS or UCSB) to stream some of the data. Strong-motion data centers have later acquired processed for engineering use and primarily stored earthquake response data. In other words, strong-motion data centers currently do not handle real-time data streaming, buffering or long-term storage.

As Table 16.2 shows, with today's feasible storage capacity, a 3–4 TB storage capacity can be handled with 3–4 1-TB arrays of disks at reasonable costs.

---

[2]Until recently, the validity of measurements using GPS was limited to long-period structures (T>1 s) because GPS systems readily available were limited to 10–20 samples per seconds (sps) capability. Presently however, up to 50 sps differential GPS systems are available on the market and have been successfully used to monitor drift ratios ([9], Restrepo, Personal Communication 2007) – thus enabling future usefulness of GPS to all types of structures.

[3]For wind monitoring of tall buildings, GPS have been deployed on the roofs of 5 buildings in Chicago, IL. (Kijewski-Correa and Kareem [8]).

Table 16.1 Instrumented Structures with Real-time Data Streaming Capability

Structure/ Year Installed	Real-time System	Channels/sps	Data Streamed to	Event Data Stored	COMMENTS
STRUCTURES IN THE US WITH REAL - TIME STREAMING					
Factor Building (UCLA, LA, CA) [2000](+4FF+IDH)	*KMI /Dolomite*	*72/200 or 500*	*IRIS*	*NESMP*	*Windowed event data FF data to USGS/Caltech*
Millikan Library, (Pasadena, CA) [200]	*Digitexx + Mt. Whitney*	*36/200 + 2 Quanteras*	*Digitexx*		*(partially to SGS/Caltech Quantera (for rotation sensor)*
Naval Facilities Hospital (NAVFACNW) Brem erton, WA [2009]	*KMI/Oasis*	*12/200*	*locally*		*Local recording of trigered data/no exchange of data*
Naval Facilities Command Northwest (NAVFAC NW) San Diego, CA [2008]	*KMI/Oasis*	*36 [partially complete]*			*Local recording of triggered data/no exchange of data*
CALTRANS (Dist. 4 H Q Building, SF, CA [2008]	*KMI/Oasis*	*33/200*	*Locally/and consultand*		*No exchange of data*
Atwood Bldg. (Free Field only) [2006]	*KMI/ Q330 (FF)*	*FF(21)/200*	*UCSB FF only*	*NESMP*	*Bladg 33 ch bldg array (phone) (FF-windowed event data)*
Cape Girarde au Bridge (MO) [2006]	*KMI/ Q330*	*84/200*	*IRIS*	*NESMP*	*Event Data to NESMP*
FDIC Bldg (SF) [2003]	Digitexx	30/200	USGS and consultant		Streaming on demand
Transamerica Building (SF) [2004]	Digitexx	21 /200	USGS	NESMP	Streaming on demand
Wells Fargo Bank (155 5th St. & 420 Montgomery St. Bldgs. California, USA. [2006]	KMI/ Q330	42 and 33 channels at 200 sps	Private/on demand		Locally and to consulting engineer only
PG&E Bldg. (SF, CA) [2006]	Digitexx	23/200	Locally only		No outside data transmittal
Castille Bldg. Mayaguez, PR [2008]	KMI/ Q330	24/100		NESMP	Locally and on demand remotely

Table 16.1 (continued)

Structure/ Year Installed	Real-time System	Channels/sps	Data Streamed to	Event Data Stored	COMMENTS
STRUCTURES IN US WITH REAL-TIMES TREAMING NOT ACTICATED					
US Federal Building San SF, CA [GSA&USGS] 2007	KMI/Q330	36/200	Not continuous	NESMP	Locally and on demand remotely
UC-Irvine It2 Bldg Irvine, CA [2004]	KMI/ K2	12 ch/200	Locally		Cannot stream and dial-up at the same time
STRUCTURES IN THE US – IN PROGRESS – TO HAVE REAL – TIME STREAMING					
VA Hospital, Palo Alto, CA [VA & USGS][2009]	*KMI/Oasis*	*36 ch (2 bldgs)*		NESMP	
Catholic Healthcare West Marian Medical Center, Santa Maria, CA [2010]					
PROJECTS OUTSIDE OF US					
Univ. of Po it o, Portugal [2007] Infante D. Herrique Bridge	KMI/Oasis	12/200	Locally and to Univ.		Local/remote continuous in blocks of 15 min each
UNA M, Institute of Engineering [Portable], Mexico [2007]					
Central Gov. Adm. Bldg. Tortola Island, BVI			UPR		3-4 story bldg. on reclaimed soft soil
Control Tower, Tortola, BVI	12 ch K 2		UPR only		Access to USGS
Isbank Towers, Istanbul, Turkey	4 Geosig	12/200	Kand illi		
Ayasofya Museum, Istanbul	9 Guralps	27/200	Kandilli		
Fatih Suspension Bridge, Istanbul	5 Guralps	15/200	Kandilli		
Atakoy Bldg, Istanbul	20 dataloggers	60/200			Transmittal to a German Univ.
90-meter Windmill Turbine, Denmark	16ch Digitexx	16/200			

Table 16.2 Only Publicly Owned Structures, Data Streaming and 30-day Storage/Buffering (*32 bit for storage. Normal digitization at 24 bit, ** increased by 10 % for telemetry and other factors).

			Total Bandwidth (Per Second)**		30 Day Buffering/ Storage Capacity	
Publicly Owned (channels)	sps	bits	megabits	megabytes	terabits	terabytes
~300 (current)	200	32*	1.92-2.1	.24-.26	5.0-5.4	.62-.68
~400 (2010 estimated)	200	32	2.56-2.6	.32-.35	6.6-7.3	.83-.90

16.3 Two Cases of Structures

16.3.1 FDIC Building, San Francisco, CA

FDIC Building in San Francisco is one of the first buildings where a real-time seismic health monitoring system was recently installed to assess the performance of the building [6]. A general schematic of the building and state-of-the-art monitoring system is seen in Fig. 16.1.

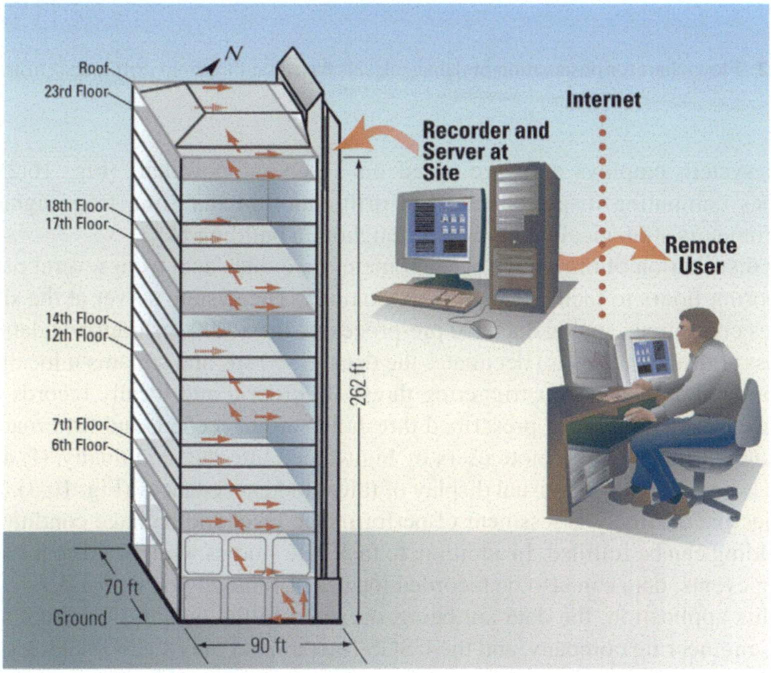

Fig. 16.1 Schematic of real-time seismic monitoring of the building [6]

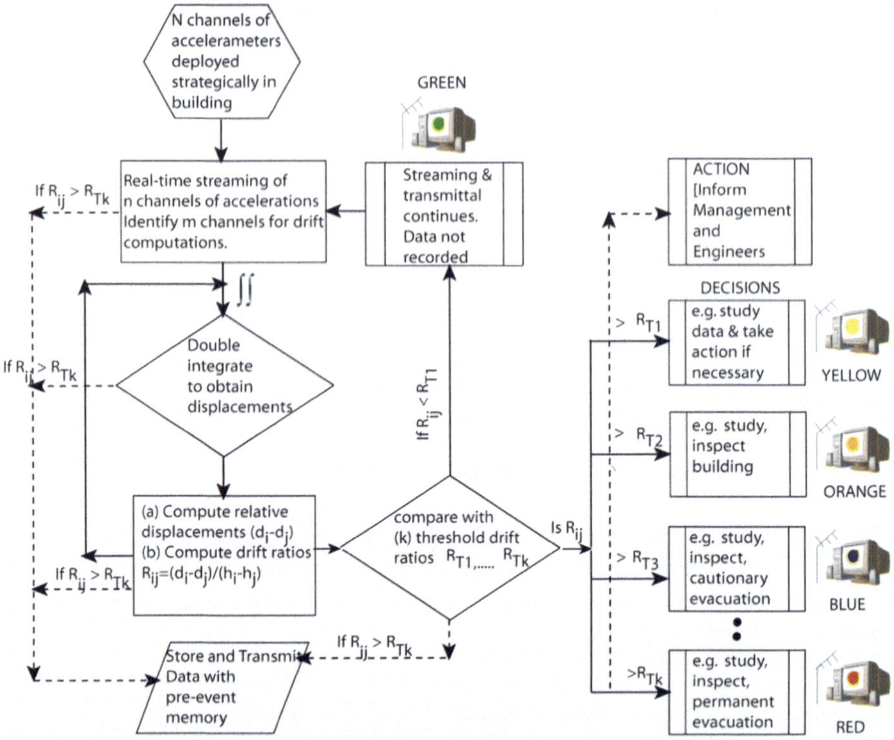

Fig. 16.2 Flow-chart for observation of damage levels based on threshold drift ratios (from [4])

The system employs software based on a general flowchart (Fig. 16.2) that describes computing displacements and drift ratios in real-time from signals of accelerometers strategically deployed throughout a building [4].

The distribution of thirty (30) accelerometers provides data from several pairs of neighboring floors to facilitate drift computations. The system server at the site (a) digitizes continuous analog data, (b) pre-processes the 1,000 sps digitized data with low-pass anti-alias filters (c) decimates the data to 200 sps and streams it locally, (d) monitors and applies server triggering threshold criteria and locally records (with a pre-event memory) when prescribed thresholds are exceeded, and (e) broadcasts the data continuously to remote users by high-speed internet, and finally, (f) allows remote access to data and visual display of threshold exceedances (Fig. 16.3). Thus, the objective of timely assessment of performance level and damage conditions of the building can be fulfilled. In addition, to facilitate studies while waiting for strong shaking events, data can also be recorded locally or remotely on demand.

In this application, the data are being streamed to the owner; the owner's consulting engineering company, and the USGS (on demand only), USGS has access to event data as well. The streaming of data is not open to the public, and is controlled by remotely utilized software that is not available to the public.

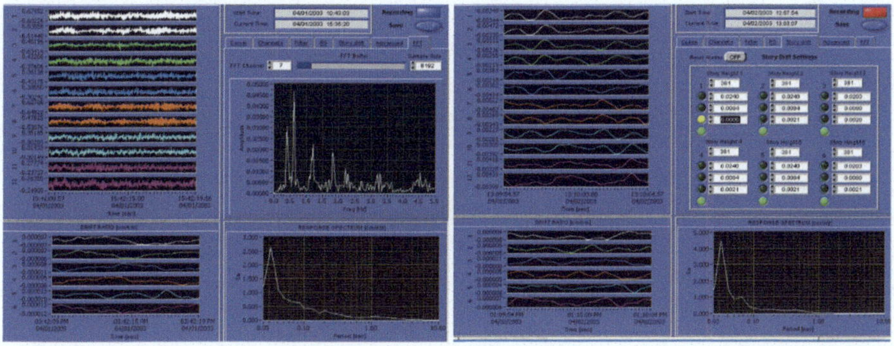

Fig. 16.3 Screen snapshots of sample client software displays: (*left*) acceleration streams and computed amplitude and response spectra, and (*right*) displacement and corresponding drift ratios and alarm systems corresponding to thresholds

16.3.2 Cape Girardeau (Bill Emerson) Bridge, MO

Figure 16.4 exhibits the cable-stayed Bill Emerson Memorial Bridge (also known as Cape Girardeau Bridge) in Cape Girardeau (Missouri, USA) which is instrumented with a monitoring system capable of streaming real-time acceleration response data that can be configured to establish performance indicators using the sensor data at the deck center or tops of towers or other instrumented locations for which data are readily available.

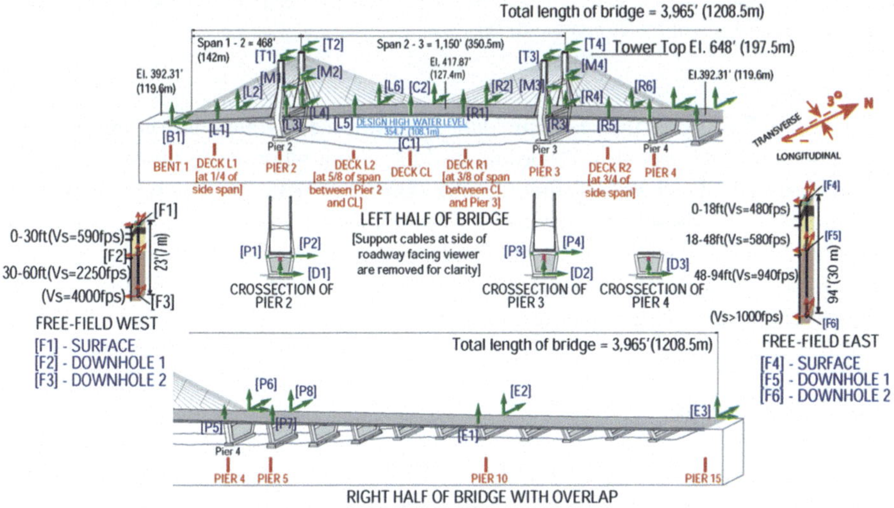

Fig. 16.4 Example of extensively instrumented cable-stayed Bill Emerson Memorial Bridge (also known as Cape Girardeau Bridge) (Missouri, USA) that has real-time streaming of responses [3]

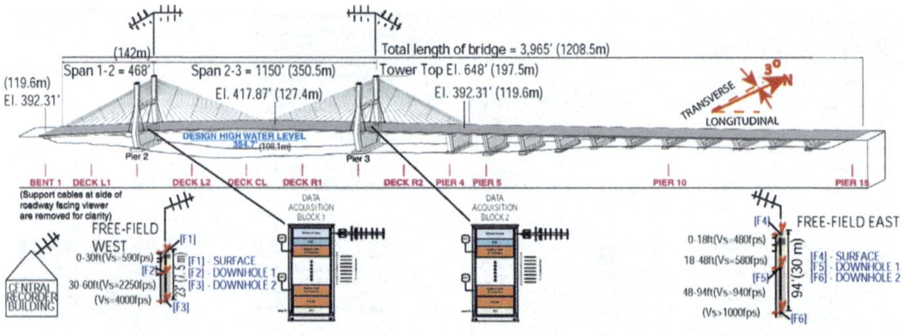

Fig. 16.5 Data Acquisition units at different locations of the bridge and local telemetry set-up transmitting data to central recording building [3]

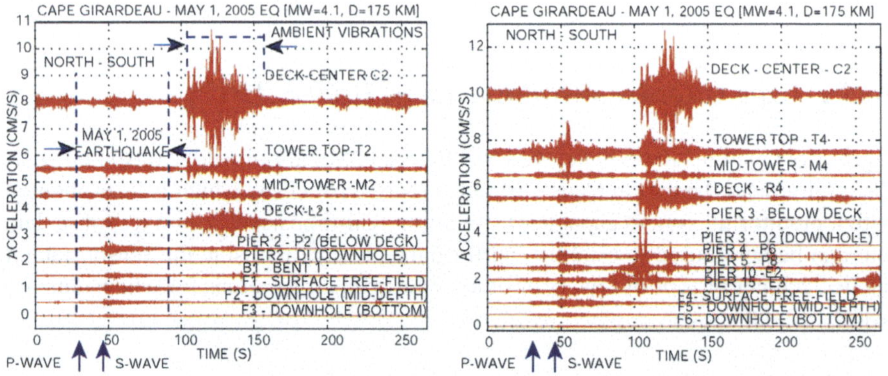

Fig. 16.6 Re-plotted from streamed data displaying response of the bridge to May 1, 2005 event (between 30 and 130 s of the 267 s record) [3]

Figure 16.5 displays the real-time data acquisition and transmission design with local telemetry to the central recording building from where, via internet, data are transmitted to the Data Management Center (DMC) of IRIS (Incorporated Research Institutions for Seismology [www.iris.washington.edu]). In this case, streaming data stored at buffer of DMC can be obtained by anyone. Figure 16.6 shows a 267 s window of streamed data that was obtained from the DMC and re-plotted to show an event and a sizeable ambient response of the bridge.

16.4 Conclusions

A list of known structures with real-time seismic monitoring capability and mainly located in the United States is presented. Data from these structures are handled in a variety of ways. For publicly owned structures, data handling and storage issues are easier than for those privately owned. It appears that a dedicated server can be used to provide a central base for streaming and buffering considerable amount of data from almost all of the publicly owned structures.

The number of structures with real-time monitoring capability is growing. Two specific examples presented demonstrate that such systems are useful and are being demanded by owners. It should be remembered that data recorded by such systems help public safety. With little cost, data from many of these systems can be centrally collected and made available to the owners and user community for future engineering applications as well as developing additional methods to assess the functionality of structures.

Acknowledgements The author gratefully acknowledges constructive reviews by Chris D. Stephens and Roger Borcherdt of USGS. Larry Baker provided input in establishing numerical computations of Table 16.2.

References

1. Applied Technology Council [ATC] (1989) Procedures for Post-Earthquake Safety Evaluation of Buildings. ATC-20, Redwood City, CA
2. Building Occupancy Resumption Program [BORP] (2001) City and County of San Francisco, Department of Building Inspection, Emergency Operation Plan (Rev. 2001). [www.seaonc.org/member/committees/des_build.html]
3. Çelebi M (2006, August) Real-time seismic monitoring of the new Cape Girardeau (MO) Bridge and preliminary analyses of recorded data: an Overview. Earthq Spectra 22(3):609–630
4. Çelebi M (2008) Real-Time Monitoring of Drift for occupancy Resumption, PROC. 14WCEE, Beijing, China, Oct 13–17, 2008.
5. Çelebi M, Sanli A (2002) GPS in pioneering dynamic monitoring of long-period structures. Earthq Spectra J EERI 18(1):47–61. February 2002
6. Çelebi M, Sanli A, Sinclair M, Gallant S, Radulescu D (2004) Real-time seismic monitoring needs of a building owner and the solution – a cooperative effort. Earthq Spectra J EERI 19(1): 1–23
7. FEMA-352 (2002) Recommended Post-earthquake Evaluation and Repair Criteria for Welded Steel Moment-Frame Buildings (also SAC Joint Venture 2000 prepared for FEMA), Washington, DC.
8. Kijewski-Correa T, Kareem A (2004) The Height of Precision: New Perspectives in Structural Monitoring, Proceedings of Earth & Space: 9th Aerospace Division International Conference on Engineering, Construction and Operations Challenging Environments, 7–10 March, Houston.
9. Panagitou M, Restrepo JI, Conte JP, Englekirk RE (2006) Seismic Response of Reinforced Concrete Wall Buildings, 8NCEE (paper no. 1,494), San Francisco, CA. April 18–22, 2006.

Chapter 17
Strong-Motion Networks in Romania and Their Efficient Use in the Structural Engineering Applications

I.G. Craifaleanu, I.S. Borcia, and I.C. Praun

Abstract Created in 1967 for strong motion recording, the seismic network of the National Institute for Building Research, INCERC, is, at present, the largest in Romania, consisting of almost 100 instruments, distributed all over the country. Other seismic networks operating in Romania are the network of the National Institute for Earth Physics, INFP, and the network of the National Center for Seismic Risk Reduction, CNRRS. Data sharing among these networks is performed under permanent collaboration agreements. A presentation of the INCERC strong motion network and of the use of recorded data in some of the most recent earthquake engineering studies performed at INCERC is made, with focus on topics as: attenuation and propagation of seismic waves during strong Vrancea earthquakes, evaluation and mapping of the spatial distribution of various quantities of interest for structural design, studies on seismic demands on buildings, assessment of the distribution of damage indices.

17.1 Introduction

The seismicity of Romania is generated mainly by the Vrancea subcrustal source. This source caused, over time, several destructive seismic events. According to the "Romplus" earthquake catalog of the National Institute for Earth Physics in Romania [11], during the twentieth century Romania experienced 32 earthquakes with a moment magnitude $M_W \geq 6$, originating from the Vrancea source.

Several historical testimonies are available from earthquakes in the past centuries. The strongest historical event is considered the October 26, 1802 earthquake, with an estimated moment magnitude Mw = 7.9 [11]. The earthquake was felt in many countries in the area (Poland, Bulgaria, Turkey and Russia). Minor damage in buildings was reported even in distant cities, as Moscow and Lvov [9].

I.G. Craifaleanu (✉)
National Institute for Building Research (INCERC), Technical University of Civil Engineering (UTCB), Bucharest, Romania
e-mail: i.craifaleanu@gmail.com

S. Akkar et al. (eds.), *Earthquake Data in Engineering Seismology*, Geotechnical, Geological, and Earthquake Engineering 14, DOI 10.1007/978-94-007-0152-6_17,

During the twentieth century, the strongest seismic event was the November 10, 1940 earthquake, with a magnitude $M_W = 7.7$, generated by the Vrancea source. Besides other less significant damage of buildings, this earthquake was remembered for the catastrophic collapse of the modern high-rise Carlton building, situated in the center of Bucharest [2, 9, 10].

However, the most destructive Vrancea earthquake in modern times occurred on March 4, 1977 ($M_w = 7.4$, $h = 94$ km), causing 1,578 deaths and about 11,300 people injured. About 35,000 families were left homeless and the total damage was estimated to more than two billion dollars. In Bucharest, about 33 high-rise buildings collapsed and 1,424 deaths were reported [1, 3].

Since then, the Vrancea source generated four more earthquakes with $M_W \geq 6$, i.e. those on August 30, 1986 ($M_w = 7.1$, focal depth $h = 131$ km), May 30, 1990 ($M_w = 6.9$, $h = 91$ km), May 31, 1990 ($M_w = 6.4$, $h = 87$ km) and October 27, 2004 ($M_w = 6$, $h = 99$ km). The 1986 event caused light and moderate damage to some buildings in Romania; 2 deaths and 558 injured people were reported. The earthquake produced stronger effects in the neighboring Republic of Moldova, where losses of about 1 billion USD were reported [20] and 4 buildings collapsed in the capital city, Chisinau. The other events caused little or no damage to the building stock in Romania.

Seismic sources in Romania are also located in Banat (in the southwest of the country, with a recent sequence of crustal events during 1991, having a maximum magnitude $M_W = 5.6$), in the Fagaras region (in the central part of the country), on the continental platform of the Black Sea and in other areas.

17.2 Strong-Motion Networks in Romania

The first strong-motion recording instruments were installed in Romania in 1967, by scientists from the National Institute for Building Research, INCERC. This year marks the creation of the INCERC seismic network.

At the time of the March 4, 1977 earthquake, nine strong-motion accelerographs (of types MO-2, SMAC-B and RMT-280) and two Wilmot seismoscopes were installed in Romania [3]. However, due to various recording and processing malfunctions, only two accelerograms from Bucharest and two seismoscope records were available after the event. Of the two accelerograms, one was recorded at the top of a ten-story building (the "Balta Alba E5" record) and the other at the basement of a one-story reinforced concrete building at INCERC.

These specific circumstances gave the second record, obtained on a three-component Japanese-built SMAC-B accelerograph, an exceptional value, as it is the single accelerographic record obtained during the 1977 earthquake in similar to free-field conditions. As, during the years that followed, no comparable seismic event occurred, the record is used until nowadays as a reference for the Romanian seismic code provisions and for seismology and earthquake engineering studies.

After the March 4, 1977 earthquake, essential support was provided by the United States Agency for International Development (USAID) to the seismic network of INCERC, through an important donation of 75 SMA-1 analog strong-motion accelerographs, together with other equipments, such as Ranger velocity transducers, signal conditioners, triaxial transducers for boreholes, signal attenuators and a PDP-11/34 minicomputer.

Thanks to this equipment, during the next significant seismic event, the earthquake of August 30, 1986, ground motion data were obtained from multiple stations. This allowed for the first studies on the spatial distribution of recorded ground motion parameters to be carried out. Most of the records of this earthquake were provided by the INCERC seismic network. Records were also provided by the network of the National Institute for Earth Physics, INFP.

Starting with 1996, the seismic network of INCERC underwent a major upgrade. New systems for data acquisition, processing and display, together with measurement and storage equipment were purchased. A number of 26 Romanian-built ADS digital accelerometers and of three SSS8 digital stations for continuous real-time monitoring were installed with the financial support of the Romanian Ministry of Transportation, Construction and Tourism.

The next major enlargement of the INCERC seismic network occurred in 2002, when 31 Kinemetrics Etna accelerometers were purchased with funds from the State Inspectorate for Construction, ISC, and installed on different locations in Bucharest and all over the country.

Other equipments acquired during recent years include Kinemetrics triaxial force balance accelerometers, installed in boreholes and on buildings. Also, new state-of-the-art computer servers, together with network and communication equipments were acquired in 2008, under an annual-based funding programme financed by the Romanian Ministry of Regional Development and Housing (Fig. 17.1). Besides the seismic stations, INCERC also operates vibration monitoring equipments installed on various significant buildings such as the University Emergency Hospital in Bucharest, the Arnota Monastery, the National Library, and in the Bucharest subway. The strong motion data recorded by INCERC is stored in a database covering seismic events recorded since 1977. The website of the institute provides a collection of accelerometric records, seismic maps and short seismic reports.

Fig. 17.1 Views of the seismic data recording, processing and storage equipment installed at INCERC

The National Institute for Earth Physics, INFP, established in February, 1977, operates both a real-time seismic network, consisting of short-period stations, and a free-field strong-motion network, consisting of 44 stations, equipped with Kinemetrics K2-dataloggers, three-component episensors and GPS timing systems. The strong-motion stations have been installed in the framework of a joint cooperation between the Collaborative Research Center 461 "Strong Earthquakes" of Karlsruhe University and INFP [4, 13]. INFP also operates a real-time seismic network and the BURAR seismic array located in the northeastern part of the country. The INFP seismic network has been recently enlarged, by adding new stations and instruments, so that, at present, it consists of 73 digital seismic stations, equipped with Q330 and K2 digitizers, accelerometer sensors and velocity sensors (INFP website [12]).

Starting with 2003, a new strong-motion network was established, that of the National Center for Seismic Risk Reduction, CNRRS. The center was founded in 2002 by an agreement between the Romanian Government and the Japan International Cooperation Agency, its creation being financed in the framework of the Japanese Technical Cooperation Project on the Reduction of Seismic Risk for Buildings and Structures. The center works in partnership with the Technical University of Civil Engineering Bucharest and with INCERC. The CNRRS network consists of 20 instruments (Kinemetrics K2, Kinemetrics ETNA and Geosig), located in free field, in boreholes or on buildings.

A map with the location of the seismic stations of INCERC, INFP and CNRRS is shown in Fig. 17.2. The location of INFP stations is as specified on the INFP website.

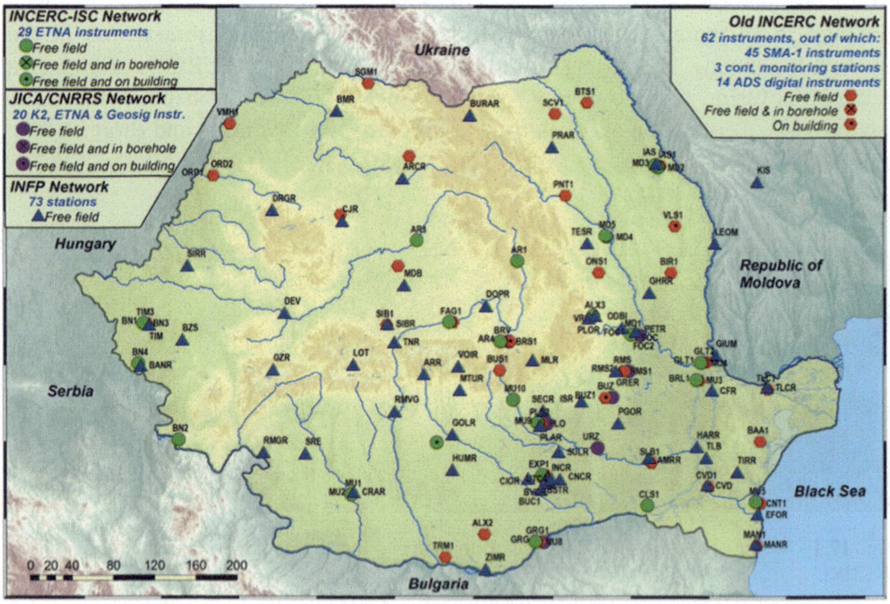

Fig. 17.2 Strong-motion networks in Romania: INCERC, INFP and CNRRS networks

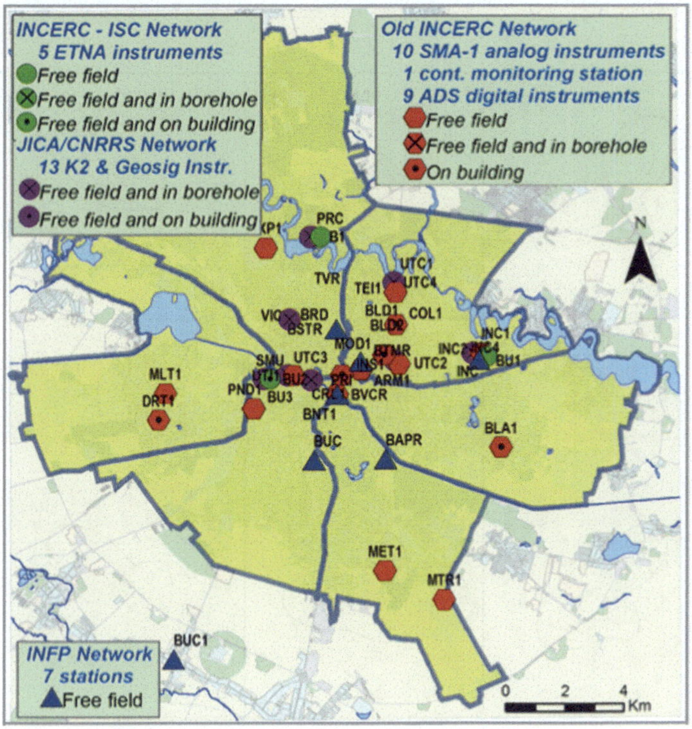

Fig. 17.3 Strong-motion instrumentation of Bucharest: INCERC, INFP and CNRRS networks

A particular attention is given to the seismic instrumentation of Bucharest, the capital city of Romania which, with a population of around two million, ranks third in the region, after Athens and Istanbul. Figure 17.3 shows the map of INCERC, INFP and CNRRS seismic stations of in Bucharest.

Data sharing among the seismic networks of the three organizations in Romania, INCERC, INFP and CNRRS is performed under permanent cooperation agreements [18].

17.3 Recent Earthquake Engineering Research at INCERC

17.3.1 Studies on Attenuation and Directivity

The attenuation analysis was performed at two levels, consisting in the analysis of attenuation in global and in directivity terms [17]. The parameters to which attenuation analysis was related were: PGA, PGV and PGD.

Figure 17.4 shows, for the three above-mentioned events, regression lines for PGA, PGV and PGD abscissas, irrespective of directivity, and, in parallel, results of directivity analysis. The regression lines are drawn against the non-dimensional epicentral distance, represented as ordinate. The results of directivity analysis are displayed as curves representing the epicentral distances where PGA values of 1.5,

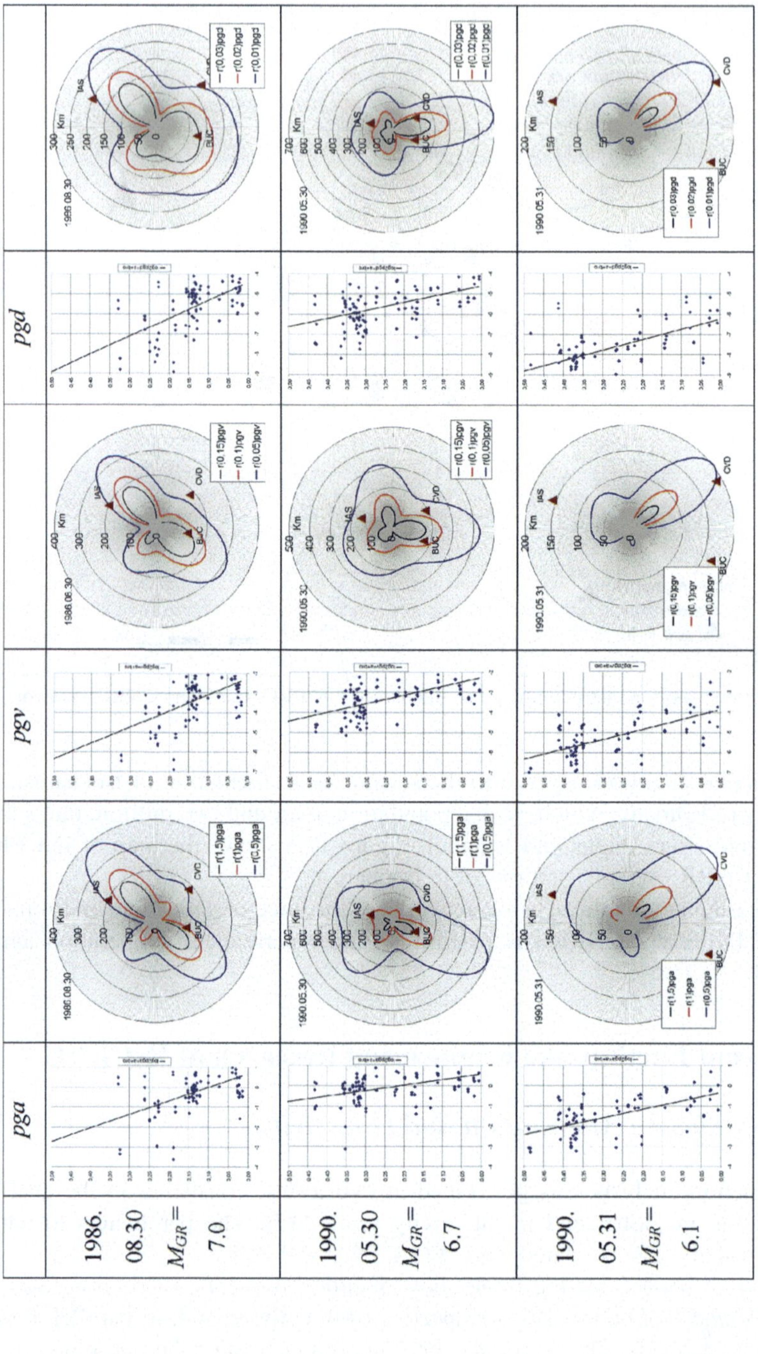

Fig. 17.4 Regression *lines* and attenuation *graphs* for three Vrancea seismic events

1.0 and 0.5 m/s^2 are expected to be reached, where PGV values of 0.15, 0.1 and 0.05 m/s are expected to be reached and, finally, where PGD values of 0.03, 0.02 and 0.01 m are expected to be reached.

The study led to the following conclusions:

- the attenuation was faster for the event of August 30, 1986 (h = 133 km) than for the events of May 30, 1990 (h = 89 km) and May 31, 1990 (h = 79 km); the result is rather unexpected, since the source was deeper for the first event;
- the directivity of radiation was strong in all cases;
- while the radiation tended to be rather symmetrical for the first two events, it was strongly non-symmetrical for the last one;
- the radiation directivity was different for the three events considered, i.e. approximately NE-SW on August 30, 1986 (as observed on macroseismic basis for the destructive events of November 10, 1940 $< M_{GR} = 7.4 >$ and of March 4, 1977 $< M_{GR} = 7.2 >$ too), N–S on May 30, 1990 and S-E on May 31, 1990.

17.3.2 Mapping of Ground Motion Parameters and of Various Spectral Quantities

During the last few years, an extensive effort has been made at INCERC in order to generate a set of reference maps displaying the spatial distribution of relevant ground-motion parameters during the four significant seismic events with $M_w > 6$ that occurred in Romania after the March 4, 1977 earthquake [7, 15]. These studies were a continuation of some previous research involving the mapping of these parameters, performed at the Technical University of Civil Engineering Bucharest by a team led by Professor Dan Lungu [14].

Initially, the peak ground motion acceleration (PGA), velocity (PGV) and displacement (PGD) were considered. In a subsequent phase, the research was extended to parameters derived from spectral values, such as effective peak ground acceleration (EPA), effective peak ground velocity (EPV) and control (corner) periods, T_C and T_D. A distinct part of the study focused on the mapping of various linear and nonlinear spectral quantities, considered as relevant for structural engineering applications. Starting with the spectral acceleration (SA) and displacement (SD) associated with specified levels of structural ductility, the research continued with the mapping of damage index spectral ordinates, as well as of strength demands on buildings.

The Vrancea events considered in the study were: August 30, 1986, May 30, 1990, May 31, 1990, and October 27, 2004. Records obtained from the seismic networks of Moldova and Bulgaria were also included, where available. Each set of maps was revisited several times during the study, either for adding data that was not available in the initial phases or for better emphasize the significant features. Detailed maps were also generated for the city of Bucharest.

It should be noted that the set of stations providing seismic data was different for each seismic event, either due to the expansion of seismic networks in Romania, or to the accidental malfunction of some older accelerometers.

All generated maps showed significant differences between the spatial distributions of the studied parameters, obtained for the seismic events considered. At the origin of this phenomenon is mainly the diversity of the source mechanisms, as well as of the wave propagation patterns of the events, both revealed by several previous studies in the field.

Figure 17.5 displays maps of the peak ground acceleration, PGA. The NE-SW orientation of the PGA map contours for the 1986 earthquake is in agreement with the conclusions presented in the previous paragraph. For the 1990 earthquakes, the contour display is less suggestive (as only the largest value of the two horizontal

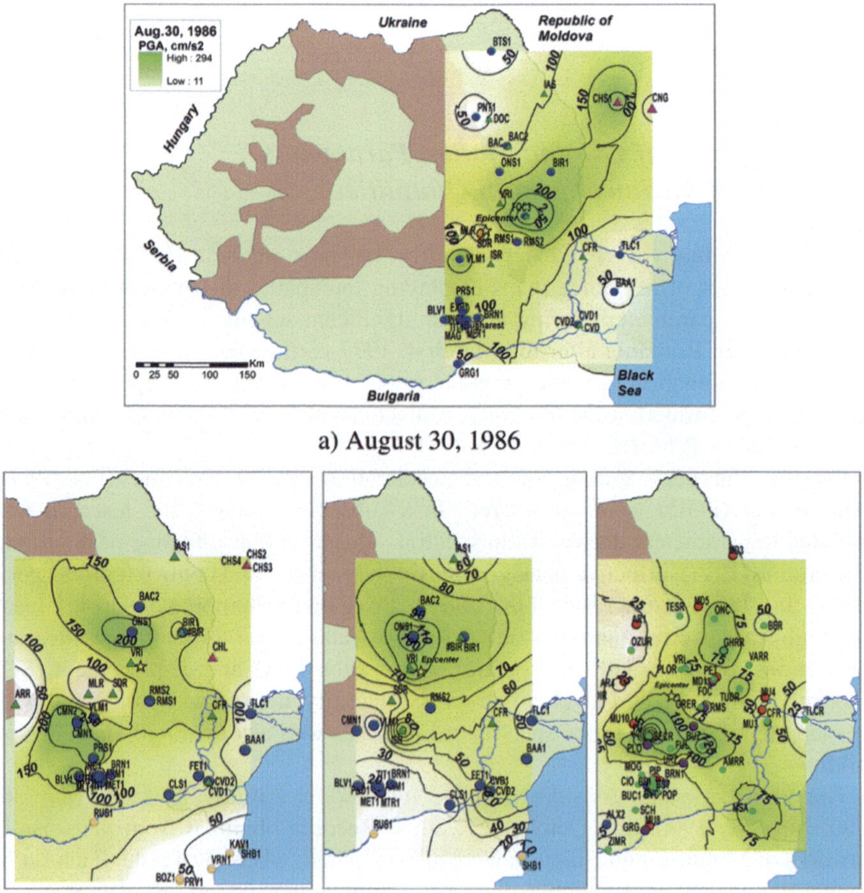

Fig. 17.5 PGA distribution for the four strongest Vrancea seismic events that occurred in Romania during the last three decades

components was mapped), but it still can be interpreted in correspondence with the results of the directivity study.

The *corner period*, T_C, was computed using the following expression:

$$T_C = 2\pi EPV/EPA, \tag{17.1}$$

where *EPV* and *EPA* are the effective peak ground velocity and effective peak ground acceleration, calculated by using the expressions proposed by Lungu et al. in 1995 [8]:

$$EPA = (SA_{averaged\ on\ 0{,}4\ s})_{\max}/2.5, EPV = (SV_{averaged\ on\ 0{,}4\ s})_{\max}/2.5 \tag{17.2}$$

In Equations (17.2), $(SA_{\text{averaged on 0.4s}})_{\max}$ and $(SV_{\text{averaged on 0.4s}})_{\max}$ denote the maximum values of the acceleration and velocity, respectively, response spectrum, averaged on a 0.4 s period mobile window.

The maps of the corner period, T_C (Fig. 17.6), also show substantial differences between the seismic events considered. Moreover, the T_C values calculated at one station for each seismic event can differ considerably. While in most cases the highest T_C values occur for the 1986 earthquake, this is not, however, a general rule. This observation needs further research in order to correlate the results with seismic source and travel path effects, as well as with the influence of local site conditions.

The maps of elastic and inelastic spectral accelerations for the 1986 earthquake are shown in Fig. 17.7, for a vibration period T = 0.5 s. Several similar maps were generated for different values of T and of the displacement ductility, μ, in order to express the strength demands of this seismic event on structures with different characteristics. A nonlinear elastic-perfectly plastic hysteretic model was considered in the calculations.

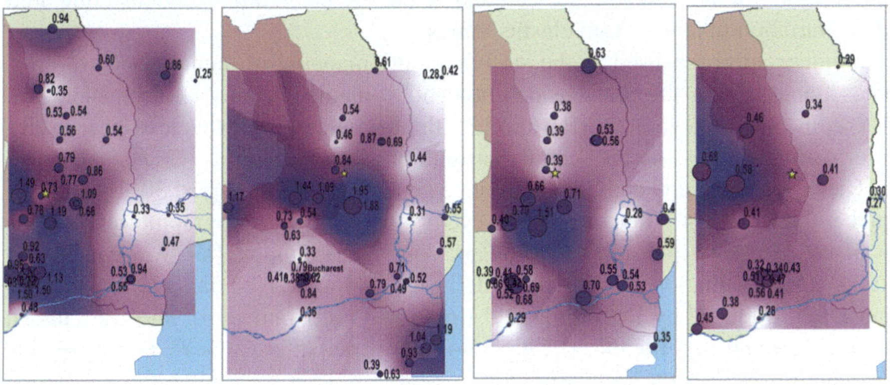

a) August 30, 1986 b) May 30, 1990 c) May 31, 1990 d) Oct. 27, 2004

Fig. 17.6 T_C distribution for the four Vrancea seismic events considered

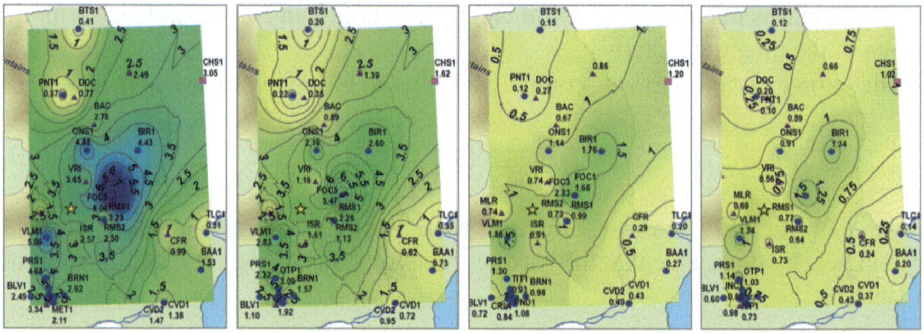

a) μ = 1 (elastic) b) μ = 2 c) μ = 4 d) μ = 6

Fig. 17.7 August 30, 1986 earthquake. Distribution of elastic ($\mu = 1$) and inelastic ($\mu = 2, 4, 6$) spectral acceleration for vibration period $T = 0.5$ s

A general observation, made for all spectral acceleration maps, was that, as ductility increases, the interpolation surfaces become smoother, simultaneously with the decrease of their ordinates. Consequently, the spatial distribution of spectral accelerations is more uniform at higher ductilities (Fig. 17.7d). It should be noticed however that, as the rates of variation of spectral ordinates with ductility for the specified period differ from one ground motion record to another, the interpolation surfaces do not "flatten" uniformly and the shapes of equal value contours on the maps are changing with ductility.

One of the most significant consequences of the first observation is that the influence of inelastic behavior is very significant and even appears to prevail, for larger values of ductility, over the influence of other factors affecting the spatial distribution of spectral ordinates (e. g. factors related to seismic waves characteristics and propagation). In engineering terms, this would mean that, for common building structures for which inelastic behavior is allowed during strong earthquakes, the spatial variation of seismic strength demands could be much attenuated, as compared to the variation obtained from elastic values.

Another approach used in the study was the mapping of damage spectra ordinates for assessing the damage potential of Romanian Vrancea earthquakes [6]. According to Bozorgnia and Bertero [5], this technique has proven to give useful information in assessing the spatial distribution of damage.

Damage spectra and damage maps were generated for the August 30, 1986 earthquake, the strongest Vrancea seismic event for which accelerographic records are available from multiple stations. The Park-Ang damage index, *DM*, was used in the study, some evaluation being also made based on the derived damage indices DI_1 and DI_2 introduced by Bozorgnia and Bertero. As a parameter of spectral curves, the yield strength coefficient, C_y, was chosen, expressed as $C_y = F_y/G$, where G is the weight and F_y is the yield strength.

According to current literature [16, 19], DM values greater than or equal to 1 correspond to collapse, $DM = 0.4$ (0.5) is the upper limit of repairable damage, while $DM = 0.2$ is the superior threshold of insignificant damage.

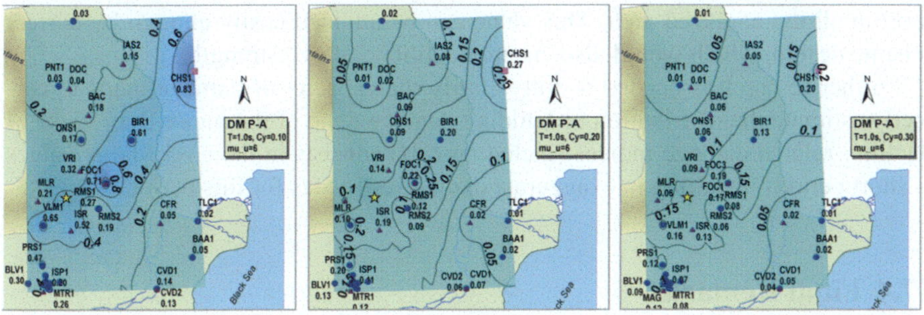

Cy=0.10 b) Cy=0.20 c) Cy=0.30

Fig. 17.8 August 30, 1986 earthquake: spatial distribution of the Park-Ang damage index, *DM*, for $\mu_u = 6$ and $T = 1$ s

Some recently developed damage maps are presented in Fig. 17.8, for vibration period $T = 1$ s and ultimate monotonic ductility $\mu u = 6$. As it results from the figure, the largest DM values occur along the NE-SW line that connects stations Valenii de Munte (VLM1) and Chisinau (CHS1). For $C_y = 0.10$, the values in Focsani (FOC2) and Chisinau exceed $DM = 0.8$. The mentioned area (delimited in Fig. 17.8 by the $DM = 0.4$ contour) largely corresponds to that where the most significant damage was reported at the time. However, the actually reported damage needs to be correlated more thoroughly with the computed DM values, based on the effective characteristics of damaged buildings.

In order to assess the building strength capacity required to limit damage at certain levels, maps of C_y were generated for different values of period and for the key values of *DM*, i.e. 0.2, 0.4 and 1. Figure 17.9 shows maps obtained for $T = 1$ s and $\mu_u = 6$.

According to Fig. 17.9, for buildings with the specified characteristics, a Cy value of 0.15 would have been sufficient, for example, to limit damage to the "repairable"

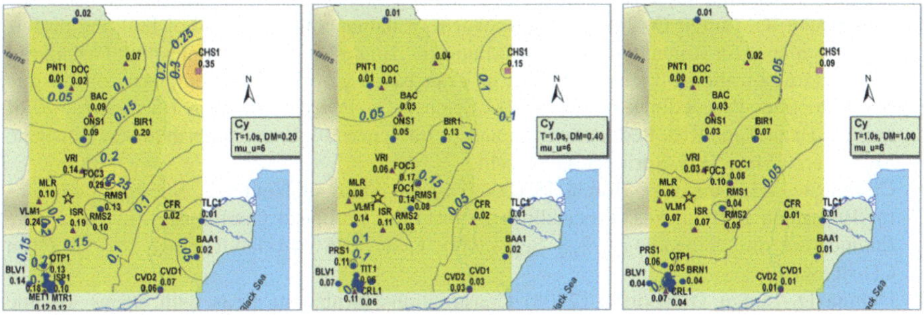

a) $DM = 0.2$ b) $DM = 0.4$ c) $DM = 1$

Fig. 17.9 Spatial distribution of yield strength demands (C_y) for three specified damage levels. August 30, 1986 earthquake. $\mu_u = 6$, $T = 1$ s level in all the analyzed area. This value can be attained easily enough by correct seismic design, which would also ensure ductility and overstrength.

Further research is needed to fully clarify the capacity of damage maps to correctly estimate the distribution of building damage. One of the major drawbacks of ductility-related damage indices (including *DM*) consists in their very large values at short periods, resulting in exaggerate damage estimates for this period range.

17.4 Final Remarks

Strong-motion networks are essential infrastructures in seismically prone countries such as Romania. The paper presented some of the most recent research performed at INCERC on the basis of data recorded in Romanian seismic networks.

Acknowledgments Part of the research reported in this work has been conducted under the NATO Project SfP 980468 "Harmonization of Seismic Hazard and Risk Reduction in Countries Influenced by Vrancea Earthquakes".

References

1. Balan St, Cristescu V, Cornea I (coordinators) (1982) The Romania earthquake of 4 March 1977 (in Romanian). Editura Academiei, Bucharest
2. Beles AA (1941) Le tremblement de terre du 10 Novembre 1940 et les bâtiments. In: Comptes rendus des séances de lÁcadémie des Sciences de Roumanie. Tome V, 3:270–288. Cartea Romaneasca, Bucharest
3. Berg GV, Bolt BA, Sozen MA, Rojahn C (1980) Earthquake in Romania, March 4, 1977: an engineering report. National Academy Press, Washington, DC
4. Bonjer KP (2004) The digital strong-motion network of the CRC461-B3 and INFP in Romania. Karlsruhe Univ. http://www-sfb461.physik.uni-karlsruhe.de/pub/B3/web/b3_research_en.htm Accessed 09/19/09
5. Bozorgnia Y, Bertero W (2001) Improved shaking and damage parameters for post-earthquake applications. Proceedings of the SMIP01 Seminar on Utilization of Strong-Motion Data, Los Angeles, California, September 12, 1–22
6. Craifaleanu IG, Lungu D (2008) An assessment of damage potential and of building performance demands for Romanian Vrancea earthquakes. In: Zaicenco A, Craifaleanu I; Paskaleva I (Eds), Harmonization of seismic hazard in Vrancea zone with special emphasis on seismic risk reduction. Proceedings of the NATO Science for Peace Project on Harmonization of Seismic Hazard and Risk Reduction in Countries Influenced by Vrancea Earthquakes Chisinau, Moldova 20 May 2008. Springer, Dordrecht
7. Craifaleanu IG, Lungu D, Borcia IS (2006) Shake Maps of Nonlinear Spectral Ordinates for Romanian Earthquakes. Proceedings of the First European Conference on Earthquake Engineering and Seismology. 3–8 Sept 2006, Geneva, Switzerland. Paper No 1257 (on CD-ROM)
8. Dubina D, Lungu D (coords) (2003) Constructions located in areas with strong seismic motions. Orizonturi Universitare, Timisoara
9. Georgescu ES (2003) Earthquake Engineering Development before and after the March 4, 1977, Vrancea, Romania Earthquake. Proceedings of the Symposium "25 years of Research in Earth Physics", INFP, September 25–27, 2002, Bucharest. St. Cerc. Geofiz., Tome 1, 93–107, Bucharest

10. Georgescu ES (2005) The earthquake of November 10, 1940, initiating vector of the modern earthquake engineering in Romania (in Romanian). Proceedings of the 3rd National Conference of Earthquake Engineering, December 9, 2005, Bucharest, Romania, INCERC Ed., Bucharest
11. INFP (2009a) Romanian Earthquake Catalogue ROMPLUS (2009). National Institute for Earth Physics. http://www.INFP.ro/catal.php. Accessed 18 Sept 2009
12. INFP (2009b) Website. http://www.infp.ro/en/static/monitorizare-seismica. Accessed Oct 2009
13. Ionescu C (2004) Romanian Seismic Network, FDSN Report 2004. INFP http://www.fdsn.org/FDSNmeetings/2004/Romania_FDSN_2004.pdf. Accessed 18 Sept 2009
14. Lungu D (2005) Seismic risk mitigation in the Romania-Synergy from international projects. Presentation at the World Conference on Disaster Reduction, 18–22 Jan 2005, Kobe, Hyogo, Japan. http://www.unisdr.org/wcdr/thematic-sessions/presentations/session4-6/romania-dr-lungu. pdf. Accessed 27 Sept 2009
15. Lungu D, Craifaleanu IG (2008) Shake maps of strength and displacement demands for Romanian Vrancea earthquakes. In: Zaicenco A, Craifaleanu I, Paskaleva I (Eds), Harmonization of Seismic Hazard in Vrancea Zone with Special Emphasis on Seismic Risk Reduction. Proceedings of the NATO Science for Peace Project on Harmonization of Seismic Hazard and Risk Reduction in Countries Influenced by Vrancea Earthquakes, Chisinau, Moldova 20 May 2008. Springer, Dordrecht
16. Park YJ, Ang AH-S, Wen YK (1987) Damage-limiting aseismic design of buildings. Earthq Spectra 3(1):1–26
17. Sandi H, Borcia IS, Stancu M (2004) Analysis of attenuation for recent Vrancea intermediate depth earthquakes. Proc. 13-th WCEE, Vancouver
18. Seismic Database for Romanian Earthquakes (2001). INCERC and INFP. Research financed by the Romanian Ministry of Education and Research
19. Teran-Gilmore A (1997) Energy Concepts and Damage Indices. EERC-CUREe Symposium in Honor of Vitelmo V. Bertero. January 31–February 1, 1997, Berkeley, CA. http://nisee.berkeley.edu/lessons/teran-gilmore.html. Accessed 08 April 2010
20. Wenzel F, Lungu D (2000) Earthquake Risk Assessment for Romania. Proc of the EuroConference on Global Change and Catastrophe Risk Management: Earthquake Risks in Europe. Laxenburg, Austria, 6–9 July. http://www.iiasa.ac.at/Research/RMS/july2000/Papers/wenzel708.pdf. Accessed Apr 2010

Chapter 18
Strong-Motion and Structural Monitoring Networks in Istanbul and Their Use for Risk Assessment

E. Şafak

Abstract Department of Earthquake Engineering of Kandilli Observatory and Earthquake Research Institute (KOERI) has installed and been operating for some time a large number of ground and structural seismic monitoring networks in Istanbul. The ground systems include 100-station Istanbul Rapid Response Network, 10-station Early Warning Network, and one 4-level downhole array. Installation of three more similar downhole arrays, and a small-aperture 72-channel dense array at the Air-Force Academy grounds are currently in progress. The structural systems include 27-channel real-time system at the St. Sophia Museum, 42-channel real-time system at Fatih Mosque, 15-channel real-time system at Fatih Sultan Mehmet Bridge (the 2nd Bosporus bridge), 24-channel triggered system at Suleymaniye Mosque, and 15-channel triggered systems at Kanyon Building (high-rise office building), Isbank Tower (a high-rise office building), and ENRON Power Plant. Installation of more real-time structural monitoring networks are in progress or at planning stages, such as those for Sultanahmet (Blue Mosque), Bayezit, and Mihrimah Sultan Mosques, and the Marmaray Immersed Tube Tunnel under the Bosporus. The objective of the early warning ground network is to provide 5–7 s early warning of an earthquake, which can be used for automatic shutdown of some of the critical systems in Istanbul. The rapid response network provides the critical values of ground shaking (PGA, PGV, and spectral accelerations at 0.2 and 1.0 s) automatically within minutes of an earthquake, which are then transmitted to those agencies dealing with emergency response. Downhole arrays are located in the areas of Istanbul, where site amplification is expected to be critical. The objective in structural arrays is to monitor the health of the structure by continuously measuring its vibrations, and detect and locate damage after an extreme event, such as a large earthquake. A large component of structural monitoring activities is directed towards the research on developing new tools and techniques for system identification and damage detection. In addition to standard modal identification, more advanced real-time techniques are explored, such as wave the propagation approach and interferometric imaging.

E. Şafak (✉)
Kandilli Observatory and Earthquake Research Institute, Bogazici University, Istanbul, Turkey
e-mail: erdal.safak@boun.edu.tr

S. Akkar et al. (eds.), *Earthquake Data in Engineering Seismology*, Geotechnical, Geological, and Earthquake Engineering 14, DOI 10.1007/978-94-007-0152-6_18,

18.1 Introduction

Department of Earthquake Engineering of Kandilli Observatory and Earthquake Research Institute has installed and been operating for some time a large number of ground and structural seismic monitoring networks in Istanbul. The ground systems include 100-station Istanbul Rapid Response Network, 10-station Early Warning Network, and one 4-level downhole array. Installation of three more similar downhole arrays, and a small-aperture 72-channel dense array at the Air-Force Academy grounds are currently in progress. The structural systems include 27-channel real-time system at the St. Sophia Museum, 42-channel real-time system at Fatih Mosque, 15-channel real-time system at Fatih Sultan Mehmet Bridge (the 2nd Bosporus bridge), 24-channel triggered system at Suleymaniye Mosque, and 15-channel triggered systems at Kanyon Building (high-rise office building), Isbank Tower (a high-rise office building), and ENRON Power Plant. Installation of more real-time structural monitoring networks are in progress or at planning stages, such as those for Sultanahmet (Blue Mosque), Bayezit, and Mihrimah Sultan Mosques, and the Marmaray Immersed Tube Tunnel under the Bosporus.

18.2 Istanbul Rapid Response Network

Various seismic hazard studies done in recent years all agree that Istanbul is likely to have a damaging earthquake in the near future [1, 3]. The probability of having a magnitude 7 or above earthquake in Istanbul is about 65% within the next 30 years. To assist in the reduction of losses from such an earthquake, a dense strong motion network is established in Istanbul [2]. The network is composed of 100 strong motion stations, placed primarily in regions close to the Marmara Sea, as shown in Fig. 18.1. The area covered by the network is approximately 50 km $\times$ 30 km, and constitutes the most likely region of Istanbul that will suffer damage due to its soil conditions and proximity to the fault line.

The network operates on triggered mode. After triggered by an earthquake, PGA (Peak Ground Accelerations), PGV (Peak Ground Velocities), and PSA (Pseudo-Spectral Accelerations) at specified periods are sent in the form of SMS messages at every 20 s directly to the main data center through a GSM communication system, telemetry, and landlines.

Spectral displacements obtained from the SMS messages are interpolated to determine the spectral displacements at the center of each $0.01^\circ \times 0.01^\circ$ geo-cell. The seismic demand at the center of each geo-cell is computed from spectral displacements. Each cell is also assigned a structural type based on the 24 structural categories determined from the structural inventory studies done for Istanbul. Using the spectral displacements and the spectral-displacement based fragility curves (i.e., the Capacity Spectrum Procedure) for the 24 building categories, the building damage in each geo-cell is computed.

The Rapid Response Information is generated automatically in the form of shake maps for different parameters, and damage distribution maps. An example

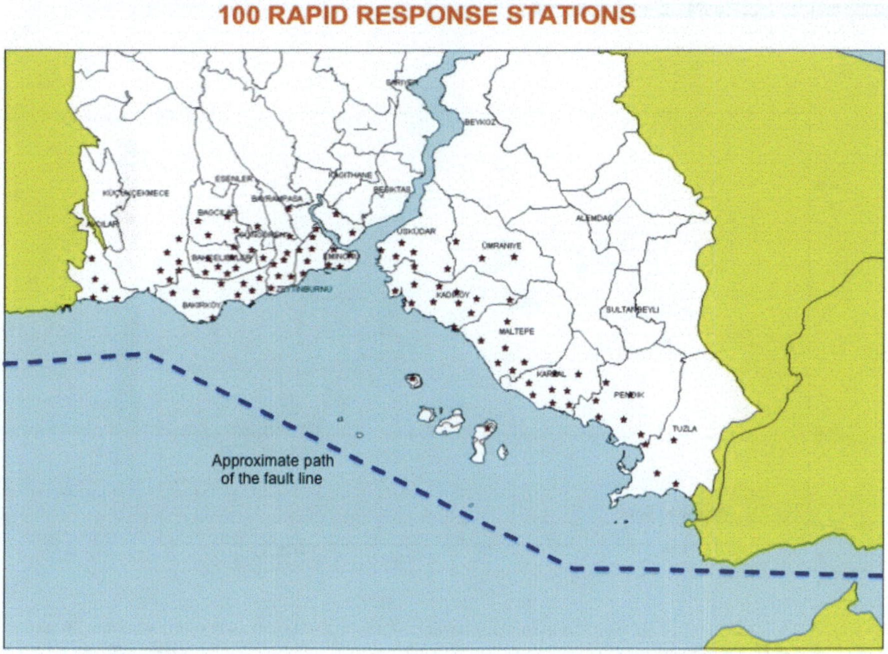

Fig. 18.1 Locations of Istanbul rapid response stations with respect to fault line

of a damage distribution map created for a simulated earthquake is shown in Fig. 18.2. These maps are transmitted immediately to the Istanbul Governorate, First Army Headquarters, and Istanbul Municipality through redundant communication channels, including digital radio modems and GPRS communication systems.

The health of the rapid response system is tested automatically every day at 10:00 am by setting off a false trigger.

18.3 Istanbul Early Warning Network

The Istanbul Early Warning network is composed of ten 24-bit broadband strong-motion stations operating in real time [2].

The stations are located as close to the Great Marmara fault zone as possible, as shown in Fig. 18.3. The data from these stations to the main data center at Kandilli are transmitted continuously via digital spread spectrum radio modem involving repeater stations, as well as satellite transmission.

Considering the complexity of fault rupture and the short fault distances involved, a simple and robust Early Warning algorithm, based on the exceedance of a specified threshold amplitude level is implemented. Two ground motion parameters are used for early warning, the PGA (peak ground acceleration) and the CAV (the cumulative integral of absolute accelerations).

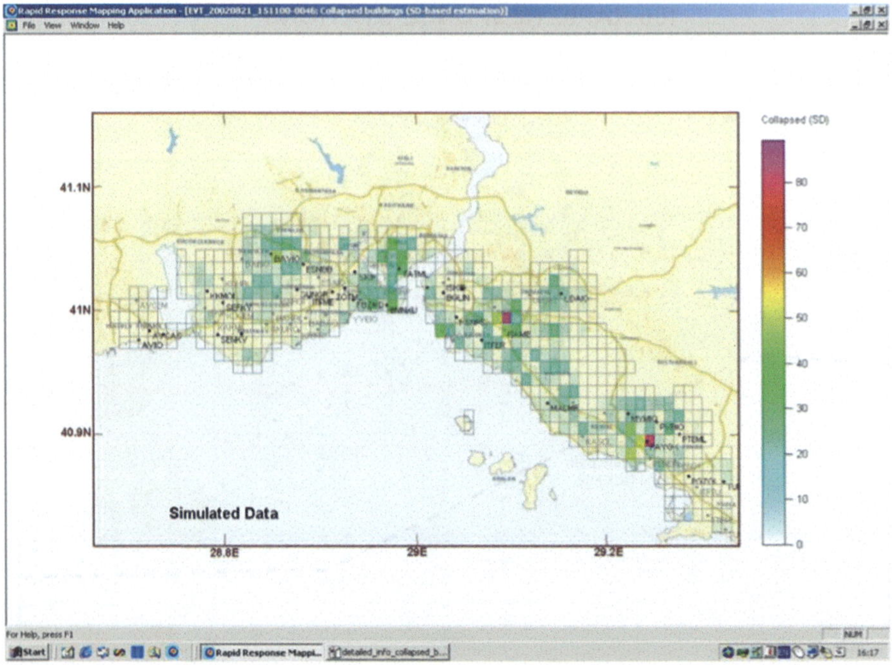

Fig. 18.2 Estimated number of collapsed buildings for a simulated earthquake

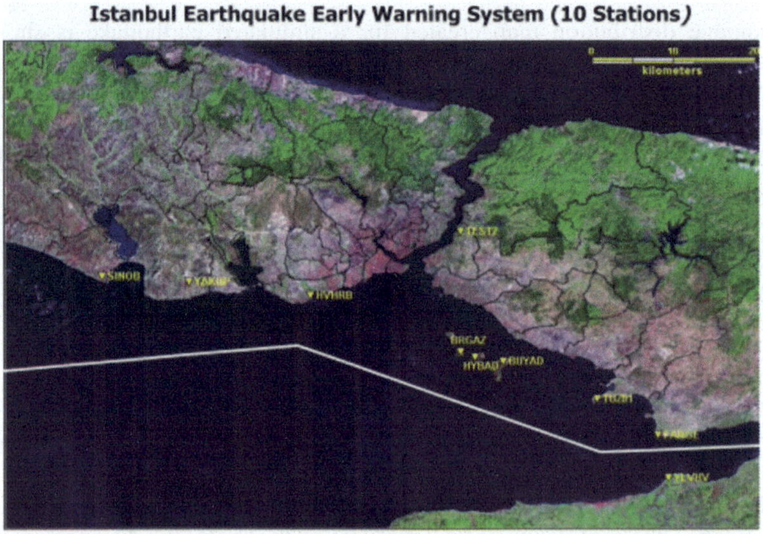

Fig. 18.3 Locations of Istanbul Early Warming stations

The PGA values are calculated after low-pass filtering the records at 12 Hz. When the acceleration in any channel exceeds 20 mg it is considered a vote. When three more stations vote within 5 s. after the first vote, it is considered the first alarm. When three more stations vote for a threshold level of 50 mg within 5 s. after the first alarm, it is considered the second alarm. When three more stations vote for a threshold level of 100 mg within 5 s. after the second alarm, it is considered the third alarm.

Similarly, the CAV values are also calculated from the low-pass filtered accelerations at 12 Hz. The CAV is computed for only those 1 s. intervals where PGA is greater than 3 mg. When CAV in any channel exceeds 20 mg-s it is considered a vote. When three more stations vote within 5 s. after the first vote, it is considered the first alarm. When three more stations vote for a CAV threshold level of 40 mg-s within 5 s. after the first alarm, it is considered the second alarm. When three more stations vote for a CAV threshold level of 70 mg-s within 5 s. after the second alarm, it is considered the third alarm.

The early warning information consisting of three alarm levels are communicated to the appropriate servo shut-down systems of the recipient facilities, which should automatically decide proper action based on the alarm level. Figure 18.4 shows the histogram of early warning times for 280 simulated earthquakes for two locations

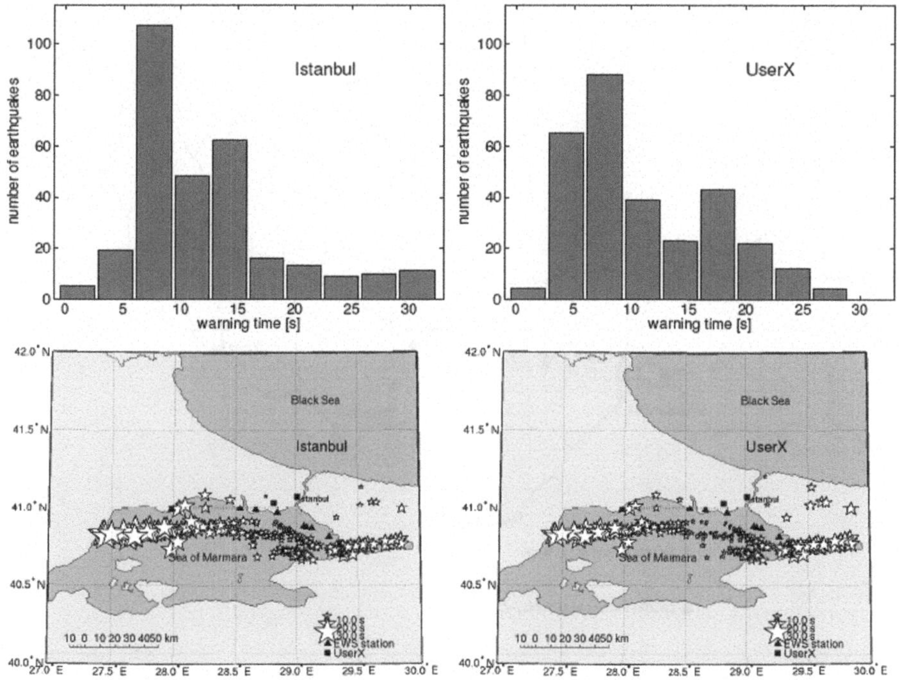

Fig. 18.4 Early warning times for two locations based on 280 simulated earthquakes

marked as Istanbul and User X in the figure. Depending on the location of the earthquake, the nucleus of fault rupture, and the recipient facility the average alarm time can be as high as about 8 s.

The early warning signals are transmitted to the end users by employing several communication companies as service providers. The encrypted early warning signals (i.e., earthquake alarms) are communicated to the respective end users by FM, UHF and satellite communication systems.

The early warning system is currently being tested for several end users (e.g., one of the İs Kule buildings in Istanbul, and the Trakya Elektrik plant in Marmara Ereğlisi).

18.4 Vertical Ground Arrays

Currently there is one vertical ground array operating in real time near the Atakoy district of Istanbul. It is composed of 3-component downhole accelerometers at three different elevations, plus a surface accelerometer. Four new downhole arrays are currently being installed. The map in Fig. 18.5 shows the locations of the old and the new downhole stations.

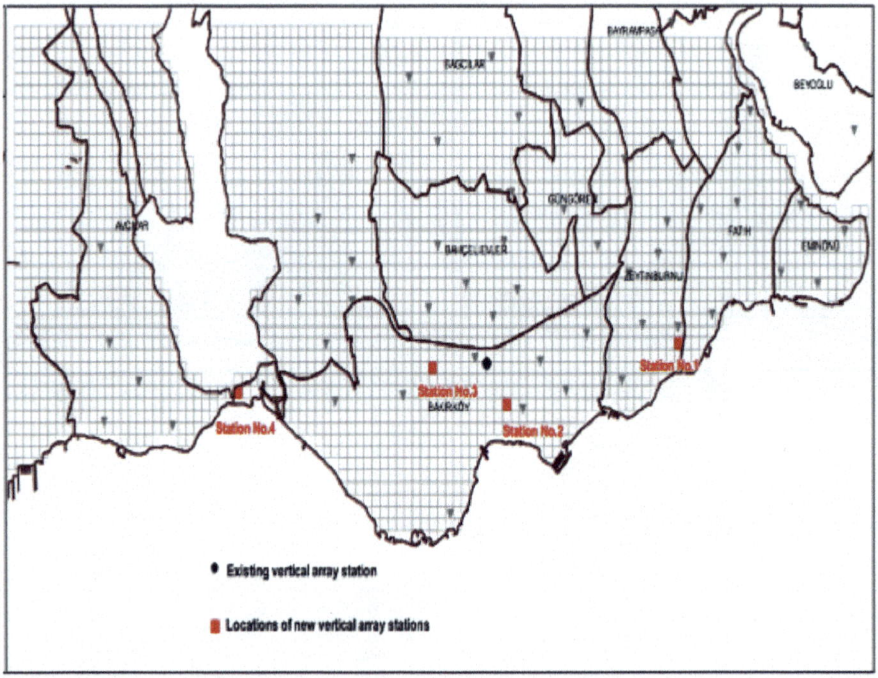

Fig. 18.5 Locations of current and planned downhole stations

18.5 Structural Monitoring Arrays

KOERI has installed and operates a significant number of structural monitoring networks in Istanbul. The majority of these structures are historical and critical structures. They include several mosques (Suleymaniye, Fatih, Mihrimah Sultan, and Sultanahmet Mosques), St. Sophia Museum, the two suspension bridges over the Bosporus (Bogazici and Fatih Sultan Mehmet Bridges), high-rise buildings, and industrial plants. Figures 18.6 and 18.7 show examples of instrumented structures and instrument layouts.

The monitoring involves mainly acceleration sensors and the accelerations are recorded continuously. Some of the structures (e.g., the suspension bridges) are being supplemented with GPS sensors to record the displacements directly.

Data from the structural monitoring networks are transmitted in real time to the monitoring center at KOERI. An n-house real-time data analysis software, KOERI_MIS, is used to process and analyze the data. The software includes data

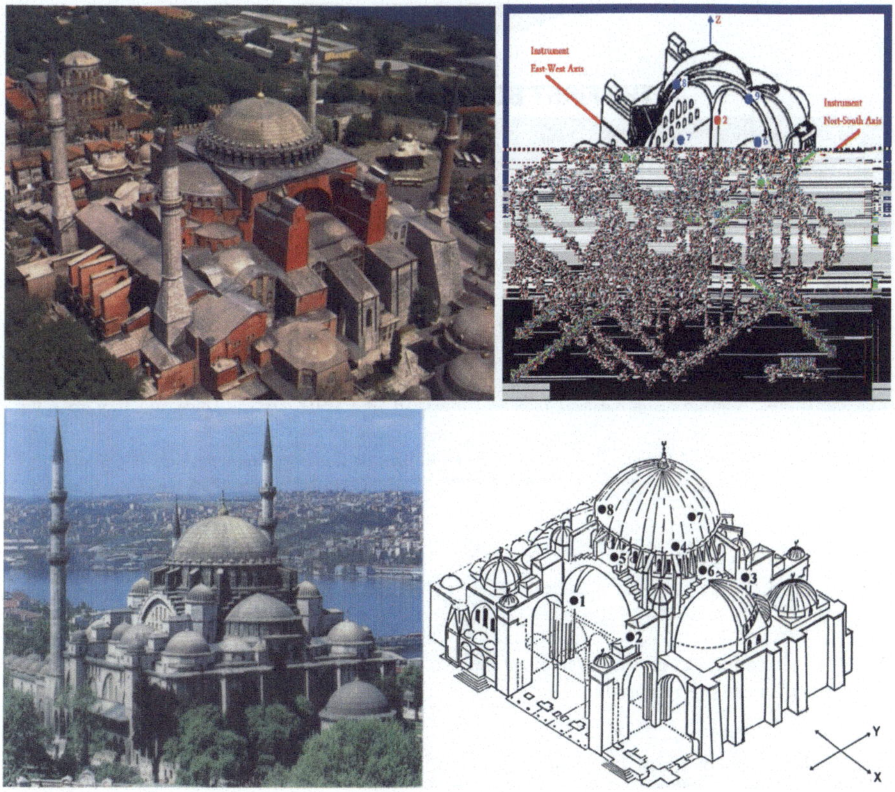

Fig. 18.6 Structural monitoring systems at St. Sophia Museum (*top*) and Suleymaniye Mosque (*bottom*)

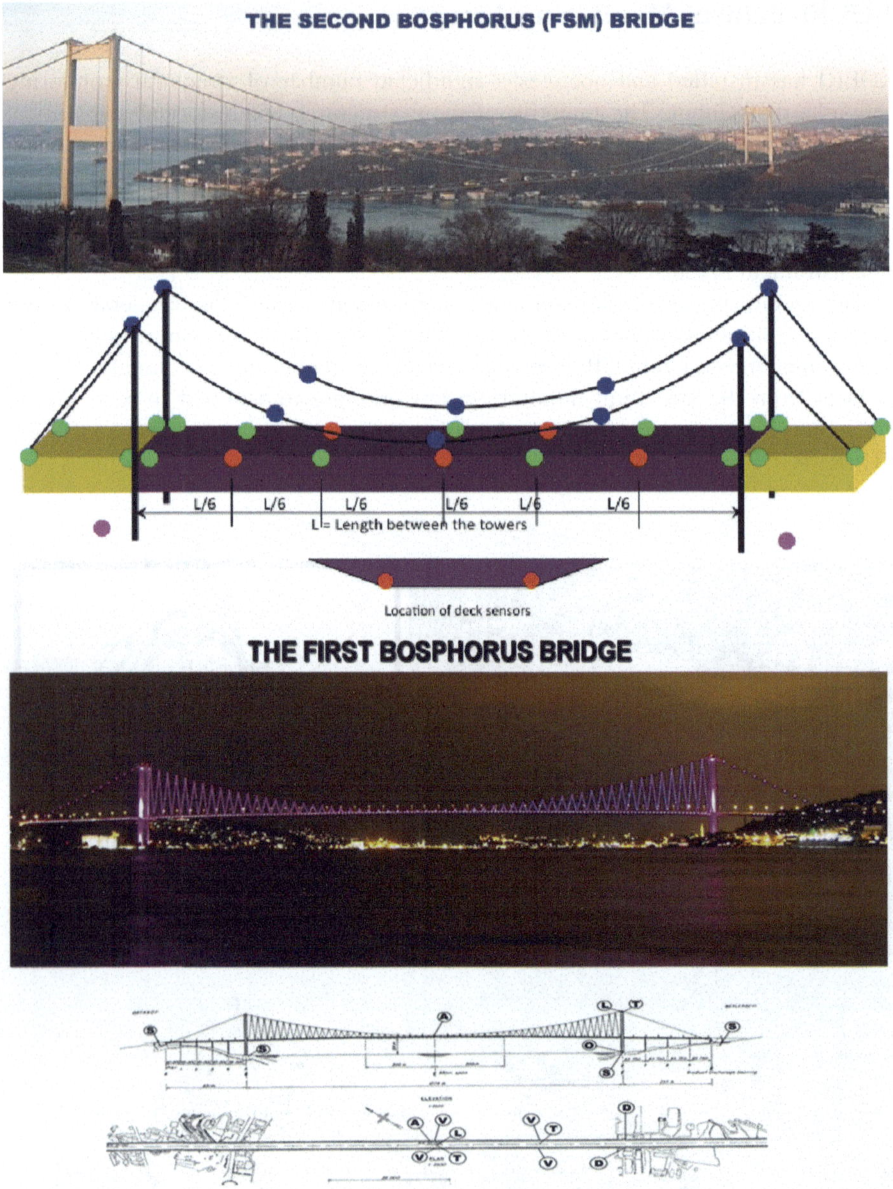

Fig. 18.7 Structural monitoring systems at the two Bosporus Bridges (the *red* sensors are already in, the others are being acquired)

processing, system identification, and animation modules. The results are displayed in real time, showing the time variations of modal properties and the structure's configuration. Figure 18.8 gives a screen-plot of KOERI_MIS for a tower structure with a heavy mass near the top.

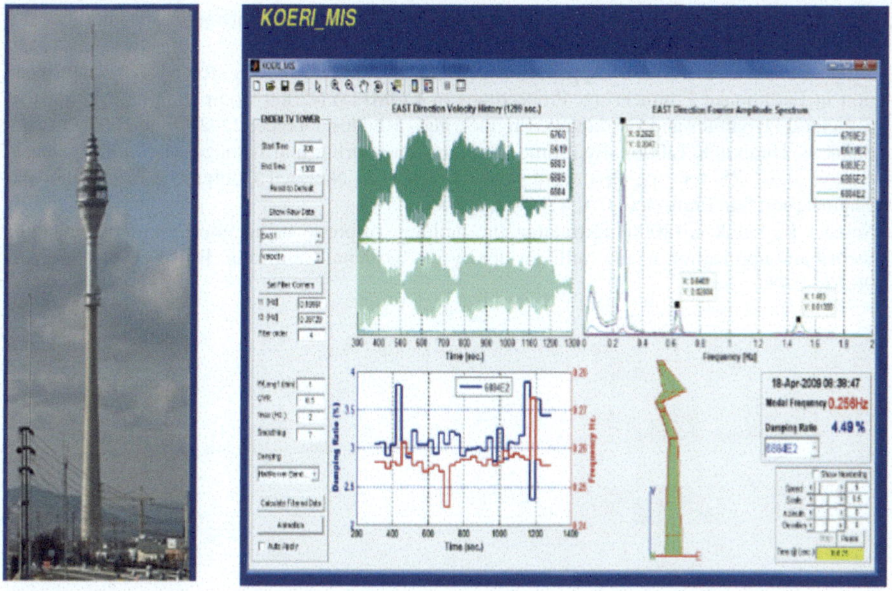

Fig. 18.8 Screen-print of the KOERI_MIS data processing and analysis software

The safety of monitored structures is assessed based on the observations of their modal properties and geometry. Changes in modal properties (e.g., a reduction in a natural frequency) can be detected from the modal identification of acceleration records. However, changes in modal properties do not always represent damage, because environmental conditions (e.g., temperature) can cause frequency changes. An additional parameter that can be used for damage detection is the changes in the structure's geometry. In most cases, the behavior of a damaged structure is non-linear. One of the characteristics of a nonlinear behavior is that the structure does not come back to its original configuration when the loading (e.g., the earthquake) stops. GPS sensors are better suited to detect changes in the geometry [6]. Another parameter that can be used for identification and damage detection is the velocity of seismic waves within the structure. Methods are being developed based on the concept of *Seismic Interferometry* to investigate the characteristics of seismic waves in structures [4, 5, 7].

References

1. Erdik M, Demircioglu M, Sesetyan K, Durukal E, Siyahi B (2004) Earthquake hazard in Marmara region, Turkey. Soil Dyn Earthq Eng 24:605–631
2. Erdik M, Fahjan Y, Ozel O, Alcik H, Mert A, Gul M (2003) Istanbul earthquake rapid response and the early warning system. Bull Earthq Eng 1:157–163
3. Parsons T, Toda S, Stein S, Barka A, Dietrich JH (2000) Heightened odds of a large earthquake near Istanbul: an interaction based probability calculation. Science 288:661–665

4. Safak E (1999) Wave propagation formulation of seismic response of multi-story buildings. J Struct Eng ASCE 125(4):426–437
5. Safak E, Cakti E (2009). Applications of seismic interferometry for system identification in Earthquake Engineering, Proceedings of COMPDYN2009-Computational Methods in Structural Dynamics and Earthquake Engineering – Rhodes, Greece, 22–24 June 2009.
6. Safak E, Hudnut K (2006). Real-time structural monitoring and damage detection by acceleration and GPS sensors, Proceedings of the 8th U.S. National Conference on Earthquake Engineering, San Francisco, CA, 18–22 Apr 2006.
7. Snieder R, Safak E (2006) Extracting the building response using interferometric imaging; theory and application to the millikan library in Pasadena, California. Bull Seismol Soc Am 96:586–598

Addresses of Principal Contributors

S. Aoi National Research Institute for Earth Science and Disaster Prevention, 3-1 Tennodai, Tsukuba, Ibaraki 303-0006, Japan, aoi@bosai.go.jp

J.J. Bommer Civil and Environmental Engineering Department, Imperial College London, London SW7 2AZ, UK, j.bommer@imperial.ac.uk

D.M. Boore U.S. Geological Survey (MS977), Menlo Park, CA 94025, USA, boore@usgs.gov

Z. Çağnan Department of Civil Engineering, Middle East Technical University, Northern Cyprus Campus, Kalkanlı, Güzelyurt, KKTC, Mersin 10, Turkey, cagnan@metu.edu.tr

K.W. Campbell EQECAT, Inc., Beaverton, OR, USA, kcampbell@eqecat.com

M. Çelebi U.S. Geological Survey (MS977), Menlo Park, CA 94025, USA, celebi@usgs.gov

J. Clinton Swiss Seismological Service (SED – ETHZ), Sonneggstrasse 5, 8092 Zürich, Switzerland; Swiss Federal Institute of Technology, Zurich, Switzerland, j.clinton@sed.ethz.ch

I.G. Craifaleanu National Institute for Building Research (INCERC), Technical University of Civil Engineering (UTCB), Bucharest, Romania, i.craifaleanu@gmail.com

J. Douglas BRGM-RNSC/RIS, 3 avenue C. Guillemin, BP 36009, 45060 Orleans Cedex 2, France, j.douglas@brgm.fr

P. Gülkan Department of Civil Engineering, Earthquake Engineering Research Center, Middle East Technical University, 06531 Ankara, Turkey, a03516@metu.edu.tr

L. Luzi Istituto Nazionale di Geofisica e Vulcanologia, Milano, Italy, luzi@mi.ingv.it

B. Margaris ITSAK – Institute of Engineering Seismology and Earthquake Engineering, 55102 Thessaloniki, Greece, margaris@itsak.gr

S. Akkar et al. (eds.), *Earthquake Data in Engineering Seismology*, Geotechnical, Geological, and Earthquake Engineering 14, DOI 10.1007/978-94-007-0152-6,

R. Paolucci Department of Structural Engineering, Politecnico di Milano, Milano, Italy, paolucci@stru.polimi.it

C. Péquegnat LGIT/CNRS/LCPC – Joseph Fourier University, Grenoble, France, pequegna@obs.ujf-grenoble.fr

K. Pitilakis Department of Civil Engineering, Aristotle University of Thessaloniki, GR-54124 Thessaloniki, Greece, kpitilak@civil.auth.gr

A. Roca Institut Geològic de Catalunya, Barcelona, Spain, aroca@igc.cat

E. Şafak Kandilli Observatory and Earthquake Research Institute, Bogazici University, Istanbul, Turkey, erdal.safak@boun.edu.tr

T. van Eck ORFEUS c/o Seismology Division, Royal Netherlands Meteorological Institute, 3730 AE De Bilt, The Netherlands, torild.van.eck@knmi.nl

Index

The manufacturer's authorised representative in the EU is Springer
Nature Customer Service Centre GmbH, Europaplatz 3, 69115 Heidelberg,
Germany. If you have any concerns regarding our products, please
contact ProductSafety@springernature.com

Printed and bound by CPI Group (UK) Ltd, Croydon, CR0 4YY
01/10/2026
02226299-0007